WOOD HICKS AND BARK PEELERS

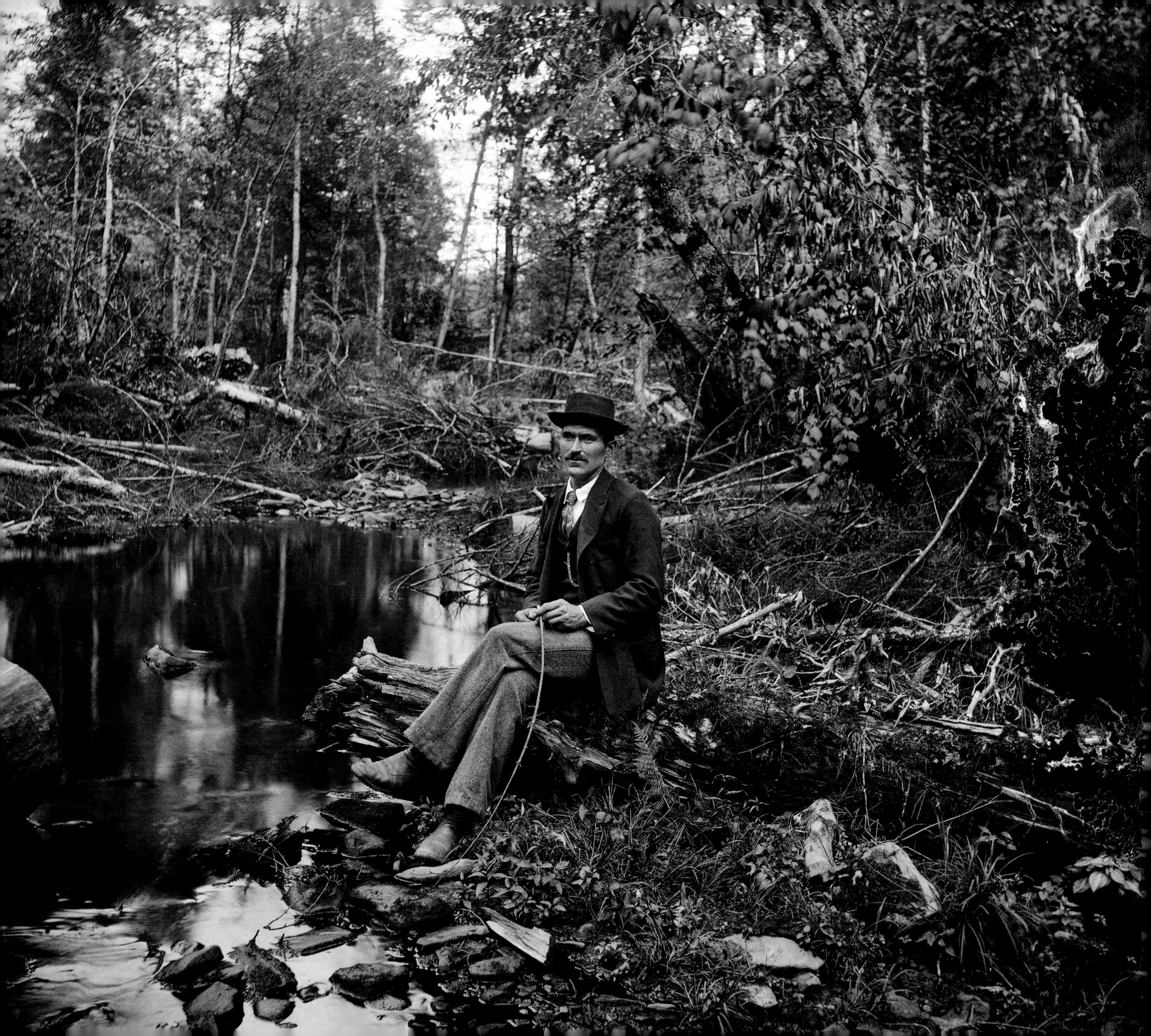

THE PHOTOGRAPHIC LEGACY OF WILLIAM T. CLARKE

WOOD HICKS AND BARK PEELERS

A VISUAL HISTORY OF PENNSYLVANIA'S RAILROAD LUMBERING COMMUNITIES

Ronald E. Ostman and Harry Littell

With an introduction by Linda A. Ries

THE PENNSYLVANIA STATE UNIVERSITY PRESS

UNIVERSITY PARK, PENNSYLVANIA

Published in collaboration with the Lumber Heritage Region of Pennsylvania and the Pennsylvania Historical and Museum Commission

A KEYSTONE BOOK®

Keystone Books are intended to serve the citizens of Pennsylvania. They are accessible, well-researched explorations into the history, culture, society, and environment of the Keystone State as part of the Middle Atlantic region.

FRONTISPIECE: This may be north-central Pennsylvania photographer William T. Clarke (1859–1930). Folklorist Henry Wharton Shoemaker wrote that Clarke and U.S. General of the Armies John Joseph Pershing were look-alikes (see fig. 10). Courtesy of the Lois Barden Photograph Collection, Candor, New York, BC 21.

Typeset by REGINA STARACE | *Printed and bound by* ASIA PACIFIC OFFSET | *Composed in* BASKERVILLE 10 PRO | *Printed on* 115 GSM KASADAKA (NATURAL) | *Bound in* JHT

Library of Congress Cataloging-in-Publication Data

Names: Ostman, Ronald Elroy, author. | Littell, Harry, author. | Clarke, W. T., photographer.

Title: Wood hicks and bark peelers : A Visual History of Pennsylvania's Railroad Lumbering Communities; The Photographic Legacy of William T. Clarke / Ronald E. Ostman and Harry Littell ; with an introduction by Linda A. Ries.

Description: University Park, Pennsylvania : The Pennsylvania State University Press, [2016] | "A keystone book." | "Published in collaboration with the Lumber Heritage Region of Pennsylvania and the Pennsylvania Historical and Museum Commission." | Includes bibliographical references and index.

Summary: "Through the photography of William T. Clarke, explores the impact of the logging industry on late nineteenth- and early twentieth-century north-central Pennsylvania"—Provided by Publisher.

Identifiers: LCCN 2015046215 | ISBN 9780271072074 (cloth : alk. paper)

Subjects: LCSH: Logging—Pennsylvania—History. | Logging railroads—Pennsylvania—History. | Loggers—Pennsylvania—History. | Clarke, W. T. | Landscape photography—Pennsylvania—History.

Classification: LCC SD538.2.P4 O78 2016 | DDC 634.909748—dc23

LC record available at https://lccn.loc.gov/2015046215

Printed in China
Published by The Pennsylvania State University Press,
University Park, PA 16802–1003

10 9 8 7 6 5 4

The Pennsylvania State University Press is a member of the Association of American University Presses.

It is the policy of The Pennsylvania State University Press to use acid-free paper. Publications on uncoated stock satisfy the minimum requirements of American National Standard for Information Sciences—Permanence of Paper for Printed Library Material, ANSI Z39.48–1992.

Pennsylvania Historical & Museum Commission

LUMBER HERITAGE REGION

To Lois and Bob Barden.
Where others saw a crate of old window panes,
you saw a historical treasure trove.

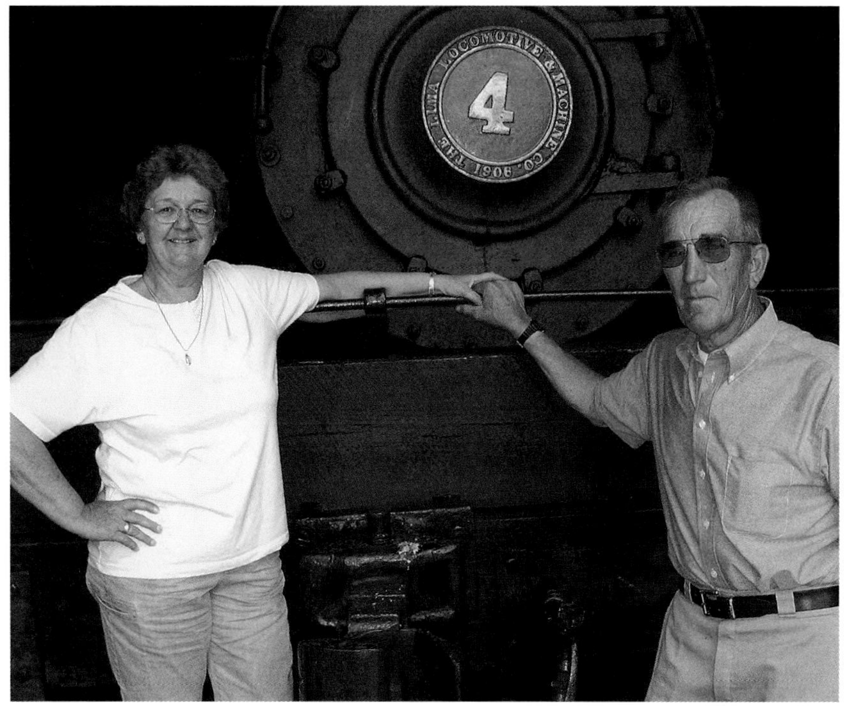

CONTENTS

ACKNOWLEDGMENTS

To all these individuals, our deep gratitude for sharing enthusiasm for the photographs and for supporting the long effort to reach a larger audience: Lois and Bob Barden, Kurt R. Bell, Matt Braun, Simon Bronner, Delores Buchsen, Dawn E. Caudell, Dale R. Corson, Robert K. Currin, Judy Gamon, Nicholas Graver, Jeffrey Hoone, Laura Ishner, Karl S. Kabelac, Alan Littell, Dawn Hull Murray, Romaine Perin, Charlee Redman, Margaret B. Redmond, Kay Brownell Reed, Laura Reed-Morrisson, Julie L. Schoelles, Nicholas Tonelli, Linn Underhill, Anne E. Wenger, Michael Wennin, Kathryn B. Yahner, our families, and our friends.

PROLOGUE

DISCOVERY AND PROCEDURES

Who were those people? What were they doing? Who was the photographer? Where were the photographs taken? When? How? And why are their pictures here, of all places? Lois Barden asked these questions more than forty years ago. This book is an effort to answer her questions.

Lois and her husband, Bob Barden, were browsing through a toolshed near Honeoye Lake, south of Rochester, New York, in 1974. Bob's other relatives were sorting through household goods and antiques in the nearby cottage. In the dark shed, Lois spied two wooden crates on the dirt floor in a damp, gloomy corner. They were filled with glass windowpanes. Maybe they'd be useful on their farm. She lifted an eight-by-ten-inch plate from a box and saw something more than dirt on it. She brushed away some of the grime, held it to the light coming through the doorway, squinted at the smudge she'd uncovered, and was shocked. A ghostly face stared back at her, a sudden and unexpected glimpse from a shadowy world beyond the grave. This was not a common windowpane! Seconds after her startling discovery, Lois replaced the glass plate, lifted the heavy crates, and moved them outside. She plucked plate after plate from the boxes and realized they were loaded with photographic negatives of a type she'd never seen before. *One thing for sure*, she thought, *these are no ordinary pictures. And they're old.*

She figured the negatives were made with one of those old-fashioned cameras on a tripod, where the photographer inserted the glass plate into the camera mechanism, huddled under a cloth that was draped over it to block out the light, and looked through the back of the camera to compose and focus the picture before asking the people in front of the lens, "Please remain as still as possible." Fascinated, Lois hauled the crates to the cottage and asked the relatives if any of them knew anything about the mystery images. No one did. Well, then, did anyone want the plates? No. Sensing the importance of her find, she stored the crates at the family cattle farm near Candor, New York. And there the negatives remained, ignored and nearly forgotten.

Some thirty years later, in early March 2004, Lois was taking a digital photography course at Tompkins Cortland Community College in Dryden, New York. She mentioned the glass plates to Harry Littell, her instructor, figuring that because he had rephotographed scenes from historic photographs from a contemporary perspective and had helped author books dealing with regional historical photographs, he might be interested in her old negatives.[1] And he was. In fact, when she brought a couple of the plates to the next class, he became excited, saying he thought she'd made an important discovery. Lois was thrilled.

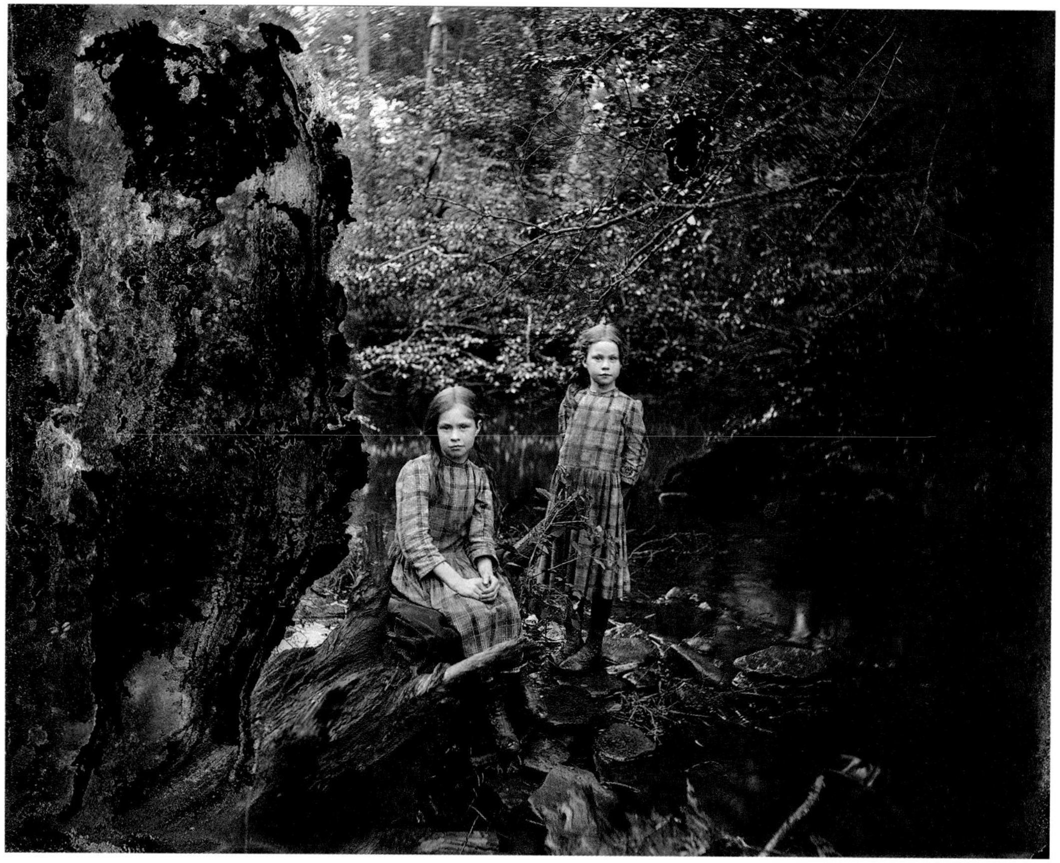

FIGURE 1
Girls near a stream in the north-central Pennsylvania forest

Upon examination of the crates' entire contents, they determined that the images depicted primarily forests, logging operations, and community life of long ago. They decided that the entire cache was remarkable and should be saved: the treasure trove consisted of 131 unique glimpses into the past.

Preserving the crates' contents was arduous. For some plates they first had to remove dirt caked on the surface. Other plates had been exposed to water and the emulsion was nearly curled off the glass. Working carefully, they saved as much of the image as they could on each plate. Then they scanned the negatives, transforming them into digital format. Through the use of modern photographic technology, they rescued the surviving images from further breakage or damage. Harry then used digital tools to remove dust and scratches. Once the negatives were scanned and retouched, prints were made,

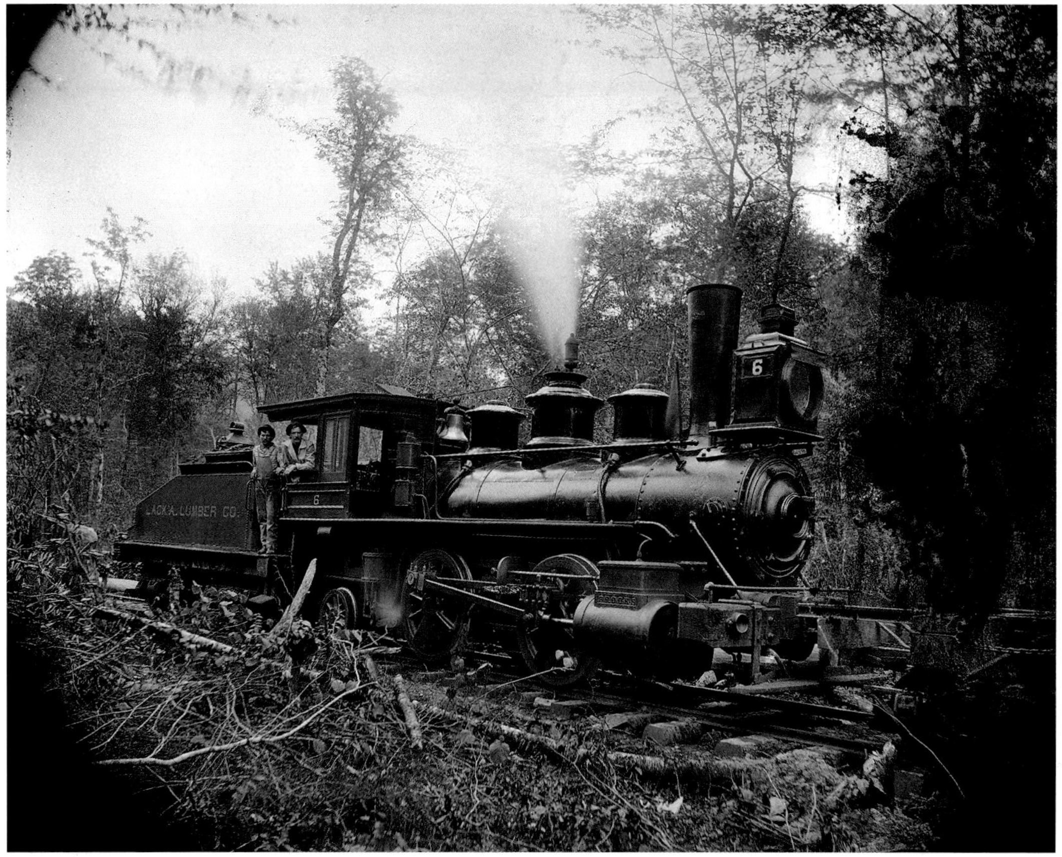

FIGURE 2
Lackawanna Lumber Company steam locomotive and tender, north-central Pennsylvania, 1898

necessitating further decisions and adjustments to create images that Harry felt would be faithful to the originals. The team decided that large areas of peeling and damaged emulsion at the margins of the images would not be altered. Subsequent prints based on this decision sometimes produced surrealistic effects, but the damaged photographs also reminded viewers that what they saw came from old, fragile negatives. During the restoration and preservation process, Lois and Harry found information scratched in the emulsion at the margins of several negatives. These clues helped them determine that the unknown photographer made the images in various transient logging camps and communities near Galeton and Port Allegany in north-central Pennsylvania during 1897 and 1898.

For example, an image with the date 1898 scratched in the emulsion, showing a steam locomotive with the words "Lack'a [Lackawanna]

Lumber Company" painted on the side of the tender, was helpful. The engineer and fireman were standing in the gangway and cab of Lackawanna Lumber Company rod locomotive No. 6, built in 1881 at the Brooks Locomotive Works, Dunkirk, New York. This locomotive and three other Lackawanna locomotives were kept busy in the 1890s, along with seventy log cars and three Barnhart log loaders. The Lackawanna Lumber Company logging railroad line operated primarily in northwestern Clinton County and extended into Potter County near Cross Fork in north-central Pennsylvania. "Lackies" transported enough logs for an estimated sawmill output of eighty thousand board feet of hemlock a day. Company headquarters were located at Cross Fork and Mina, operating from 1888 to 1903. And other images of gondola cars filled with logs and bark cars carried "Buffalo and Susquehanna Railroad" markings, providing a further lead.[2]

At that point, Lois Barden and Harry Littell did not know why these time-consuming gelatin dry plate images were made at what must have been great expense and effort under primitive and trying conditions. Nor did Lois Barden know how or why her husband's late relatives came into possession of the crates of negatives. Lois and Harry invited a third person, Ron Ostman, and, later, others to help with the historical research concerning the Pennsylvania logging industry at the end of the nineteenth century.[3] An Internet search led the trio to the website of the Pennsylvania Lumber Museum, in Ulysses, Pennsylvania. A visit with Delores Buchsen, the museum's director, and staff members and with Robert Currin, a curator and historian with the Potter County Historical Society in Coudersport, provided rich depositories of expert knowledge and stored information. The search was on.

Linda A. Ries served on a panel that evaluated the authors' successful joint application for a Pennsylvania Historical and Museum Commission (PHMC) Visiting Scholar in Residence award for 2006–7. With her intimate knowledge of the Pennsylvania State Archives photographic holdings, she strongly suspected that the authors had access to a theretofore unknown group of images made by William T. Clarke.

William T. Clarke was born in September 1859 in Rochester, New York, the son of William C. Clarke, a boat builder, and Isabella Clarke. In the 1860 U.S. census the birthplace of both parents is listed as Ireland.[4] They reported five children for the 1870 census and by then had lived at various Rochester locations.[5]

Rochester City directories indicate that William T. Clarke worked as a photographer in 1879–80 for Charles W. Woodward, a nationally prominent publisher of stereoviews, and during 1881–83 for the Union View and Copying Company, a prolific stereoview firm directed by Woodward. Woodward hired photographers for itinerant work. They traveled the country making images of interesting places for stereoview sets. It is possible Clarke did that type of work for Union View and Copying.[6] Stereoviews created a three-dimensional image when seen through a special hand-held device. They were highly popular in the latter part of the nineteenth century as parlor and group entertainment. Packaged in sets and arranged by subject, they covered topics ranging from the American West to the great cities of Europe and included a particular favorite, Niagara Falls. Clarke would later make stereoviews of the Pennsylvania lumber region to sell on his own.

William T. Clarke disappeared from the Rochester City directories between 1885 and 1917. Sometime between 1884 and 1887, he traveled to the northern tier counties of Pennsylvania, where he established a photography business, possibly seeking, as Pennsylvania journalist and folklorist Henry Wharton Shoemaker noted, "rest and outdoor life for a couple of years . . . to regain . . . lost health and ambition."[7] It has been difficult to learn of Clarke's biographical details. Only a few of his letters survive that we know of and we have found no diaries or similar sources of intimate data about his life and character. Some information is unfortunately clouded with embellishment in Shoemaker's tales. The curious and intriguing relationship between Clarke and Shoemaker is explored in detail in the introduction.

In his three decades in the lumber region, Clarke appears to have lived in several different locations within McKean and Potter Counties. Clarke bought property on Main Street in Austin, Potter County, Pennsylvania, from E. O. Austin on November 22, 1888. In the deed, he listed his residence as the Borough of Port Allegany, McKean County, Pennsylvania. He sold the Austin property to George B. Sharp in 1900, according to Potter County deed books. He also lived and had a studio at Joseph Hull's settlement (now Conrad) at the East Fork Valley of the Sinnemahoning in Potter County as indicated in the 1900 and 1910 census. He was a registered voter, listed in the Democratic poll book for Eulalia Township in Conrad in 1904. Some of his stereoviews at the Potter County Historical Society bear a rubber stamp impression, "W. T. Clarke, Photographer, Conrad, Potter Co.," on the reverse. He also apparently lived for a time at Cross Fork in the same county. His last Pennsylvania residence was in Betula, McKean County.

The Pennsylvania State Archives has nearly 470 images made by Clarke in the north-central part of the state. They were acquired in 1923 from the photographer by the Pennsylvania Department of Forests and Waters with the assistance of Governor Gifford Pinchot, state forester Joseph S. Illick, State Forest Commission member Henry Wharton Shoemaker, and W. W. Thompson of the then newly formed Potter County Historical Society. Thompson, a founder and former owner of the Coudersport newspaper the *Potter Enterprise*, was treasurer of the historical society. A page-one article appearing in the *Enterprise* on February 8, 1923, gave the story, with happy and sad details.

Linda Ries, working from the State Archives, initiated genealogical inquiries about William T. Clarke. She received enough information to ask Lois Barden, "Was there a 'Bessie Mayo' in the Barden family?" There was. Lois and Bob Barden knew her as "Grandma Mayo." Lois Barden's husband, Bob, was related to William T. Clarke. The photographer was distant family! Clarke likely left Pennsylvania in late 1917 or early 1918. By the time he returned to New York State, his youngest sister, Isabelle, probably had inherited the parental home at 5 Madison Square South in Rochester. City directories dated from 1918 to 1930 show him as living at the family home. Possibly Clarke's sister Isabelle needed his help and company after their mother died in 1916. Isabelle perhaps was in poor health. Although a decade younger than Clarke, she died more than two years before he did. Perhaps, too, his age and health were factors.[8]

We can only speculate on why Clarke left Pennsylvania. By the second decade of the twentieth century, the Goodyear Lumber Company had completed logging their lands in Potter and McKean Counties and the Goodyear brothers had moved their lumber business to Louisiana and the state of Washington. Some of Clarke's last

Pennsylvania images, dated November 1917, are of wood chemical plants in Betula and sawmills at Norwich. The loss of the area's largest employer and subsequent declines in population meant less work for supporting industries such as photography.

Initially, Clarke's occupation after moving back to Rochester was given in city directories simply as "helper." However, the 1930 census stated he was a "camera developer" and "photographer," suggesting he was again active in a business related to his life's career.

Grandma Mayo and her husband had a cottage near Honeoye Lake, south of Rochester. This may answer the question of how the crates of negatives were located there where Lois Barden found them in a shed. We can speculate that Clarke picked out some of his eight-by-ten-inch negatives from the thousands he'd made and took them with him to Rochester when he moved north from Pennsylvania. He chose to store the remainder of his life's photographs, the five-by-seven-inch negatives, in a Betula barn rather than take all the heavy glass plates with him. The negatives would have been awkward to transport. Clarke's storage choice was ill fated. The barn had a leaky roof. Eventually, the great majority of Clarke's negatives were destroyed. The Pennsylvania Department of Forest and Waters rescued the remaining undamaged five-by-seven-inch negatives, most dating from 1910 to 1915 (later becoming part of the Pennsylvania State Archives collection). This story is elaborated in the introduction.

Additional contacts with individuals, institutions, and libraries helped round out the information presented in this book. A 2006 article by the authors published in *Smithsonian* magazine led to many inquiries from readers, some asking for our judgment about whether their old photographs were by Clarke.[9] Whenever possible, we examined their vintage prints to make that assessment, using judgment and criteria developed over several years of examining and studying known Clarke photographs. We found one that we felt Clarke had made; with the owner's permission, this image was scanned from the original print and is included in a later chapter, along with additional information and comments the owner provided.

Seeing Clarke's original vintage prints for the first time at the Potter County Historical Society, Harry realized that Clarke had used an all-but-forgotten method to print the negatives. Clarke printed directly from his glass plates using "printing-out paper," commonly known by its abbreviation, POP. It was available in two formats, a gelatin chloride or a collodion chloride paper, both sold ready made by manufacturers. Clarke placed a sheet of the paper in an appropriately sized frame behind the negative. He then positioned the frame in natural light, where exposure took place. He could check the progress of an exposure without damaging the print because of the frame's construction, which kept the paper in exact registration with the negative while he peeked at a portion of the exposure in dim light to see if it was completed. If it wasn't, he closed the frame and returned it to brighter light until subsequent checks showed the print was exposed properly. The print then was rinsed, toned, fixed in hypo, washed, dried, and mounted on stiff card stock. Finally, it was ready for delivery to his customer.[10] The knowledge of this process, plus the opportunity to see original Clarke vintage prints, helped Harry reconstruct prints with the appearance he believed Clarke had achieved.

Harry, in a wistful mood one day, reflected on the Clarke photographs: "I wish I could describe the emotions these images evoke

. . . the faces so solemn and haunting . . . the bodies so rugged and fit . . . some hillsides so completely ravaged, while other images show a nearly pristine environment. As we printed these remarkable photographs, we became more and more curious about the stories that connect the people in them to each other, and through time, to us."

Lois Barden now devotes her time to fine arts photography, specializing in portrayals of working people in everyday environments and during special occasions, much as William T. Clarke did in his era. She is exuberant and obviously proud of her Clarke collection and the fact that she was responsible for saving his images from an almost certain trip to the landfill way back when. "I've saved something that needs to be remembered—photographs for people, some not even born yet, to gaze at and marvel over."

And marvel they have. Several local and regional photo exhibitions featured huge enlargements of Clarke's images of loggers and their communities.[11] The Pennsylvania State Archives hosted a major exhibition titled *Wood on Glass: The Lumber Industry Photographs of William T. Clarke* from November 1, 2010 to May 1, 2011. After it closed, the Pennsylvania Lumber Heritage Region acquired the display for a traveling exhibition, showing first at the Jefferson County Historical Society in Brookville in June 2012 and then at the Pennsylvania Lumber Museum near Galeton in Potter County. The Lumber Museum will receive the exhibit as an accession of its permanent collection. Critical reviews of the exhibitions have been heartening.[12] Several regional publications brought a small selection of Clarke's photographs to focused audiences, who responded positively.[13] As with the *Smithsonian* article, regional viewers and readers contacted the authors with questions and with family album images from the same era and same places. Others shared letters and memoirs that had been exclusively in their families and never published. Two such memoirs, one by Aden Byron Albertson, writing of the Nordmont, Sullivan County, area, and the other by Leo Bailey, who grew up in Leetonia, were poignant, entertaining, and informative. We quote anecdotes and observations from the memoirs with their descendants' permission in chapters that follow.

Prints of Clarke's images from Harry Littell's digital files are sharp and precise, depicting details such as musical instruments, children's toys, specialized tools, and the spikes in the loggers' calked boots. The inherent beauty of modern digital prints from fragile negatives more than a century old is a tribute to Harry, Lois Barden, and the photographer who made the originals.

As an itinerant photographer, Clarke lugged his heavy photographic equipment and glass plates from place to place throughout north-central Pennsylvania for nearly thirty years in a quest to document the logging and lumber industry and to earn a living. He would no doubt be amazed and pleased that his work is still being viewed and appreciated. Through the diligent efforts of many people who have brought his images of the logging and lumber industry to appreciative audiences, the ghostly faces that stared back at Lois Barden in a damp shed on a cold, wet day decades ago have achieved an immortality of sorts.

Scottish poet and essayist Alexander Smith's time on earth (1830–1867) briefly overlapped William T. Clarke's (1859–1930). In his posthumously published *Dreamthorp*, Smith addressed the importance of human beings to themselves, arguing that in literature and art, it would be well if a person were known by something other than his or her name alone. In art, for example, an individual should be known for his or her personal outcomes, "the only thing truly valuable," because individuality produces works that are peculiar to the person, not something that is interchangeable with or held in common with other artists. If an artwork has such a peculiarity, it has the elements of and potential for endurance because it will express "a human interior."[14]

William T. Clarke would not be remembered today had he not produced such art through his photographs. Indeed, in his selection of subject matter and place, he was working in a documentary genre and in a locale that few other photographers chose. Moreover, Clarke's surviving work is of value to today's viewer, not only for the window he presents on a world that no longer exists but also for the individuality brought to his task and the expressions of documentary consciousness and realism revealed.

What type of people were loggers? What were their jobs? When did they work? What did they do when they weren't working? Where did they spend their time? Why? How? With what effect? Clarke's photographs provide at least partial answers to all those questions. To some extent, each chapter addresses all the questions raised. For example, all the chapters examine a particular time period and motivating impulse (*when, why*). However, some chapters focus more than others on certain questions. Chapter 1, "The Black Forest," addresses the natural environment and *where*. Chapter 2, "The Machine in the Garden," gives information on *how*. Chapter 3, "Wood Hicks, Bark Peelers, and Other Woods Workers," gives special attention to *who* and *what*. Chapter 4, "Camp Life," describes *where* and *how* the loggers spent time when not working in the woods, thus examining facets of psychology and sociology. Chapter 5, "Community Life," discusses residential and business lives in nearby communities and the occasional forays of loggers into those communities (*who, where*). Chapter 6, "The Pennsylvania Desert," describes the aftermath of decades of clear-cut logging (*with what effect?*). And Chapter 7, "A Mighty Transformation," examines responses to the situation explored in Chapter 6. Thus, the reader will be exposed to the boom-to-bust days of the Black Forest to the Pennsylvania desert disaster and then will visit contemporary developments to celebrate the regeneration and transformation that provides a new lease on life in north-central Pennsylvania today (*when, with what effect?*).

We believe that William T. Clarke's imagery, produced over the evolution of boom, bust, and early response to disaster, provides illustrations leading to insights into the fate and fortune of north-central Pennsylvania better than words alone could convey.

Automobile innovator and industrial magnate Henry Ford (1863–1947) was quoted as saying, "History is more or less bunk. It's tradition. We don't want tradition. We want to live in the present, and the only history that is worth a tinker's damn is the history we make today."[15] However, Henry Ford was fascinated with everyday lives as lived in the past. Even though he professed a belief that history (as described in the history books of his time) was humbug, he was farsighted enough to assemble and reconstruct past architecture and historic sites of human ingenuity in Dearborn, Michigan. Ford's reconstructed eighty-one-acre Greenfield Village contains such gems as Thomas Edison's Menlo Park, New Jersey, laboratory and the Wright brothers' Dayton, Ohio, home and bicycle shop—where much of the design, testing, and manufacturing of the fledgling airplane concept and prototypes were done. Greenfield is today a major tourist attraction that celebrates American history, as does the adjacent Henry Ford Museum. Robert F. Bruner, dean of the University of Virginia's Darden School of Business, lauded those Ford creations as among "the best repositories of industrial history in the world."

Clarke and Ford were but two late nineteenth- and early twentieth-century contemporaries among people true to themselves who lived lives of peculiar individuality and who left "things truly valuable" after their deaths. Their works allow us to see and remember the past as we examine their accomplishments and those of others. That past, of course, is neither a wholly accurate and true record nor a clairvoyant blueprint to explain the world of today or to predict the future. However, that doesn't mean history doesn't matter. Mark Twain (1835–1910) reportedly said, "History does not repeat itself, but it does rhyme." No direct source for the quote has been located, but it does convey a middle ground that makes sense.[16] Historical occurrences and achievements reveal a glimpse into and a sense of what led to our own lives and perhaps may inspire us to shun or emulate the lessons we derive from history as we explore our own special vision and endeavors.

William T. Clarke remains an enigma. However, through his photographic legacy we get a view of logging in north-central Pennsylvania during the late nineteenth century and the early twentieth that suggests what his world was like and adds texture to our understanding and appreciation of the area and its residents past, present, and future.

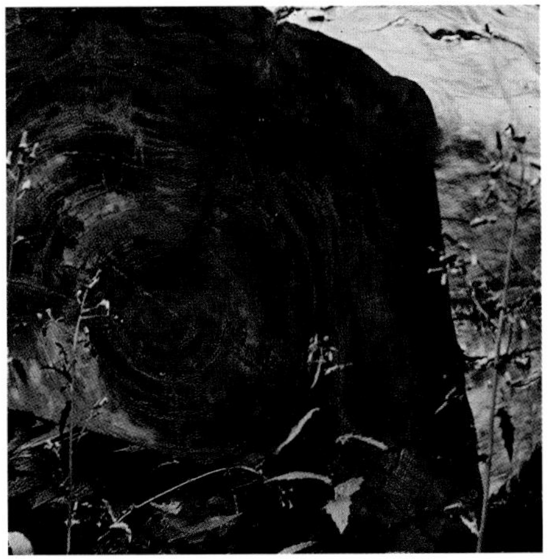

INTRODUCTION

The Salvation of William T. Clarke

LINDA A. RIES

We know very little about him. What we do know has been teased from civil records and contemporary publications. Only a handful of his letters survive. With such meager data, we can at least evince a slight figure, suffering from heart problems, arriving around 1887 from his native New York to ply his trade in the Pennsylvania woods. William T. Clarke stopped time and place for us before and after the turn of the twentieth century in north-central Pennsylvania. He created a precious visual legacy, enabling for us an appreciation of the landscape and people of that era. Yet he himself is a mystery.[1] To gain further perspective on his life, it is useful to describe Clarke's world and photography's place in it. Also, contemporary use of his images, his fascinating relationship with Henry Wharton Shoemaker, and the ultimate salvation of his works warrant additional study.

Like any technology, photography serves to enhance our lives. It is a tool, like a toothbrush or computer, to make existence secure and efficient. Photography's particular technological purpose is to capture and retain graphic information for later use: a means of creating a record. We now accept image making as a given, a ubiquitous means of documenting ourselves and our surroundings. In the nineteenth century it was a miracle and a revelation. Photography's commercial introduction in 1839 permanently altered the way humans perceived themselves and their world. For the first time in history, people could see what they looked like without gazing at a reflection or having their portrait painted, the latter something normally reserved for the wealthy. And for the first time, they could graphically record their world: homes, landscapes, anything piquing a curious mind. Images also could be shared with family and colleagues in distant locations. This is photography's greatest gift: people who cannot readily see a place or person for themselves can appreciate them through the recorded image. This was obviously useful in the immediate sense for the Victorian mindset: friends or relatives plying the gold fields of California or gone soldiering to the Civil War could share their likenesses and surroundings with family and friends back home. There is a deeper value as well to this sharing of visual information: the tool of photography is an extension of the human memory, enabling us to reach back many generations to times and places long past—a great and critical helpmate to the study of ourselves and our history. William T. Clarke did this for us.

By the time he entered the profession, photography was four decades old, well established worldwide as an industry, and accepted as a part of mainstream life.[2] It had been through several format and process changes, from the direct positives of the daguerreotype, ambrotype, and tintype in the 1840s and 1850s to the positive/negative process available commercially around 1855. Until the recent digital era, the photographic negative would dominate the industry. A single negative could produce an infinite number of positive prints on paper, which in turn could be widely disseminated. The earliest negatives had paper substrates, but when Clarke entered photography, they were produced on plate glass. A plate would be coated with a viscous light-sensitive collodion solution that the photographer had to mix and use immediately. While tricky enough to do in a photography studio, the process was compounded when making landscapes. Image makers prior to Clarke had to transport chemicals, the heavy and easily broken glass, and a wooden camera and tripod through the steep Pennsylvania hills often via a horse-drawn wagon; compose and make the exposure; then develop it on the spot in a small darkroom box on the wagon. After paper prints were produced from the collodion glass negatives, the glass was often scraped clean of its emulsion and reused. The degree of difficulty in execution, fragility, and reuse of glass accounts for why there are very few surviving landscape images before 1880 of the Pennsylvania lumber region.

Clarke entered the business in the late 1870s on the cusp of another revolution in photography, initiated by the gelatin dry plate negative. Beginning in April 1880, George Eastman marketed dry plates in Rochester, New York. In his factory, glass was precoated with light-sensitive chemicals suspended in a gelatin emulsion, dried, and shipped to the photographer in light-tight boxes, eliminating the need to formulate and apply collodion on the spot or to process immediately. Flexible film negatives, another technological breakthrough during Clarke's career, eliminated the need for heavy, bulky, and breakable glass, but they had an inherent problem: the base was made of cellulose nitrate film, chemically unstable and flammable. It is probable Clarke used both the earlier collodion process and later, nitrate film, but the chances of any of his negatives from these processes surviving today are unlikely.

Judging by the identified subjects and locations of his surviving negatives, Clarke made his living as an itinerant by traveling to remote farms, homesteads, sawmills, chemical factories, and lumber camps. Much of his income likely came from contracting with lumber companies, especially the Goodyear Lumber Company, based in Buffalo, New York, to make images of their logging operations in Pennsylvania's northern tier counties. The Goodyear brothers, Frank and Charles, were the last operators in that area at the turn of the twentieth century and they took down the final remaining stands of original hemlock and white pine. Clarke was likely also hired on an individual basis by local residents for various purposes; he photographed families standing before their homes, proud shopkeepers and clerks at businesses, and workers at their camps, shops, and mills. Clarke's activities in this regard were reported periodically in the local papers: for instance, the *Potter County Journal* listed under "Costello News" for February 12, 1891, that "Mr. Clark of Austin was down the other day & photographed the school of L. D. Ripple." In this era, making a photograph was a reportable event.[3]

Clarke was not the first to photograph the lumber business in Pennsylvania, nor would he be the last. Several figures deserve mention. Some of the earliest known images of logging in northern Pennsylvania are by J. W. C. Floyd (ca. 1853–1930) of Lock Haven, who made images of the G. B. Merrill Lumber Company operations in Penfield, Clearfield County, in the 1880s. Godfrey Hess (1844–1903) and his apprentice Albert F. Zimmerman (1863–1938) often did portraiture, but they also photographed logging operations in Jersey Shore, Lock Haven, English Center, Eagles Mere, and their hometown of Williamsport, the locus of the state's lumber region. Also based in Williamsport, Francis Edwin Bradley (1870–1943) made images for the Central Pennsylvania Lumber Company and for the Susquehanna and New York Railroad, which served lumber interests in the region. Nelson Caulkins (1874–1965) photographed timbering around Pine Creek and its tributaries from studios in Galeton and Waterville in the early 1900s, producing his photographic prints primarily on postcard stock. Sadly, Caulkin's home burned in 1956 and along with it his priceless collection of negatives. George Clark (1880–1967), no relation to William T. Clarke, began his career in Austin but lost his business in the disastrous Austin Dam flood of September 30, 1911. He later established a studio in Galeton. In Bradford County, C. H. Krumm photographed the lumber operations at the now ghost towns of Laquin and Barclay.[4]

Why did these men make these images? Like Clarke, they practiced photography as a livelihood. They made images to please their clients. Today we are tempted to think of their extant work through the lens of "art," for in terms of composition they surely are in many cases. We should be mindful that the original purpose was to fulfill the terms of a sales agreement and please the sitter or contractor. Composition was secondary to the main task, but Clarke had a wonderful eye for it. A typical landscape photographer's method of filling the image frame of, for example, camp life was to show a group of people in the foreground, the buildings behind. Clarke did this as well, but he seemed to enjoy adding a vanishing-point perspective,

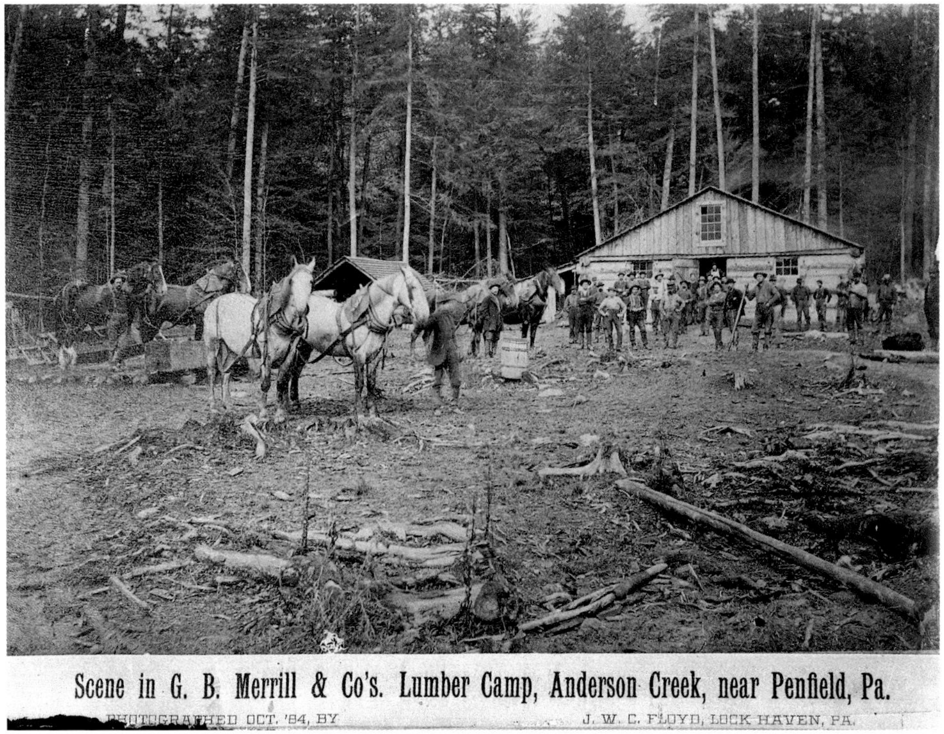

FIGURE 3
J. W. C. Floyd, "G. B. Merrill Logging Operations near Penfield, 1885"

placing his people and structures in the middle of or near to a road or a railroad track, and having it wind off into the distant right or left. This pattern can be seen through many examples in this volume (for example, plate 48), and the unidentified logging camp with a gravity slide (see plate 30).

Photographers were also proactive in selling their wares. Local events, especially disasters, whether the Battle of Gettysburg or the Johnstown Flood, were opportunities to make and sell images. In the lumber region, any logging or rafting event or train wreck or other catastrophe—notably, the Austin flood of 1911—would be fair game. Clarke was no exception. His body of work encompasses the aftermath of the flood at Austin, train wrecks along the Buffalo and Susquehanna Railroad, and the swift dramatic rise and fall of the now ghost towns of Betula and Norwich. He is different from his fellow commercial photographers in that he did not seem to specialize in studio portraiture. In the late Victorian era, photographers printed likenesses as cabinet cards, a common and popular format for portraits. Usually these bore the name of the photographer on the front or back as an early form of branding. There seem to be no such extant photographs bearing Clarke's name or logo. Why did he not do this, especially for one likely suffering from heart problems? Portraiture was an easier means of income than the rigors of photographing lumber camps. Henry Shoemaker suggested that it was on "doctor's orders" that Clarke was to get out and into the hills. We may never know.

The bulk of the surviving Clarke negatives are of landscapes showing the lumber industry. Some seem to be deliberately made as a series for the Goodyear Lumber Company or other companies, visually explaining how timber is cut, transported to sawmills, sawed into boards, and shipped to market. The fact that workers are deliberately posed, as in, for example, a tree notcher near the end of his job (plate 23), suggests that Clarke was producing images for a company publication or promotional campaign. This hypothesis is supported by the explanatory captions for a grouping of negatives in numerical sequence (figs. 5–7) showing timbering along Nine Mile Run. "After the Crash to the Ground" and "Making Logs" (WTC 3822) were made within moments of each other. "Sawing the Tree into Logs, Nine Mile, Potter County" also belongs to this series. Some of his negatives were made with a stereoview camera, a device producing two slightly offset images on one plate; when the images were printed and placed beside each other and viewed through a stereoscope, a three-dimensional effect was produced. Viewing theme-related series of stereos was a wildly popular Victorian pastime. It was lucrative for photographers to photograph far-flung scenes and put them on stereoview cards. We do know Clarke made and sold stereoview cards, and, as surviving ones at the Potter County Historical Society attest, some were part of a series on the Goodyear-owned Buffalo and Susquehanna Railroad. Whether stereo or not, all the dated negatives now at the Pennsylvania State Archives include the Goodyear operations in and around Potato Creek and between Keating Summit, Norwich, and Betula between 1910 and 1917. Other groupings include Austin before and after the 1911 dam break and logging in the Hammersley region of northwestern Clinton County. Homes and businesses in the East Fork Valley of the Sinnemahoning Creek in Potter County are well represented, especially the area around Jamison, Logue, and Hull's (Conrad), where Clarke made his home. There do not seem to be any existing Goodyear publications using these images. However, a lithograph of a locomotive on Rich Lumber Company Keystone Railroad

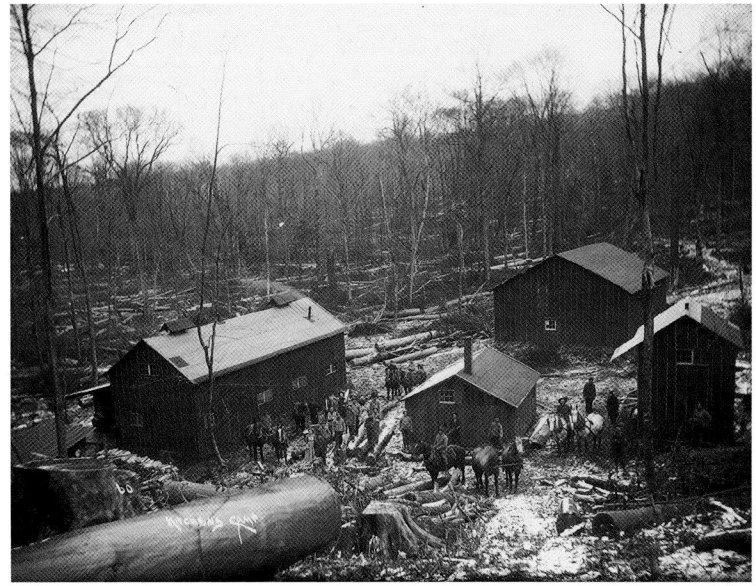

FIGURE 4
Nelson Caulkins negative no. 60, "Kitchens Camp"

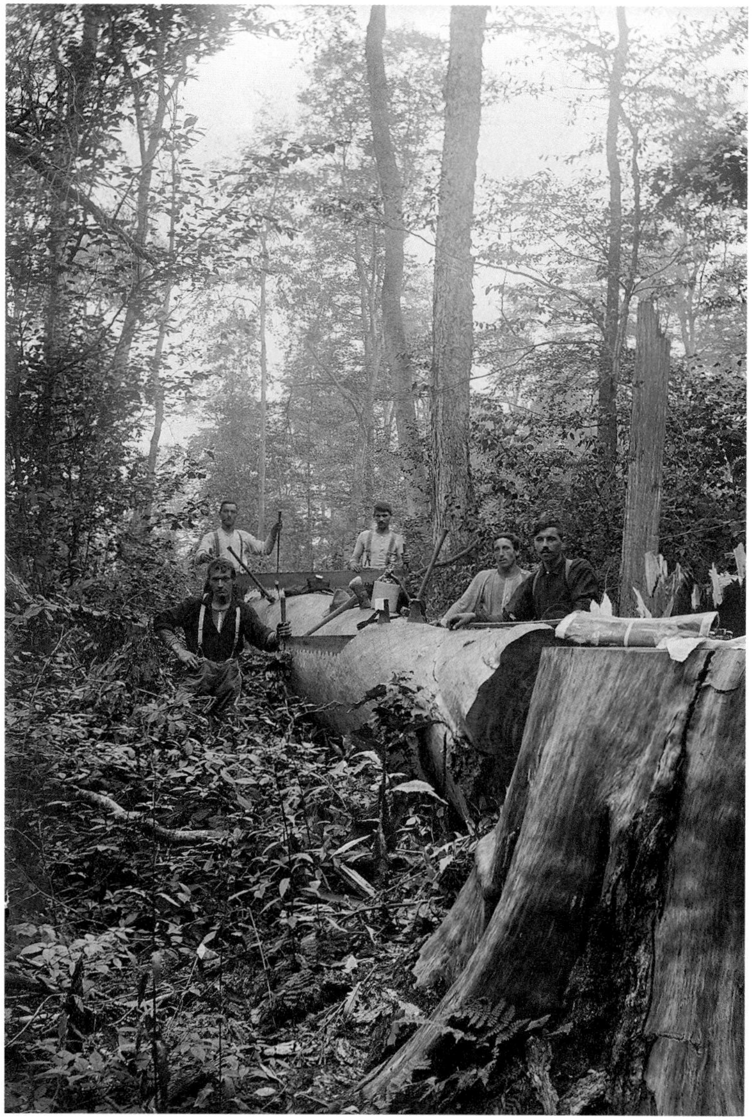

FIGURE 5
"After the Crash to the Ground,"
Nine Mile, Potter County

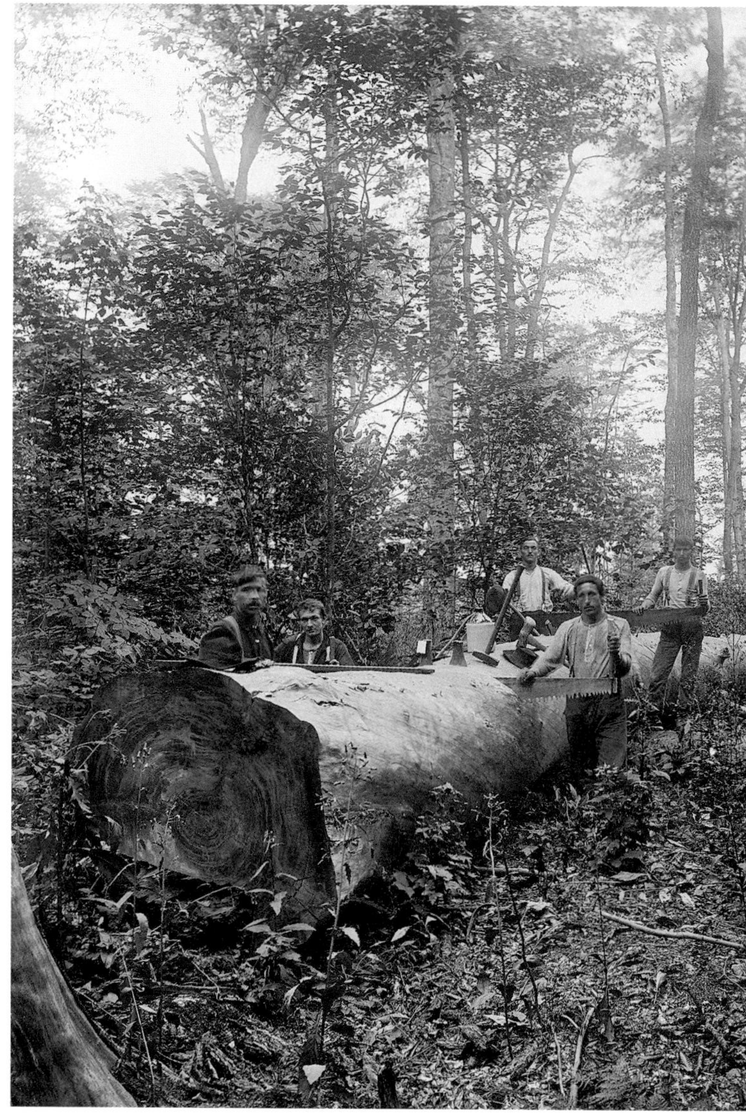

FIGURE 6
"Making Logs," Nine Mile,
Potter County

stationery (see plate 73) was copied from a Clarke photograph (see plate 10), a shining example of corporate use of his images.

Clarke also made and sold landscape images to noncommercial interests and individuals, some of the images being published in newspapers, magazines, and books. The halftone printing process, a fast and cheap method of mechanically reproducing photographic images, was beginning to be widely used in commercial publishing at the turn of the twentieth century.[5] In the 1890s American newspapers began featuring halftones relating to newsworthy people and events, for pictures on a printed page would draw the reader's attention, as they still do today. Clarke likely supplied some halftones to Potter and McKean County papers for local features and to help sell issues. In addition to such use, and worthy of note here, his timbering images also were seized on as a powerful tool for reform by Progressive Era crusaders, especially those trying to stop the destruction of the Pennsylvania forests by lumber interests. We now call these *documentary* photographs, those which dramatically underscore an author's intent to sway the reader to a specific point of view. For example, Clarke's image of a grand giant hemlock in the forest along Indian Run near Norwich (see plate 1) was captioned as "The Forest Primeval—the Black Forest" and featured as the frontispiece for a 1916 Pennsylvania Wild Life League pamphlet, which argued for increased legislative appropriation for managing the Commonwealth's forests.[6] The not-so-subtle implication that the magnificent old hemlock and its companions were doomed would not be lost on readers.

The most notable contemporary use of Clarke images as documentary tools for reform was by Henry Wharton Shoemaker (1880–1958), a dominant figure in Clarke's later history. From a wealthy family, Shoemaker spent summers at Restless Oaks, a property owned by his mother's family in McElhattan, Clinton County. There he witnessed firsthand the timbering of the breathtaking old-growth stands of hardwoods, white pine, and hemlock, and the land's subsequent abandonment by logging companies with no thought of replanting. Deeply affected by this experience, he became, among other things, a proponent of environmental conservation. Nostalgic yearning for this time, place, and a people forever gone became a

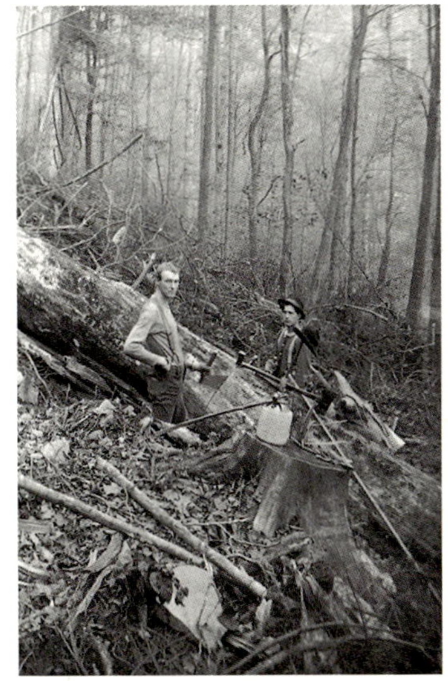

FIGURE 7
"Sawing the Tree into Logs,"
Nine Mile, Potter County

PLATE I

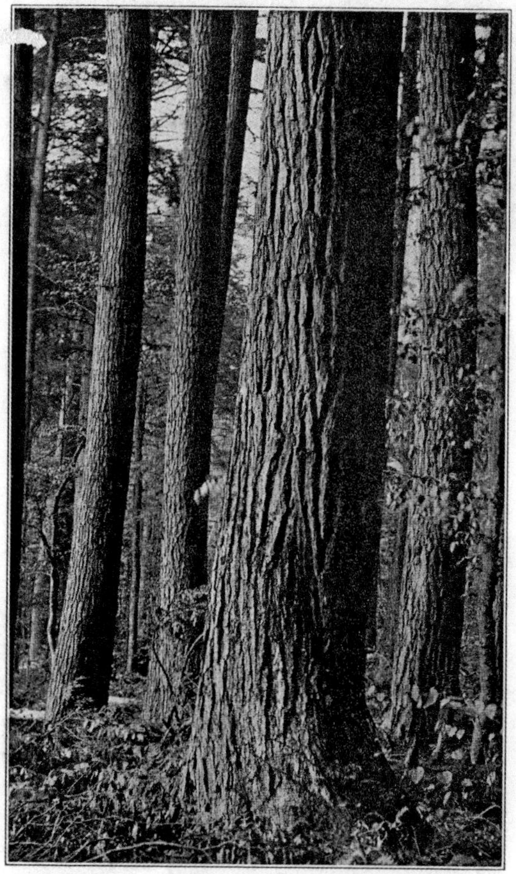

THE FOREST PRIMEVAL—THE BLACK FOREST

The lumbermen cut down the trees but intelligent forest practice will restore their successors, of equal size and quality

FIGURE 8
Frontispiece of Albert Heck's Wild Life League pamphlet, *The Black Forest of Pennsylvania* (1916)

recurring theme in Shoemaker's works. A true Renaissance man, he was owner and publisher of the *Altoona Tribune*, U.S. minister to Bulgaria, state folklorist and archivist, collector of Pennsylvania folktales, and prolific author. He promulgated myths in regard to Princess Nita-nee, the Nittany Lion, the Indian Steps in Huntingdon County, and many others as a champion for the history and culture of the Commonwealth. He also may have coined the term "Black Forest" of Pennsylvania. Scholars criticize him, and rightly so, for fabricating or altering facts to enhance the drama and passion in stories he collected or wrote. Historian and folklorist Homer T. Rosenberger stated that when Shoemaker "took pen in hand it ran away with him," going on to add, "Had Henry W. Shoemaker done more careful writing and had he let his readers know what part of his writing was fact and what part was fiction, his writings would be useful to the historian today."[7] For us, Shoemaker has unfortunately marred the historical William T. Clarke.

Shoemaker used a number of Clarke images in his books over a ten-year period.[8] The power of publishing Clarke's logging images was not lost on Shoemaker, who wisely understood photography's ability to graphically advance his Ruskin-like theme: the loss of the veritable Eden that was Penn's woods. The book *Black Forest Souvenirs*, for example, illustrated with halftones made from twenty Clarke images, depicts life in the forest, lumbering activities, and the like. Shoemaker assigned them generic captions, such as "Black Forest Camp Life" and "Bark Peelers at Work." Interestingly, and oddly, the images he chose have little or no direct bearing on the stories. Shoemaker likely needed them to break copy and decorate pages—the desire to attract the eye with an image and an inducement to read the text. The same is true for the volumes *Eldorado Found* (in this case the Eden, or Eldorado, is the Seven Mountains) and *The Indian Steps*. Shoemaker also manipulated captions for Clarke images to reflect disdain for the lumber companies, often referring to them in text as "the Despoilers." A Clarke view of lumbermen posing with logs in *Eldorado Found* is titled "The Despoilers Again. A Group of Black Forest Loggers."

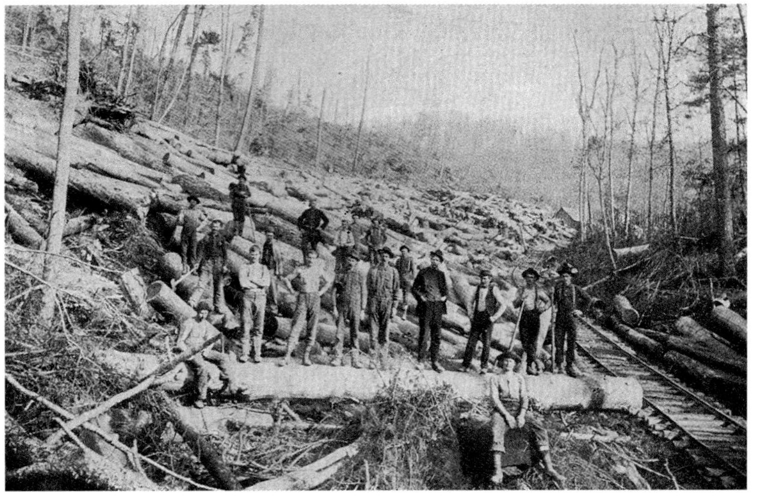

FIGURE 9
Clarke photo opposite page 78 in *Eldorado Found*, captioned by Henry W. Shoemaker as "The Despoilers Again. A Group of Black Forest Loggers." Betula, McKean County

Shoemaker and Clarke seem to have shared the same feelings about the loss of the original forests. This is demonstrated by a Clarke letter to Shoemaker, dated August 12, 1912. It discusses types of images available to Shoemaker for his books and is especially poignant, revealing bits of Clarke's life and business:

> Conrad, Potter Co. Pa. Aug 12, 1912
>
> Mr. H. W. Shoemaker
> Riverside Conn.
>
> Returning here Saturday I found your latest awaiting and note remarks of your book about ready from the publishers I should have been much pleased if you could have had others to go with those you had on hand. However I am glad that you had some prints to make use of. Perhaps later you may be able to use others that I will soon forward—Most of them may be 5x7 prints as I have been making such size of late years. The ones you have are from 8x10 negatives. Think you will be much pleased with the assortment I will send. (Negatives) subjects of rafting are about the only ones I have not on hand. Some may be smaller than 5x7 being Steroscopic & right here. If you have a scope you will want them for what may well be of more pleasure than to sit & look at those scenes just as they were? Would it be much troubl to forward me a list of the subjects you have I think I annexed title to prints when I sent them & now forget what ones. Certainly will be pleased to receive the Copy of the Work you mention and hope for you the very best of success. Before closing it no doubt will be of interest for you to know that the Goodyear Co. operations are now on Potato Creek over in McKean Co. Railroad starting at Keating Summitt crossing Penn R.R. tracks near Liberty going up Saoffel [?]

> lick through the old Liberty works and also the Riches (Gardeau) works Railroad passes at the mouth of old Gardeau Coal Mine going on down through timber until new towns of Norwich and Betula are reached about 14 miles from Keating Summitt and 6 miles from Colegrove up stream At Norwich are located Saw Mill, Kindlewood, & R.R. Shops At Betula one mile from Norwich is a Heading and Stave Plant with Wood Alcohol plant being constructed I mention these places as I think you know the location When I made first trip of the R.R. road it simply was a ride back to former days when I was much a young man haveing been on some of the runs in 1887 As you mention the last of the hill forest are about gone And this is the last of it (Around this vicinity) Even at that what a hurry for the fastest mill ever run in this country is now eating up the trees at a rate of 275,000 to 300,000 per 24 hours Why? When the Hemlock can not last there more than 7 or 8 years at most Emporium Co. expects to be done with hardwood at the Summitt by spring also at Austin in Year or two Remember I will be at your service to assist if possible such as you may wish etc
>
> Very Truly Yours
>
> W. T. Clarke[9]

The letter confirms and provides interesting details about Clarke's business: that Clarke was working in a five-by-seven-inch format and had been earlier working in the larger eight-by-ten-inch size, and that he had been in the area by at least 1887. He says he can send Shoemaker "an assortment," suggesting that he kept a stock file of his negatives to be printed on demand. He tries to interest Shoemaker in purchasing some stereo images as well. The poignant coda at the letter's end is especially notable. After describing the speed and efficiency of the new sawmill at Norwich, he questions, "Why?" It is the only written expression we have of Clarke's emotions regarding the transformation of the forests. It proves he recognized he was a witness to their destruction and lamented their passing and must have realized the images he made were more than for just immediate use; they recorded a significant event. He knew he was creating, through photography, a legacy, an extension of the human memory for this time and place.

In a 1923 letter, Shoemaker claims to have been corresponding with Clarke for twenty-five years, which would have taken their relationship to 1898.[10] While Clarke probably considered Shoemaker a customer and a friend, Shoemaker's relationship with Clarke was a bit more complicated. Shoemaker viewed Clarke as more than a photographer plying his trade and lifted him through his writings to a more dramatic literary plane. Clarke himself, and his images, were springboards for furthering Shoemaker's theme of forest and lifeways lost. The photographer's life of rambling about the Pennsylvania woods inspired Shoemaker to fictionalize him into a muse for at least two stories. In both, Shoemaker presents Clarke as the "Old Mortality" of the Pennsylvania woods, a wistful gadabout, photographing what intrigued him. This appellation is worthy of note, for Old Mortality is the nickname of Robert Paterson, a real-life eighteenth-century Scot who spent forty years wandering about the lowland hills cutting and carving gravestones for old Covenanters. Paterson is also a character in a novel of the same name by Sir Walter

FIGURE 10

Folklorist Henry Wharton Shoemaker wrote that General of the Armies John Joseph Pershing (1860–1948) and Clarke were look-alikes

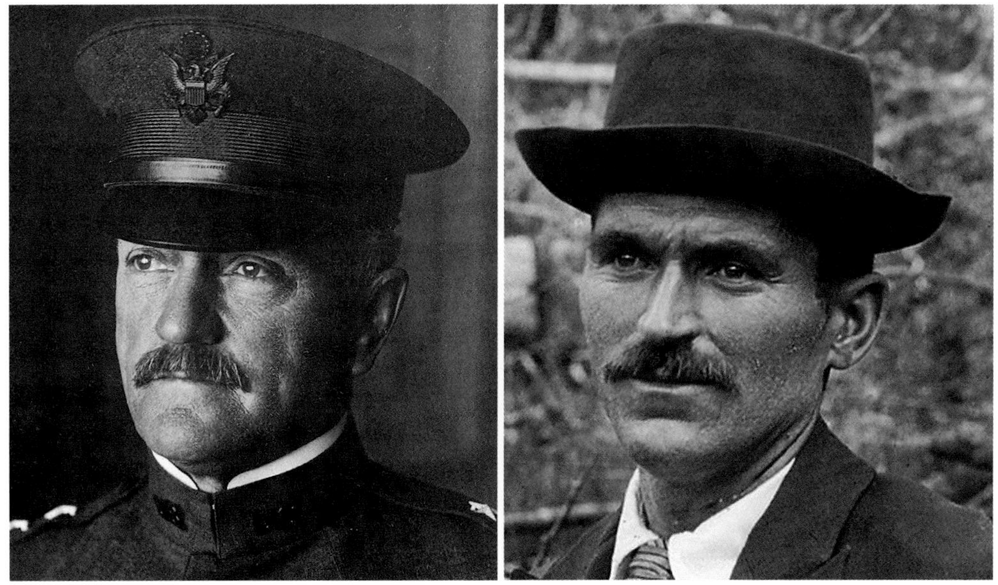

Scott.[11] Some of this romantic comparison can be considered true, for there are extant Clarke images obviously taken for sheer interest and amusement. Also, in both tales, Clarke is the narrator of a story within a story, and his photographs are key to the plot, being catalysts for advancing the tale. "An Antique Dealer's Romance" appeared in 1931, a year after Clarke's death, in *Some Old Deserted Houses in the Central Pennsylvania Mountains*. In "An Antique Dealer's Romance" Clarke is both narrator and character, telling the story of a Chinese immigrant, Charley John, who is prompted to tell his tragic story after viewing a Clarke photograph of a young woman from the lumber region. The second story, an unfinished manuscript titled "Women's Part in Pennsylvania Lumbering Days," is undated. Similarly, in "Women's Part," a young man becomes obsessed with a Clarke image of a beautiful mountain woman doing a man's job of spudding bark.

For both stories, Shoemaker exaggerated Clarke's backstory. He portrayed Clarke as being from "Rasselas," not Rochester, New York, serving in World War I in the U.S. Signal Corps ("selected by George Eastman himself"), attaining the rank of major, and being awarded the Croix de Guerre by the grateful people of France. Investigation into New York and Pennsylvania military records have found no evidence to support either his service or his receiving the Croix de Guerre. In 1917 Clarke would have been fifty-eight, too old to serve, and with probable health problems, not fit for duty. A place named Rasselas does appear in Pennsylvania, but not in New York.[12] One wonders why Shoemaker waited until Clarke's death to include him in stories, perhaps as tribute, perhaps to permit embellishment of character. Perhaps both.

Shoemaker's relationship with the photographer is paramount in the final chapter of Clarke's story, for it is one thing to create photographs and quite another to preserve them. Sometime after Clarke's return to Rochester in 1917, Shoemaker realized the significance and potential use of Clarke's body of work for the Commonwealth—it was a valuable means by which state government could illustrate the loss of the old-growth forests and further advocate the need for wiser management. By this time Shoemaker was a member of the Pennsylvania State Forest Commission, headed by Gifford Pinchot. The Forest Commission was assigned the large task of creating Pennsylvania's system of parks and forests out of land forsaken in part by lumber companies and other natural resource extractors. As America's first professionally trained forester, Pinchot had earlier established the United States Forest Service with President Theodore Roosevelt and served as its first director. No doubt Shoemaker received Pinchot's enthusiastic approbation to seek, find, and save Clarke's photographs. It is possible efforts to do this began as early as 1918, when Governor Martin Brumbaugh appointed Shoemaker to the commission. The hunt was still on in 1922, for a May 22 letter to Shoemaker from A. C. Vorse, the chief of the Office of Information, Department of Forestry, stated, "District Forester Port believes he is on the trail of W. T. Clarke."[13] Through the combined efforts of Shoemaker, other state officials, and W. W. Thompson of the newly formed Potter County Historical Society, Clarke, and his negatives, were found in early 1923. Clarke had said his negatives were left in a barn near Betula, his last Pennsylvania home. Upon discovery, however, the barn had leaked sometime between 1917 and 1923, and many of the plates were rendered unusable.

Shoemaker gave alternate accounts of how the negatives were salvaged. In "An Antique Dealer's Romance," he wrote of finally meeting Clarke in February 1922 at a barn in Norwich where the negatives were stored, three thousand in number. In "Women's Part," he states that he and Joseph S. Illick, the Department of Forestry's chief of research, met Clarke in Betula in February 1923, the negatives numbered five thousand, and fifteen hundred were in good condition.[14] A subsequent letter from Shoemaker to Clarke, of January 18, 1923, thanking Clarke for the plates, describes their meeting; this is presumably more accurate than the above accounts. They met at Richardson's Drug Store in Betula around January 14 and spent an evening talking and telling stories. This was the only time the two ever met face to face.

Two different accounts in Potter County newspapers claim that 300 and 150 plates, respectively, were salvaged, with businessman W. W. Thompson assisting in the discovery.[15]

> SEARCH FOR PHOTOGRAPHER SUCCESSFUL
> State Officials for Years Try to Locate W. T. Clark, Woods Photographer, Who Worked in Northern Pennsylvania All During Lumbering Days. W. W. Thompson of Coudersport Finds Him in Rochester. Has Many Valuable Pictures.
>
> W. W. Thompson of this place is much pleased that he has finally been able to locate W. T. Clark, who for years was a photographer in Potter County. Clark made a specialty of

pictures of lumber camps, log trains, woodsmen, etc., and since these views are no longer obtainable his old negatives have great historical value. Mr. Clark, after leaving Potter County went to Betula, McKean County, where he continued his profession until lumbering was on the decline. He then went to Rochester where all trace of him was lost until a few days ago when a letter from Mr. Thompson reached him. An appointment was made with Clark by State Officials who met him in Betula. The wanted negatives had been stored by Clark in the basement of a barn and most of them had been ruined by water coming through a leaky roof, but some 300 have been found unharmed and copies of these will be saved for the State records. Mr. Thompson for his part in the hunt for the Photographer has requested that a complete set of the pictures be presented to the Potter County Historical Society.

An additional version is from an undated newspaper clipping at the Potter County Historical Society:

A VALUABLE FIND

Colonel Shoemaker, who has been active in a search for pictures taken by W. T. Clark, a one-time photographer at Cross Fork, has at last succeeded in locating Mr. Clark at Rochester, and finds that he has the negatives of a large number of pictures taken years ago of lumber camps, log trains, woodsmen and other scenes connected with lumber operations. W. W. Thompson, who has also been active in trying to secure these pictures, received word from Mr. Shoemaker that he thinks about 150 negatives can be saved, a good many having been destroyed when stored away in the basement of a leaky barn. It is expected that a set of these pictures will be secured for the Potter County Historical Society, through the efforts of Mr. Thompson.

We will never know the exact number of original plates discovered. And what of the plates not salvaged, and left behind? The ones considered unusable? Probably Illick and Shoemaker wanted only ones directly relating to the lumbering industry. What about plates Clarke may have made of homes, businesses, and events in the region? These subjects are few in number compared with the existing logging views. This again is part of the Clarke mystery.

Illick, in the Norwich and Betula area on business for the department in January 1923, picked up the plates after the Clarke-Shoemaker meeting and took them to Forestry headquarters in Harrisburg. There they were cleaned, housed in new envelopes, captioned, printed, and added to the agency's photograph library, which also included glass negatives by the "father of Pennsylvania forestry," Joseph Trimble Rothrock (1839–1922), Illick, and others. In the 1960s they were transferred to the Pennsylvania State Archives, where they remain. About 470 Clarke plates currently are archived there. It appears the Department of Forestry made a set of prints for the Potter County Historical Society, for the society has a list of at least seventy-two Clarke prints matching captions with ones in Harrisburg.[16]

After their acquisition by the state, the surviving plates were immediately promulgated to illustrate publications regarding the northern tier counties and the lumber industry, especially by the Department of Forests and Waters.[17] In June 1923, only a few months after the plates' salvation, Pinchot, now governor of Pennsylvania, featured Clarke halftones for an article in *American Forestry* about timber depletion, among them a 1917 image of the Norwich sawmill and, once again, Clarke's giant hemlock, previously used by the Wild Life League.[18] Clarke's images later graced the pages of Victor L. Beebe's *History of Potter County*, published in 1934 by the Potter County Historical Society. Over the years the society also included Clarke's photographs in their newsletter, their journal, and a 1976 bicentennial publication.[19] His images were notably part of a series of studies done in the 1970s of the logging industry by Thomas T. Taber III, Benjamin F. J. Kline Jr., and Walter C. Casler.[20] They have also appeared in displays in historical societies and museums and orientation areas of state parks and forests of the region as educational tools about the logging era.

Despite his literary flights of fancy, Henry Shoemaker is to be praised for his part in helping to facilitate this last chapter of Clarke's life: the salvation and preservation of his legacy. Although Clarke stopped time and space for us with his images, Shoemaker ensured they would be around for us to study and enjoy. W. W. Thompson and Joseph S. Illick deserve credit as well.

We now know that Clarke took along a selection of his negatives when he returned to Rochester—over 130 glass plates. We have Lois Barden to praise for ensuring that this second grouping of Clarke's work lives on. As with the salvaging of the negatives from the leaky barn, she and Harry Littell spent much time cleaning, cataloging, and printing them. Lois generously has made them available for all to study and enjoy. Now, a total of around six hundred original negatives are therefore known to exist. Numerous vintage prints are in private hands and at historical institutions like the Potter County Historical Society. Examples are lovingly reproduced in these pages; more doubtless will be discovered in the future.

By 1920 almost all of Shoemaker's "Black Forest" of north-central Pennsylvania was gone, the lumber people had departed, and communities like Betula and Norwich had become ghost towns. The record Clarke created through the tool of photography is precious indeed, his unique images testifying to a time and place gone forever. Although Clarke himself will always remain an enigma, his photography, an extension of human memory, has enabled us to visit the distant past and see original portions of the Pennsylvania forests before, during, and after their demise. And thanks to the vision of people like Shoemaker, Rothrock, Illick, and Pinchot, who used Clarke's images as leverage for their advocacy, many of the areas Clarke photographed in the Black Forest are now reforested. As the region today again undergoes another transformation, this time for natural gas extraction, the preservation and appreciation of images of north-central Pennsylvania are even more imperative. A century ago, William T. Clarke's camera bore witness to a metamorphosis of the Pennsylvania landscape. We are the beneficiaries of his work. His legacy may be that another photographer will be inspired to take up the lens for the next transformation.

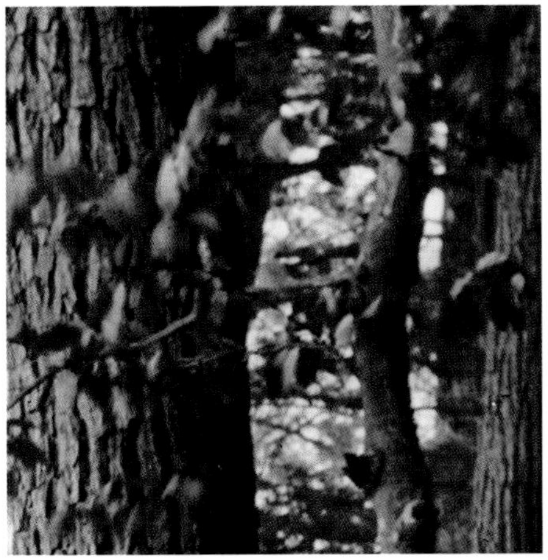

THE BLACK FOREST

CHAPTER 1

William T. Clarke's photographs of the untouched forest depict the legacy of another William—William Penn (1644–1718), the founder of Pennsylvania. Penn became a man of property and influence after his father's death in 1670. He petitioned England's King Charles II for land in the New World. Charles II had long owed a debt based on a loan from Admiral Sir William Penn, William's father. William proposed the transfer of forty-five thousand square miles of land (almost as large as England itself) from the Crown to him as a way to discharge the king's obligation. Penn, a Quaker, was keen to open a new land of refuge to believers of his persecuted faith and other religious groups. He wanted to name the land Sylvania, a Latin-derived choice suggesting a grove or woods.[1] Charles II agreed, but on granting the charter on March 4, 1681, the king insisted that the name Sylvania be prefaced with "Penn" to honor Penn's father. Pennsylvania (literally, "Penn's woods") became a colony with William Penn as proprietor.

Penn's woods were huge. The original forest covered between 25.8 million and 27.2 million acres, largely untouched by ax or saw. The Allegheny Plateau region of north-central Pennsylvania showed no sign of aboriginal burning or clearing.[2] In William Penn's time, trees covered 90 to 95 percent of Pennsylvania, an estimated 500 billion board feet of lumber, a supply thought inexhaustible.[3]

Although the land had not been properly surveyed, Penn sold huge sections of rural land to settlers, mostly farmers, for but a few pence an acre.[4] Generally, the early pioneers were wary of primeval forest. Many feared the solitude and loneliness, the dangers, the unknown, and the diminished human control over the environment. Early European settlers came from countries where agriculture was the dominant occupation and where most forests had long since been cut. The settlers generally intended to become farmers in the New World. Rather than seeing the New World's forests as a valuable resource or an economic opportunity, they thought trees were obstacles to be eradicated. One new settler proclaimed of Pennsylvania, "It is a very fine Country, if it were not so overgrown with Woods."[5] Clearing away trees; eliminating brush; and removing boulders, rocks, and stones were necessary before agriculture could be practiced. Their response to being confronted by the forests was to work hard to get rid of them and to use the wood, which they thought necessary in order to survive and to build a life for themselves and future generations. Most wanted a pastoral, rural, self-sufficient, and community-based existence. They wanted to lessen their labor, to enhance their quality of life, and to improve their security. They labored to tame, cultivate, control, and (in their estimation) improve upon the wilderness they faced.

Henry Wharton Shoemaker coined the name Black Forest to refer to the dense oak, white pine, and hemlock of the north-central Pennsylvania forest. He claimed that the term originated with early German settlers who were reminded of their native Schwarz Wald, or Black Forest.[6] Pennsylvania logging began in the 1680s on a modest basis after Penn and new settlers arrived. As more and more forest passed into private hands during the eighteenth and nineteenth centuries, sounds of the ax and crosscut saw resonated through Pennsylvania's mountains and valleys. Logs were needed to build and heat houses. Outbuildings and fences were needed for crop storage and for sheltering animals. When it was not convenient to cut trees with ax and saw, the land was cleared for efficient farming by burning.[7]

Pioneer families soon realized that logs had use and value beyond their small family farms. The growing population led to numerous small local sawmills that converted logs to lumber. For example, it is said that all "forest-farmers" on the upper reaches of the West Branch of the Susquehanna supplemented agricultural income by cutting logs from their Allegheny Plateau land in the winter. Logs were sawed into lumber by the farm's own water-powered sawmill or moved to a larger river and floated by one or more timber rafts to local sawmills. "Every farmer ran one timber raft at least. . . . It was the timber trade that made the pioneer prosperous."[8] Landowners found that livelihoods and, indeed, a good deal of money could be made from the forest. The flow of lumber, boards, shingles, and ship timber down the Susquehanna River as early as the 1790s made possible an upriver return flow of store goods and supplies. An important market for logs was the iron industry, the first large Pennsylvania business in the eighteenth and nineteenth centuries. To feed the fires of the iron forges, a huge section of the forest was cut down, especially in the southeastern part of the state. By the mid-nineteenth century, large numbers of people were settling Pennsylvania, mostly German and Irish immigrants. Urban hubs were expanding rapidly. Those population centers required wood in vast amounts. Logging timber to be used in large-scale building construction and manufacturing began in the 1840s. The era of impressive lumbering operations was in full swing by 1860, when historian George Korson estimated there were fifty thousand lumbermen.[9] By that time, Pennsylvania led the nation in lumbering.

In the early 1880s, observers described the pine belt of northern Pennsylvania with awe. The area included Susquehanna, Bradford, Tioga, Potter, Cameron, Elk, and Clearfield Counties. A vast primeval forest, dense with white pine and hemlock trees, existed throughout the twenty-five-thousand-square-mile drainage area of the Susquehanna and its tributaries.[10] The white pine forest contained trees that were from two hundred to four hundred years old. A single big tree produced enough lumber to build a six-room house.[11] Trees were said to tower as high as two hundred feet and to have a diameter of six feet. The most abundant area was on Pine Creek in southwestern Tioga County. In north-central Pennsylvania, the first lumber boom of commercial harvests concentrated on the magnificent white pines, the "aristocrats of the forest," located near rivers, tributaries, and creeks, among them the Allegheny, Clarion, Cross Fork, Driftwood, First Fork, Kettle, Oswayo, Potato, Pine, Tionesta, Sinnemahoning, and West Branch. Throughout the hills and valleys the lumbermen's axes struck thudding blows against tree trunks to form a notch, followed by the deceptively quiet, rhythmic *sh-sh-sh-sh-sh-sh* of their saws, followed in turn by the sudden and quick loud crackling, swishing, and thumping of falling timber. It was hard work to harvest trees from those woods, which often grew on steep inclines or in deep ravines. However, it was worth the effort. White pine stands typically yielded twenty thousand to forty thousand board feet an acre, with some acres producing as much as one hundred thousand board feet.[12]

Inevitably, the loggers thinned out the white pine stands, particularly those close to waterways. Remaining stands of the old-growth forest increasingly were isolated and confined to remote and inaccessible places. Despite the obstacles faced by the hardy woods workers, the white pines were exhausted in Tioga County by 1865 and in Potter County by 1880. Michael Williams noted that much of the white pine cut in Pennsylvania, beginning in 1805, and especially during the peak years for harvest (1832–40) was sent by raft as far as New Orleans via the Allegheny, Ohio, and Mississippi Rivers.[13]

Once the white pine forest was gone, the hemlock forest, which had been largely ignored, from the same areas provided a natural resource for a second lumber boom. Hemlock bark, a source of tannin, was used in the production of leather. In the early days of harvesting bark for tanneries—for an industry that by the 1870s numbered hundreds of small businesses—the wood itself was often left behind to rot, since the trees, which produced "a relatively brittle, coarse, knotty, tough and splintery wood," resisted the carpenter's forged wrought-iron nails of earlier times.[14] Those nails simply were not up to the task of easily penetrating hemlock lumber.

The majesty of a Black Forest hemlock grove was suggested by Aden Byron Albertson (1887–1979) in his description of the forest near Nordmont in Sullivan County that provided logs for his father's sawmill: "I would that I might give you a mental picture of a dark, dense, hemlock forest—midwinter—when the temperature was around 0° F and the snow a foot or more deep. You would walk into one of these hemlock cathedrals, dark and awe-inspiring, on a carpet of fronds and hemlock burrs, the snow having been captured and held above you by the density of the forest, and a kind of mist arising from the warm earth beneath. To do so was certainly a thrill. The wild life often gathered in these places for warmth and shelter during the night while they would feed upon the shrubbery of the valley midday."[15]

Setting aside its beauty, why had hemlock been ignored during the first white pine logging boom? The answer is that carpenters preferred to work with pine, which originally had been so plentiful. Robert Currin, late curator of the Potter County Historical Society, noted that "woodworkers thought of the pine as the Lord's wood, soft and easy to work, and hemlock was the Devil's wood, full of Hell-fired knots." Another authority noted, "Due to its strong and durable, yet easily worked nature, white pine was the preferred species for most buildings, i.e., for timber frame construction, for panels, for clapboards, for doors, windows, sashes, and molding."[16] In fact, lumber companies considered hemlock a pariah of the early nineteenth-century forest. However, after the white pines were depleted, loggers in the late 1800s turned their attention to hemlock, historically more abundant than pine in Pennsylvania, particularly in the northern region, which held the largest block of hemlock in the world.[17] No wonder, then, that the eastern hemlock is the state tree of Pennsylvania, the tree most readily identified with the Commonwealth. In addition to hemlock bark being valued as a key ingredient in tanning hides for leather, the peeled hemlock logs increasingly were sawed for lumber. Hemlock had become more useful to carpenters after the mass production of stronger round wire nails that replaced weaker square-cut nails. Once driven in, moreover, the newer nails held well.[18] Once made into boards, hemlock was manufactured into a thousand and one useful products—to name just a few, barrels, boxes, shingles, laths, pickets, staves, mangle rollers, rolling pins, gears, clothespins, tools, rifle stocks, toys, furniture, bed slats, wagon hubs, veneer, mop handles, hoes, brooms, toothpicks, kindling wood, wood alcohol, medicines, and matchsticks (for America's wood-burning stoves). The U.S. population by the late 1800s had an insatiable demand for the wood that the forest provided. And the woodsmen were more than pleased to cut and sell the trees they sometimes called "greentop gold."

Logging in Pennsylvania in the nineteenth and early twentieth century was an arduous and dangerous occupation. Steep mountain slopes, unpredictable and often inhospitable weather, the remoteness of civilization, and lack of modern and relatively safe tools and machinery were among many obstacles faced by early woodsmen who chose to make their livelihoods from the forest. Pioneer women associated with logging generally did not do the actual woods work, but their lives were no easier. They served in other ways, working as camp cooks and bookkeepers or tending farms while the men were logging.

The next chapter will consider the fate of the vast Black Forest hemlock stands that William T. Clarke knew when he first traveled to north-central Pennsylvania in the early to mid-1880s. Like the white pine forests, the hemlock stands would be but a memory, and images on Clarke's negatives, by 1920. The decimation of the forest resulted from human persistence, ingenuity, and greed, with the increasing use and adaptation of efficient machines to speed logging operations and no thought of replacing what was cut.

When Clarke photographed children, he often chose a bucolic location of great tranquility and beauty. It's known from the rare letter discussed in the introduction, from Clarke to Shoemaker, that Clarke did not approve of the wanton wholesale destruction of the forest under way in Pennsylvania during his lifetime. If those children remained in north-central Pennsylvania until middle adulthood, they would have witnessed the near decimation of the original forest.

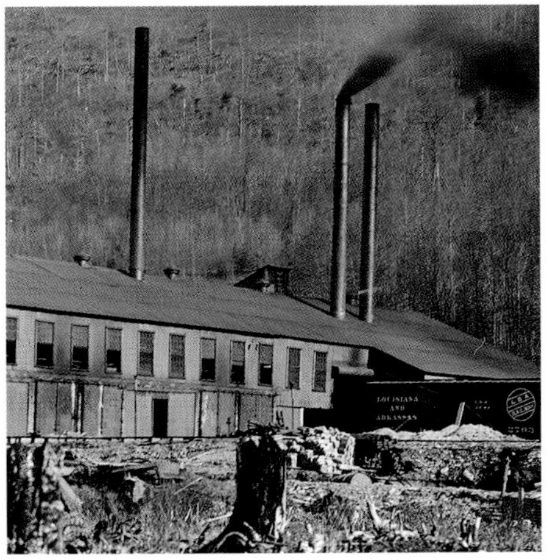

CHAPTER 2

THE MACHINE IN THE GARDEN

William T. Clarke's images of the timber and bark industries before and after the turn of the twentieth century represent an era on the knife edge of momentous upheaval. He did not see or record the earliest logging in Pennsylvania, although he did find virgin forests all around him when he moved to the Commonwealth. His photographs concentrated on what historians consider the third phase of U.S. logging and lumbering, the use of steam power and machines.

The earliest traditional logging and lumbering operations were done by hand. A tree could be, and often was, used in the round to build log buildings and fences. Or, if squared timbers were desired, a man with an ax could hew a log to a rough square while it was on the ground. An alternative was to dig a pit, roll a log that was longer than the pit over the center, and send one man down into the pit while another man stood above on a platform attached to the log, usually on either side so that he had better balance with his legs spread apart.

The two men, one on each end of a long saw, cut the tree to square or to lumber of a desired thickness. Either method was hard and dangerous work. When using the two-man pit sawing method, no one wanted to be the pit man and get sawdust in his eyes, nose, and mouth and on his clothes. The man above had to keep his balance as he moved forward or backward as sawing progressed.[1] As the scope of logging and lumbering increased, improvements were made, using buildings containing water-powered machinery. Such sawmills produced lumber for use in immediate neighborhoods. These were operated by pioneer farmers who had streams or rivers on their land. At first, perhaps, they operated their small sawmills to produce lumber for their own use. Farmers could not work at agriculture year-round, so many turned to woods and sawmill work during seasons when agriculture was dormant. No doubt some eventually gave up farming and earned a living through barter or cash by cutting timber for others, using additional logs from their own land or logs supplied by other farmers. As time went on, farmers and sawmill owner-operators might have continued logging and lumbering to sell surplus logs and lumber to others downstream as inhabitants continued to clear land for agriculture or used their forests as a primary or supplementary source of income.

Over time, a second phase of logging and lumber production came into existence. The size and output of sawmills increased in response to greater consumer demands from distant populations for lumber. Such sawmills usually were supplied with large numbers of logs by rafts that were piloted down rivers and streams or with logs that were transported via horse-drawn sleds, drays, or wagons overland to the mills. Yet another method of log delivery relied on moving large numbers of logs to upstream riverbanks, and, when conditions were right (for example, spring thaws and flooding), dumping them en masse into the water. Downstream "drivers" kept a watchful eye on logs to prevent foundering or jamming. It was a dangerous and strenuous job. The drivers used peaveys to guide individual logs or to break up logs jammed together, so as to keep them floating downstream until they were corralled in booms and eventually delivered to sawmill holding ponds to await sawing.[2]

In time, however, Pennsylvania logging and lumbering moved into a third phase in response to large-scale U.S. industrial innovations and transportation expansions during the second half of the nineteenth century, thus becoming part of the Industrial Revolution.[3] The delivery of logs to sawmills and subsequent lumber-to-market transportation began to rely on steam power and railroads to supply and service sawmills that were growing in number, size, and speed in order to sell lumber to regional centers and urban markets. Builders needed vast amounts of lumber to create housing for far-flung expanding urban populations. Manufacturers made more and more products from wood in meeting the needs and wants of increasingly affluent consumers. Those industrial changes had profound implications for local and regional rural life.

Leo Marx (1919–), a professor at the Massachusetts Institute of Technology and a noted twentieth-century American thinker, published a landmark book in 1964 and coined a compelling phrase to describe the catalyst in that social and cultural cataclysm—"the machine in the garden."[4] His simple phrase encapsulated a momentous transition in human life and civilization. The traditional ideals of pastoralism, espousing a simple and innocent way of life in harmony with nature and a predictable, comparatively slow pace of life, were much affected by the Industrial Revolution, with its growing emphasis on machines, bureaucracy, urbanization, the centralization of power and influence, and the notions of speed and "progress."

The railroads were to play a huge part in that transition. The first commercial steam locomotive to be operated in the United States, in 1829, was imported from England to haul coal in northeastern Pennsylvania.[5] The next year, a locomotive named Tom Thumb was in service on the thirteen-mile Baltimore and Ohio line from Baltimore to Ellicott's Mill in Maryland. That same year, the West Point Foundry of New York City finished an American-built locomotive, the Best Friend of Charleston, for the South Carolina Railroad.[6]

At first, steam locomotives, pulling their water- and fuel-carrying tenders and following rolling stock, were depicted as relatively benign and romantic additions to the landscape. They were slow, colorful, dependable, and powerful. The rate of technological improvements to railroads grew rapidly, as did the construction of rail beds and routes. By the mid-1830s, "railroad fever" was in full swing in the United States. Enthusiasts preached the gospel that railroads would open up lanes of mobility to the western frontier and would quickly transport people, goods, and materials from here to there and from there to here, from somewhere to anywhere and from anywhere to somewhere.[7] All-weather railroads would act as spurs to development and wondrous opportunities. They would generate great, unheard-of revenues and wealth. Trains would transform society completely; would eliminate uncivilized, poor, backwater communities; and would foster wealthy cities and cultures never before experienced. That, at least, was the hope and vision of many early to mid-nineteenth-century futurists.

On a practical level, Pennsylvania's abundant resources of coal, iron, oil, and timber would be used to create and sustain the railroads and generate wealth beyond anything ever contemplated in pre-rail thinking and dreaming. The 1844 revised edition of *A Practical System of Modern Geography; Or, A View of the Present State of the World* listed the principal railroads finished or in progress within the United States. Of seventy-one listed, fourteen were associated with Pennsylvania. Eleven of those existed solely within Pennsylvania. The Commonwealth shared two railroads with Maryland and one with New Jersey. New York State, the second-most-active railroad location, had thirteen, of which one was shared with Massachusetts. Twenty states and the District of Columbia had one or more railroads finished or in progress in 1844. The railroad farthest to the northwest was in Michigan and the farthest to the southwest was in Louisiana. Lengths varied. The shortest, west of Philadelphia, was nine miles long and connected West Chester and Columbia, Pennsylvania. The longest was five hundred miles, connecting New Orleans and Nashville.[8] By 1844, the Philadelphia and Reading Railroad was the first U.S. line to carry one million tons of freight in a year. The Industrial Revolution in the United States had begun in earnest. Pennsylvania became the nation's leading railroad developer from the 1840s to the 1860s, the same period in which the Commonwealth was the nation's largest timber producer, harvesting pine. By 1860, the Commonwealth led all U.S. states in the total number of railroad track miles (2,660), a position it held into the early twentieth century as Pennsylvania

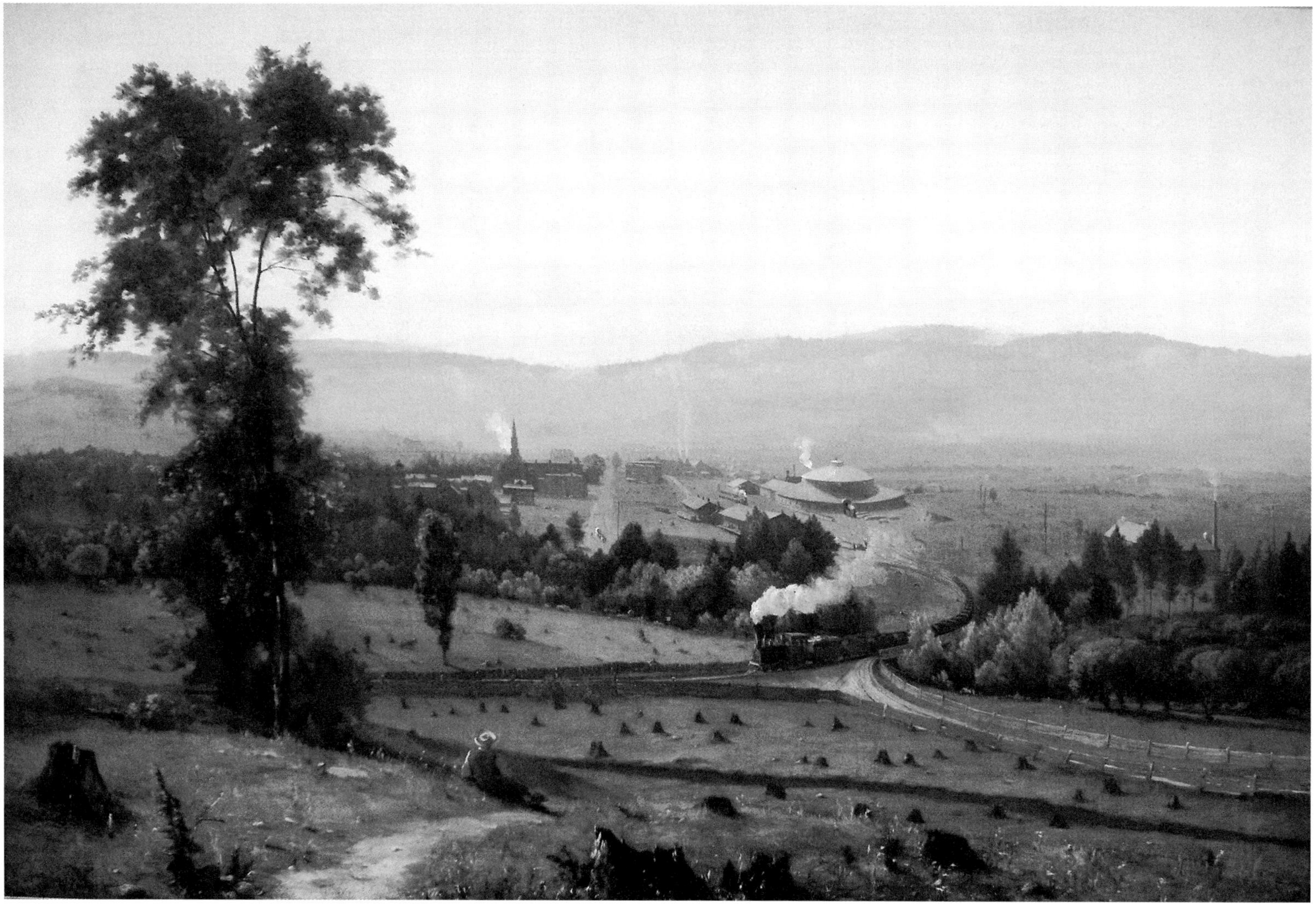

FIGURE 11

George Inness, *The Lackawanna Valley*, ca. 1856

continued to dominate the railroad industry.[9] Pennsylvania's maximum rail track mileage was more than 11,500 miles, achieved in 1920.[10]

Artists, bellwethers of sociocultural shifts and trends, were quick to sense and reflect upon the importance of the new machine. The locomotive's massive scale, its moving parts, the firebox glow, the smell of smoke, the way the puffs bellowed in concert with the pace of the machine, the changing pitch and tempo accompanying its movement, the bending of rails and ties under its weight that made the very earth tremble below its wheels—all these would have been remarkable, especially to artists, but also to the general public. The fact that this new machine was practical and useful must have made it an easy sell. Artists, sensitive to iconic images, were among the first to realize that the locomotive and the railroads were worthy of status as *the* symbol of the technologically driven changes that would transform U.S. society from top to bottom.[11]

To celebrate the prevailing spirit of optimism and enthusiasm, John Jay Phelps (1810–1869), founder and first president of the Delaware, Lackawanna and Western Railroad (D, L & W) commissioned American landscape artist George Inness (1825–1894) to create a celebratory vision of the D, L & W. Inness finished a pastoral landscape painting around 1856 titled *The Lackawanna Valley*. Viewed from the top of a large hill overlooking Scranton, Pennsylvania, a steam engine chugs along in the distance, climbing the hill, with white steam roiling behind. From the distance, the machine seems innocent and benevolent. D, L & W's new roundhouse for maintenance and repair of its engines and other equipment dominates the city far below. Other far-distant trains send up minuscule smoke plumes, suggesting a busy environment. Tree stumps form part of the painting's foreground. A reclining viewer near a dirt path leisurely watches as the unfolding parade passes through the bucolic scene. Art historian Adrienne Baxter Bell perceived an implication in the scene that the recumbent individual in the foreground was contemplating the woes that coming industrialization would visit on the Edenic American wilderness.[12] John Jay Phelps rejected the initial painting. He insisted that all four of the railroad's trains be shown and that the name D, L & W be included on the locomotive. Inness was incensed, but his precarious financial situation during his early career and his wife's urging to accede to the request led him to comply. For his efforts, Inness was paid $75, or slightly more than $1,877 in 2014 U.S. dollars.[13]

The early railroad activities depicted in Inness's *Lackawanna Valley* were to become commonplace in many parts of the nation.[14] Railroad speculators and entrepreneurs were busy during the pre–Civil War era raising money to establish effective coal and other natural resource carrying capacity in Pennsylvania. For example, in 1857 in Jefferson County, the Great Western Coal and Transportation Company offered a thousand $500 mortgage bonds at generous incentives (7 percent annual interest). The stock offering price of $500 in 1857 was equivalent to slightly more than $12,522 in 2014 U.S. dollars.[15]

The United States boasted 30,000 miles of railroad tracks in 1860. By 1890, trains ran on 160,000 rail miles. By 1916, tracks covered 254,000 miles. By this time, the nation was on a twenty-four-hour-a-day work schedule. Trains ran faster and farther, night and day, almost always unhindered by weather. While the earliest steam locomotives were wood fired, coal became the dominant fuel in the 1860s, creating black sooty smoke and subsequent grime.[16]

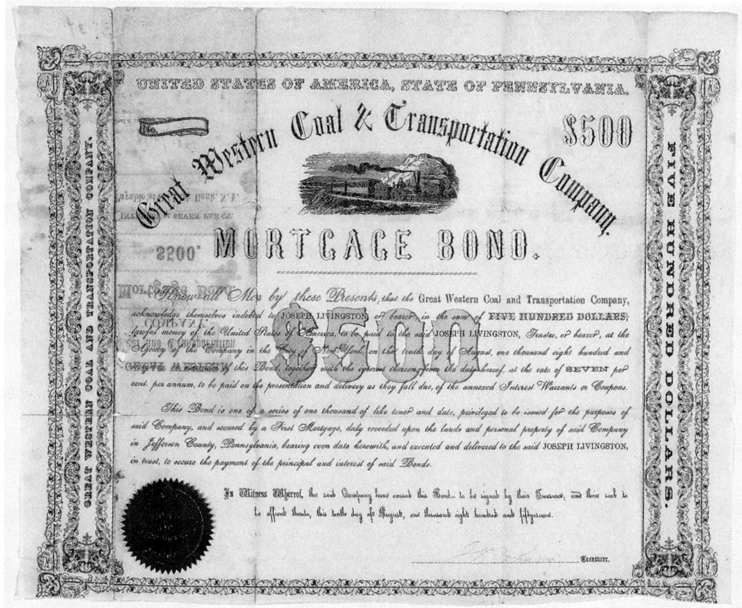

UNITED STATES OF AMERICA, STATE OF PENNSYLVANIA.

No.

Great Western Coal & Transportation Company.

$500

MORTGAGE BOND.

GREAT WESTERN COAL AND TRANSPORTATION COMPANY.

FIVE HUNDRED DOLLARS.

Know all Men by these Presents, that the Great Western Coal and Transportation Company, acknowledge themselves indebted to JOSEPH LIVINGSTON, or bearer, in the sum of FIVE HUNDRED DOLLARS, lawful money of the United States of America, to be paid to the said JOSEPH LIVINGSTON, Trustee, or bearer, at the Agency of the Company in the City of New York, on the tenth day of August, one thousand eight hundred and ... of this Bond, together with the interest thereon from the date hereof, at the rate of SEVEN per cent. per annum, to be paid on the presentation and delivery as they fall due, of the annexed Interest Warrants or Coupons.

This Bond is one of a series of one thousand of like tenor and date, privileged to be issued for the purposes of said Company, and secured by a First Mortgage, duly recorded upon the lands and personal property of said Company in Jefferson County, Pennsylvania, bearing even date herewith, and executed and delivered to the said JOSEPH LIVINGSTON, in trust, to secure the payment of the principal and interest of said Bonds.

In Witness Whereof, the said Company have caused this Bond to be signed by their Treasurer, and their seal to be affixed thereto, this tenth day of August, one thousand eight hundred and fifty-seven.

Treasurer.

FIGURE 12

Great Western Coal and Transportation Company mortgage bond, 1857

Commerce increased and lives changed. For example, agricultural products moved across the country, changing dietary expectations and habits. The number of women in the workforce increased. Ambitious young people boosted their educational qualifications to ensure jobs in the white-collar occupations that industrialization required.[17]

A later generation of urban artists working in the 1920s through the 1940s, among them Reginald Marsh (1898–1954), saw railroading differently from how it had been depicted by Inness in *The Lackawanna Valley*.[18] Marsh frequently visited the railroad theme as an artist committed to social realism. Works such as *Locomotive Watering* (lithograph, 1927), *Tank Car Rail* (etching, 1929), *Erie R.R. and Factories* (etching, 1930), *Locomotives, Jersey City* (oil painting, 1934), *Jersey City Landscape* (etching, 1939), and *Switch Engines, Erie R.R. Yards, Jersey City* (lithograph, 1948) are examples that progressively took much of the romance out of railroading by capturing not only the raw power but also the mundane workaday world of the smoke, soot, oil, and grime of the iron horse, railroad cars, railway workers, and environs.[19] Marsh's interest in railroading persisted his whole life. Norman Sasaowsky remarked that Reginald Marsh fixated on the locomotive "almost as a leitmotif throughout his career." The artist's feelings about railroading evolved, however. An early lithograph, *Loco Watering*, of 1927, was printed with a generous amount of white space. By contrast, his 1934 etching *Erie R.R. Locos Watering* and his painting *Locomotives, Jersey City* were dark and smoky. Later art featured cramped environs and congested designs. While his locomotives could be "stately, powerful, and detailed," he showed railroad workmen as "secondary [and] miniscule, mere functionaries for the machine."[20]

William T. Clarke's images of locomotives and railroad men lie somewhere between the early romanticism of Inness and the social realism of Marsh. Clarke was not a romantic. He saw Pennsylvania industrialization with unvarnished clarity but, like Marsh, retained an aura of monumental scale in his photographs of the railroads, mills, and factories that changed the face of north-central Pennsylvania, depleting in a generation the forest resources once thought inexhaustible. The railroads he photographed, however, were located in the rural wilderness and therefore never suggested the dirt and

FIGURE 13
Reginald Marsh, *Locomotives, Jersey City*, 1934

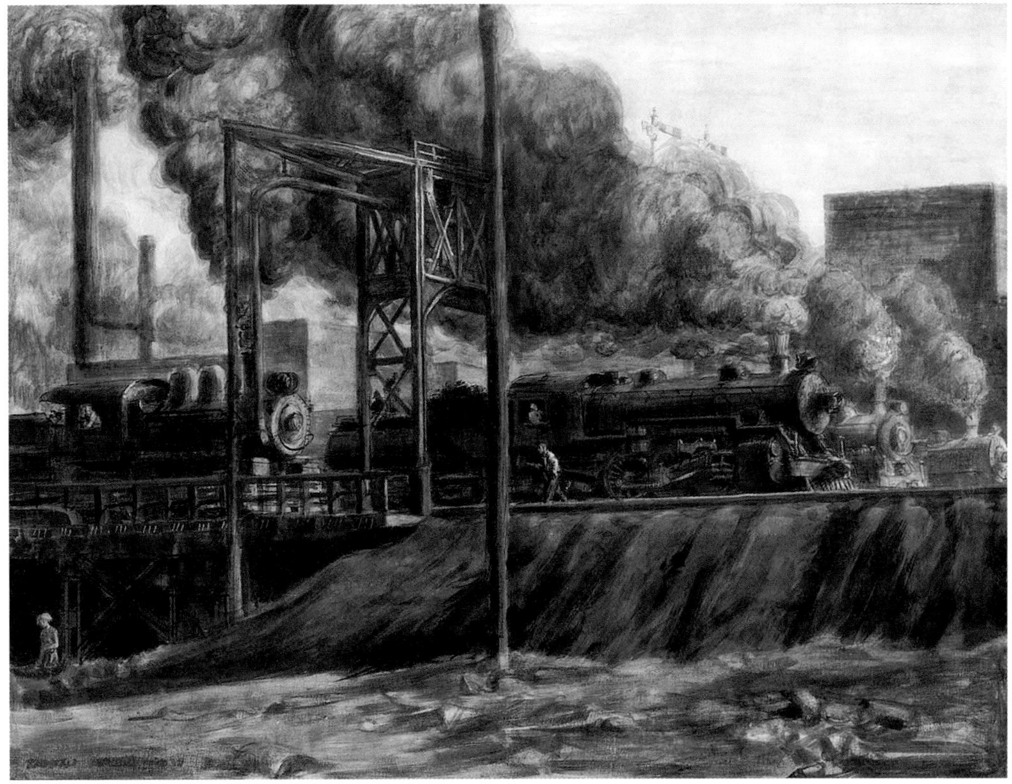

grime characteristic of the urban surroundings Marsh captured in lithographs, engravings, and etchings. Clarke's images of railroads and their related lumber operations were likely made as a work-for-hire by the Goodyear operations. For example, plate 11 is a record shot of the pristine Goodyear Shay locomotive No. 5. Probably the Goodyear brothers wanted an image of their new acquisition to proudly display in their offices. Like Inness, Clarke was producing work on contract and wanted to flatter his clients by showing them in a positive light.

Although the railroad had existed in the United States from 1829 and had linked the Atlantic and Pacific Oceans in 1869, it didn't arrive in north-central Pennsylvania with a major Class I railroad until the Buffalo and Susquehanna (B&S) was formed in 1893 through the merging and consolidation of smaller railroads.[21] The vast timberlands of the Allegheny Plateau, long thought unreachable, were about to be harvested en masse and transported as lumber and wood products throughout the country (see plate 3). Pennsylvania's timber production had declined in the 1870s after most of the pine forests were cut, but a second boom began in the 1880s when hemlock increased in value. With this boom came steadily increasing profits until the forests were mostly depleted around 1910. The boom would not have been possible without machines that increasingly enabled people to go places and to accomplish work that previously would have been very difficult and uneconomic.[22]

Railroads drastically altered the old ways of producing lumber. Traditional methods used gravity whenever possible, required the abilities of muscular men and animals, and used simple technologies to wrestle logs through the forest to natural waterways where they were formed into rafts or placed individually in rivers for transportation to sawmills downstream. Of course, trees that grew in close proximity to rivers and streams were the first to be cut. The practical limit of transporting trees to the waterways was about two miles. If the trees were farther inland from rivers and streams than that, it proved unprofitable. Rafting logs required selectivity in length and straightness. Driving logs was less selective. Any tree that could produce a sixteen-foot log was cut for river drives. When railroad transportation and steam log loaders reached into the uncut forests of the Allegheny Plateau far from waterways, all logs were cut.[23] After 1849, the West Branch of the Susquehanna River, running some two hundred miles from its headwaters in Cambria County, was a major channel for floating logs, either loose or in rafts, to be captured behind the Williamsport "boom." The river also permitted loggers from Clinton, Centre, Clearfield, Cameron, Elk, Potter, Lycoming, and McKean Counties to use the boom.

Upstream owners "branded" their logs with symbols or initials stamped with heavy hammers into log ends. Each owner's brand was registered in offices downstream at Lock Haven and Williamsport, as well as in county courthouses. "Drivers" were stationed along the river to keep loose logs from jamming or foundering in stagnant water. Downstream at Williamsport, huge logs were chained together and stretched from stone crib to stone crib to form a six-mile boom across the middle of the Susquehanna to capture floating logs. The Susquehanna Boom Company, ready for operations in late 1849, corralled logs that piled up as far as seven miles behind the boom, representing as much as 300 million board feet at a time.[24] From the Susquehanna boom, logs were sent to sawmills of southern Lycoming County and elsewhere downriver as far as Harrisburg. "Boom rats" sorted them and kept careful records of ownership once the logs were stopped at Williamsport (or by other Susquehanna booms). Then, logs were routed by rafts into sawmill holding ponds for processing. There were several major boom breaks caused by spring floods, but each time the Williamsport boom was rebuilt and reinforced. By the mid-1800s, Williamsport had become the nation's greatest lumber producer. Thirty-three sawmills consumed logs at a horrific pace.[25] The last raft drive through the Pine Creek Gorge in the early 1900s was an occasion for great regret and nostalgia, since rafters were legendary in their ability to guide the huge rafts through dangerous waters and they were equally well known for their bravado, boastfulness, and cheekiness on land.[26]

However, the coming of the railroads to north-central Pennsylvania meant there was no longer a need to move logs by water for great distances. Rail transportation meant logs could be moved shorter distances to modern local or regional steam- and electricity-powered sawmills. Owners and investors favored local and regional control over the transportation process. Accordingly, local and regional operators built large sawmills close to the forest supply. Before the railroads, lumbermen not only had to worry about their logs being lost on the river but also had to contend with unscrupulous pirates and mill owners who stole logs or who altered brands to change their ownership. Railroads and local waterways provided an opportunity to saw logs into lumber near the harvest sites, prevented losses, and eliminated the traditional downstream middlemen. The bottom line? Profits increased.[27]

In the early 1880s, entrepreneurial logging activities of well-heeled magnates established a small cadre of Pennsylvania lumber barons. Frank Henry Goodyear (1849–1907) worked as a bookkeeper to a successful New York businessman who owned several different businesses, including a sawmill. Frank married the boss's daughter in 1871. The newlyweds were given a large tract of Pennsylvania timber land as his wife's dowry. "F. H.," as he often was called, went into the lumber and coal business in Buffalo, New York. He became very active in 1884–87 in north-central Pennsylvania, buying huge tracts of timber land, some eighty thousand acres, and also speculated in western Pennsylvania coal. The time of his arrival on the scene coincides nicely with William T. Clarke's coming to north-central Pennsylvania, suggesting the idea that Clarke, from nearby Rochester, was sent there as an employee or contractor of Goodyear. Frank's brother, Charles Waterhouse Goodyear (1846–1911), joined him in 1887, forming the F. H. and C. W. Goodyear partnership. For the most part, their vast forests were in the remote, uninhabited interior far from waterways. Their acres were filled with untouched eastern white pine, eastern hemlock, and hardwoods (maple, birch, beech, and lesser amounts of ash and cherry) that would yield many thousands of board feet per acre, plus the bark, once harvested and processed. The Goodyears' holdings were estimated as containing an average of fifteen thousand board feet of hemlock and two thousand board feet of hardwoods an acre. The better tracts of hemlock held twenty-five thousand board feet of lumber.[28] With an eye on those old-growth forests, with wealth to invest in order to accrue ever greater fortunes, and with knowledge of emerging technologies, the Goodyears established a personal empire based in Buffalo that they maintained under their personal control until they died. Their Pennsylvania woodlands tracts were located on the headwaters of the Pine and Sinnemahoning Creeks. The Goodyears built a modern sawmill in Austin (Clarke bought property there in 1888), ran other sawmills at Galeton and elsewhere, and all the while extended rail transportation into the wild woods, a vivid example of the machine in the garden in action. The Goodyear operation depended upon technology. Foremost were the steam locomotives, their tenders, the cars they pulled, and the log-loading machines that loaded and unloaded the cars (see plate 4).

The first step in improving the transportation of logs to sawmills was to firm up the railroad grades begun in 1883 for use by the Jersey Shore, Pine Creek, and Buffalo Railway. Upgrades were needed to support the steel tracks required by the Class I railroad and its loads.[29] Main rail lines were upgraded and constructed without much difficulty because they followed the comparatively level bottomlands created by rivers.[30] Eventually, the Pine Creek Gorge route linked the New York Central Railroad to the north with the Clearfield Coalfield to the southwest and with New York City–allied lines in Williamsport to the southeast. William T. Clarke's bird's-eye view of a portion of the "Grand Canyon of Pennsylvania" reveals the railroad main line situated in the forty-seven-mile-long valley eight hundred feet below, paralleling Pine Creek (see plate 3).[31]

It was a much more difficult job to build "spurs," or temporary offshoot tracks that branched off the main lines into the mountainous hinterlands, in order to gain access to the old-growth forests far from waterways and the railroad's main line. It took tremendous labor and ingenuity in rail bed building to allow a Shay locomotive to gain access to the vantage point at Colton Point atop the Pine Creek Gorge's west rim, for example. But that is what is shown in another photo very similar to that depicted in plate 3. The image shows the front of Leetonia Lumber Railroad's Shay locomotive No. 3 high above a nearly clear-cut Pine Creek Gorge. The scene was recorded, perhaps by Nelson Caulkins, before 1921 when the Leetonia mill closed and the railroad was torn up.[32]

"Swampers" carved out logging spur routes, or paths for new rail grades, that clawed and crawled their way up hills and mountains and down into remote valleys. The spurs required numerous switchbacks with primitive hairpin turns to permit the heavy locomotives, tenders, and rolling stock to climb steep slopes and to descend with incredibly heavy loads without gaining enormous speed (see plate 5). When necessary, the workmen fashioned wobbly-looking but dependable bridges over streams or gullies. The spurs typically were as short as three miles or as long as thirty. Spurs were notoriously crooked, twisted, and dangerous. All too often, spurs were hastily, crudely, and sometimes shoddily constructed because it was assumed they'd be abandoned in a year or two after all the trees had been cut and removed from a particular area. Materials used to build spurs were cheaper than those for main lines. Comparatively flimsy lighter-gauge steel rails were used, for example. Period photographs of spurs sometimes show bent, crooked rails resting on widely spaced ties made of various-size rough-hewn logs that had been cut nearby, and no underlying earthen bed (see plate 6).

Thomas T. Taber III and Walter Casler stated that "[spur] grading was universally poor and followed the natural terrain as much as possible."[33] Spurs had grades of 10 percent or more and sharp curves (a thirty- to fifty-foot radius was common). Taber and Casler also noted the extreme dangers of spur railroading. Units of rolling stock, often in poor condition, were connected by link-and-pin hook couplers. There were no air brakes and sometimes not even hand brakes. Only the engine's steam brake was used to control the train most of the time. "Runaways and wrecks were frequent and often men were maimed or killed, particularly when load after load of logs started to cascade over a wrecked engine." Clarke photographed a typical derailment in Clinton County (see plate 7).

The Goodyear brothers contracted with about fifteen "jobbers" to harvest specified quotas of logs during a given season. The jobbers, in turn, also might sublet parts of their contracts to smaller-scale operators and families. Typically, jobbers in the Goodyears' operating area for a given season employed about twenty-four hundred laborers (including fifteen hundred seasonal bark peelers). Most lived in temporary logging camps. Loggers from families and independent seasonal workers often were settled in the area being cut and operated independently of the camps. Once the logs were stripped of bark, they were moved to skidways near the rails for loading, often in a huge jumble of logs known as a "rough and tumble" (see plates 8–11).

On an average day, five of the Goodyears' eight Shay locomotives were shuttling to and from the main line and traversing about seventy-five miles of spur lines. Shay locomotives were designed to handle rough, hurriedly constructed rail grades and were adept at climbing and descending from steep terrain and traversing switchbacks and hairpin curves (see plates 12 and 13). Goodyear locomotives hauled from eight to eleven steam-operated Barnhart mechanical log loaders that loaded logs from the skidway landings. The Goodyear lumbering operations worked with the Marion Power Shovel Company of Ohio to adapt the steam shovel to work as a mechanical log loader, according to historian Keith Haddock. The Goodyear-Marion collaboration led to an innovation of great utility and value in the lumbering business. The log loader's success can be inferred from the fact that many log loaders were sold to other companies. The coal-fired log loader worked from rails fastened to beds of gondola cars, allowing it to move from car to car as it loaded each, working toward the train's rear. Log loaders could be detached from the train, remaining at the skidway while the loaded gondola cars departed for the journey up and down hills to the mills and factories. One locomotive operated by an engineer, fireman, and brakeman typically tended two log loaders and their crews (see plates 14 and 15). Prior to the mechanical log loader, gondola rail cars were loaded by hand, a slow, labor-intensive, and extremely dangerous operation. The machine-in-the-garden approach to logging represented a 15 percent savings over the older method of hand loading logs onto the cars, even when the costs of manufacturing the loader ($3,250) and coal were included.[34]

In a year or two, after an area had been clear-cut and the tree trunks and bark moved out, the section of land would be abandoned. The temporary camps that housed the woods workers would be dismantled and moved out. The railroad tracks and any wood ties that were still serviceable would be taken up and hauled out during the final trips. The spur would be abandoned. The salvaged rough boards and equipment of the "old" camps would be relocated to the next tract after another railroad spur was built to provide access to a fresh logging area of forests to be clear-cut. The "new" logging camp would be built mostly from the recycled materials.

The Lackawanna Lumber Company changed some of that camp-rebuilding inefficiency by creating a mobile logging camp, known as "the shacks." The components of the shacks were built on railroad flatcar platforms and provided cooking, eating, lounging, and sleeping areas for the men and a "barn" for their horses (see plate 16). The railroad easily moved the camp crew and animals close to where the forests remained to be clear-cut. Several major benefits could be attributed to this innovation. The inefficient need to constantly tear down and rebuild camps was eliminated. Travel time to and from the work sites was reduced because locomotives moved the shacks' men and horses closer to their job sites as needed. Moreover, the process permitted shorter distances from the sites of felled trees to the loading landings. Logs no longer needed to be concentrated in single-site rough-and-tumble landings, which were inherently unstable and extremely dangerous. The safety of woods workers, railroad men, and animals was enhanced, because smaller log piles positioned along the rail tracks close to where they were cut were less prone to avalanche-type catastrophes. However, the main reason for the shacks probably was to enhance productivity. Getting the men to the cutting sites quickly without the need to build and rebuild fixed camps was a step toward greater efficiency, as was the adoption of mechanical log loading to replace hand log loading.

A 1907 report on northwestern Pennsylvania logging operations estimated the Goodyears' lumber empire was processing 200 million board feet of hemlock annually, making it the world's largest manufacturer of hemlock. The business processed nearly as much in hardwood, some of which was sold to other companies. Goodyear trains also delivered hemlock bark to leather tanneries in Costello, Coudersport, Galeton, Harrison Valley, Roulette, and Shinglehouse in Potter County, and elsewhere. Hemlock was established in the tanbark trade before it became a mainstay in the lumber industry. Tanbark was a key ingredient in the leather-making industry and hemlock was the primary source.[35] The tannins extracted from the bark acted on the proteins in animal hides to produce strong, supple, water-resistant leather, essential to the manufacture of harnesses, saddles, belts, industrial applications, shoes, boots, and so on.[36] Pennsylvania was at the center of hemlock tanbark and leather industries until the exhaustion of hemlock in north-central Pennsylvania during the second decade of the twentieth century.[37]

The Goodyears achieved a virtual monopoly on many aspects of logging and lumbering from their portion of north-central Pennsylvania. They achieved vertical integration within their business by controlling the entire transition of the forest-to-lumber market during a period when logging in northern Pennsylvania represented the "highest degree of forest utilization that the world has ever seen in any commercial lumbering era."[38] They owned the Buffalo and Susquehanna Railroad, which serviced the Goodyear Lumber Company. Their railroad transported the logs to Goodyear mills or to other companies that manufactured lumber and other wood products (see plates 17–21).

When the lumber and products were ready for shipping, the Goodyears' B&S Railroad stood ready to move the lumber, leather from tanneries, coal from their Buffalo and Susquehanna Coal and Coke Company, and other manufactured products to market.[39] At its zenith, Goodyear's B&S consisted of more than 382 miles of first-class standard-gauge track and rail bed, running northeast from Sagamore near Kittanning, Pennsylvania, to Addison, New York, and from Ansonia, Pennsylvania, northwest to Buffalo, New York, where the Goodyear brothers had ready access to Great Lakes shipping via two boats within their Buffalo and Susquehanna Steamship Company. The *Frank H. Goodyear* and the *Stephen M. Clement* normally carried iron ore from the brothers' mines in Minnesota and Michigan east on the Great Lakes to be smelted in the blast furnaces of their Buffalo and Susquehanna Iron Works (begun in 1902) on Lake Erie near Buffalo. The ships made the return trip west on the Great Lakes filled with coal or other raw materials and goods transported from Pennsylvania by the B&S. When necessary and cost effective, the Goodyears reworked and even relocated their main rail line. For example, four problematic switchbacks over the Hogback Mountains around Cherry Springs were bypassed, making the journey less hazardous and more rapidly traversed. Clarke photographed these tracks around 1910.

Goodyear locomotives chugged day and night, year after year through the north-central Pennsylvania valleys with smoke billowing from their stacks and whistles blowing. On its way into New York State, north to Buffalo, the B&S went through Genesee, Wellsville,

Belvidere, Pixley Summit, Arcade, Yorkshire, Springville, Big Tree, and many other burgs and hamlets. The hills, farmlands, villages, and towns near B&S tracks echoed constantly with the rumble of steel wheels on steel rails, as they squealed and clacked in an efficient, industrial cadence as the trains clamored along with heavy loads from 1893 throughout the early decades of the twentieth century.[40]

William T. Clarke's photographs over the decades are an ongoing document of the evolution of logging. In Clarke's early photographs, we see images of people logging as a small enterprise based on individual decisions to satisfy needs of self-sufficiency (such as buildings to live in and to house animals, to clear ground for farming and fencing, for heating, and so on) and secondarily to create a commodity for community barter or cash. For them, logging was part of an independent life and had no large-scale impact on community life or the environment.

Clarke didn't begin his Pennsylvania photography with the intention of contributing to a history of logging, or as a personal artistic endeavor, or as a personal protest about the denuding of the countryside by cutting all the trees, but as a means of making a living. Similarly, artists such as Inness probably were not producing their art as social or economic criticism. However, as "fine" artists, they indirectly addressed impacts of the growth of technology. For instance, Inness featured a picturesque train in the far landscape distance in his D, L & W *Lackawanna Valley* commission. Inness seemed to accept small-scale logging, as evidenced by tree stumps, as benign and normal. Such logging appeared to be a natural part of reaping the bounty of the good earth. Marsh's art was produced later and he focused predominantly on New York City. He was employed by newspapers and magazines and later taught at art schools. He therefore was not encumbered by contracts to produce work flattering to his employers. He was free to pursue his own artistic vision. When Marsh ventured afield to depict railroads and rail workers, he made a quite different comment on technology and its impact. Machines in his images are seen up close. They seem potentially malignant because of the undesirable by-products of their hearty consumption of natural resources as they assisted in harvesting wood, coal, iron, and other resources on a much more ambitious, economically motivated scale. In the view of late nineteenth- and early twentieth-century industrialists, the earth's natural resources, such as trees, existed wholly for the use of human beings and were subject to human domination and exploitation. Marsh's art, however, suggested that increased use and dependence upon machines diminished the stature of humans and led to environmental changes that could get out of control and be detrimental to life. Fewer and fewer individuals remained in control of machines and their subsequent effects on society and environment. Entrepreneurs such as the Goodyear brothers and their vertically integrated railroad-based logging empire tended to replace individual and shared decisions by the population. Masses in society often were reduced to roles as workers dependent upon carrying out the bosses' mandates or as consumers of the goods thereby produced.

However, woods workers, when callously trod upon, resorted to violence. When a Goodyear jobber absconded with money that was due his workers, they appealed to Goodyear to make the debt good. Goodyear refused and told them they should seek redress from the jobber. The result? "Fire broke out all over his slashing in Big Moore

Run in Potter Co., Pa. In vain Goodyear offered four dollars per day for firefighters. The men jeered him. Hopelessly, he watched a million dollars go up in smoke." Similarly, men who worked in Williamsport sawmills collectively challenged mill owners through early union demands calling for fewer hours and more pay. Workers participating in the country's first major lumber strike, in July 1872, demanded that hours of work per day be reduced from twelve to fourteen to ten; they also pushed for more pay. Violence erupted when mill owners tried to reopen their mills with strikebreaker workers after three weeks. The conflict finally was quelled through the intervention of guardsmen and local militia. In the settlement, workers gained the right to an eleven-and-a-half-hour day and a pay raise of twenty-five cents a day.[41]

But despite the enthusiastic adoption of machines that did much of the labor in the timber industry, logging still depended upon human and animal muscle, know-how, and grit.[42] Machines alone were not sufficient in the garden. Machines could not be invented, operated, maintained, and improved without human ingenuity and determination. And as time passed and the forests were decimated, much of the romance of railroading faded and people began to see that machines were not exactly running on the road to paradise.

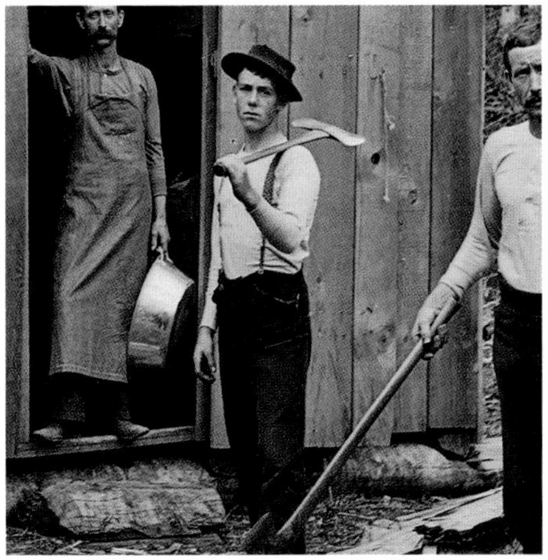

CHAPTER 3

WOOD HICKS, BARK PEELERS, AND OTHER WOODS WORKERS

William T. Clarke moved to Pennsylvania from Rochester in the late nineteenth century, reportedly for the sake of his health and also for employment. He perhaps experienced culture shock. As a young man who'd worked to make stereoscopic images for the general public, he no doubt was acclimatized to an urban environment and the then-modern conveniences that a relatively large upstate New York city afforded. His surprises after his move to north-central Pennsylvania probably were both pleasant and distressing.

The pleasures would have been the mostly pristine environment, with huge forests, beautiful mountain terrain, sparkling waters, and clean air. The things he no doubt missed about city life would have included paved streets, solid, comfortable houses with the amenities of the era, a developed transportation system, established businesses that catered to almost any need or whim a person might experience, ample human companionship, and plentiful choices regarding arts, culture, and entertainment.

His probable distresses may have been caused by north-central Pennsylvania's limited employment options, crude or nonexistent roads, elemental shopping choices, sparse population, fewer educated individuals, and lack of educational amenities such as libraries. Accommodations were sparse and often bitterly cold in winter. Exciting pastimes were few. Even the silence and the immensity of bright stars covering the heavens at night sometimes could make a person feel lonely, insignificant, or even bewildered.

Clarke needed to make a living. He must have examined the situation, reviewed his prospects, and decided he could eke out an existence through the field he'd pursued to that point. Photography was a possible occupation, if for no other reason than there were few photographers, at the time, in north-central Pennsylvania. The promise of photography work with firms such as the Goodyear Company to make images of their facilities and operations also made it enticing. Many people in that part of the world had never had their photograph taken and might be interested in having a memento of themselves or others for which they'd be willing to pay. He had to have faith in himself and in his experience as a photographer as well as the gumption needed to talk to people and convince them he could capture their images using a medium that would yield permanent keepsakes, plus the equipment and supplies needed to convince people he could capture their portraits and landscapes. In order to succeed, though, he had to travel widely to meet people and show them examples of what he was capable of producing. Then, if they agreed to his services, he had to keep records of names, places, prices, and payments, so that after he'd developed the prints from his negatives, he would be able to return and deliver the merchandise. As noted elsewhere, Clarke did not necessarily specialize in individual studio portraiture (there is a lack of evidence of cabinet cards bearing his name and few existing negatives of single individuals). However, Clarke made many outdoor group portraits. In spite of health concerns, he probably realized that he needed to get out and about among people congregated in widely dispersed small communities and logging camps. To sit in a studio and wait for people to come to him was not an option.

Thus, William T. Clarke became an itinerant photographer in a sparsely populated area. He was one of a special breed. While there were thousands of photographers around the nation scratching out a living during that era, few chose to work under the trying circumstances of the wilderness or barren areas. Four examples of other itinerant photographers illustrate this old-time class of photographers.[1]

Henry M. Beach (1863–1943), from the early 1880s until his death, specialized in scenic postcards and panoramic views of the Adirondack region as well as studio photographs of people of northern New York State. Much of his output was sold in bulk to innkeepers and small business owners, who in turn resold them to tourists.

Solomon D. Butcher (1856–1927) began to photograph the difficult pursuit of the American dream in Nebraska in 1880 for two history books he produced: *S. D. Butcher's Pioneer History of Custer County, and Short Sketches of Early Days in Nebraska*, published in 1901, and *Sod Houses; or, The Development of the Great American Plains: A Pictorial History of the Men and Means That Have Conquered This Wonderful Country*, which followed in 1904. To finance his undertaking, Butcher made subscription book contracts, mostly with farmers, whose photographs would appear in the publication. Sometimes payment was made through barter, with Butcher receiving a chicken or a hot meal. Just when it appeared that his plan to publish the first book would fail, Butcher was bailed out by a wealthy cattle rancher who saw value in his venture. Butcher continued to document Nebraska life until 1917.

Evelyn Cameron (1868–1928), a woman of high English social class, moved with her husband, Ewan, to the American badlands. They chose to live a frontier life in eastern Montana's Northern Plains, then semi-arid ranch country, after a honeymoon excursion there. Cameron and her husband, who was fifteen years her senior, were initially lured to Montana in 1893 by prospects of wealth from a thoroughbred horse-raising venture, the abundant wildlife, and excellent hunting. However, once they settled on their ranch, they struggled to survive economically. Cameron was resourceful and helped the couple earn a livelihood by taking up photography, traveling great distances to make her photos and selling her prints to cowboys and settlers. She worked as an itinerant photographer from 1894 to the end of her life in 1928.

Finally, Erwin Evans Smith (1886–1947) recorded the great changes occurring on large open-range ranches in the Southwest around the turn of the century, initially earning his way by working for ranch owners at a small salary, taking photographs while employed as a cowboy.

Independently, during the late nineteenth and early twentieth century, a number of individuals in remote U.S. frontier areas relied upon photography and an entrepreneurial spirit to make their living or supplement their finances. In each case mentioned above, the photographers identified with the people, loved the place they were in, became longtime residents, and found photography more than simply a means of making a living. Their photographic archives give witness to their personal missions to document ordinary people doing herculean work during critical times in special places. William T. Clarke was such a person, putting his photographic skills to use as a means of making a living in north-central Pennsylvania, which at that time was still a frontier area in an isolated part of the northeastern United States. Over the course of more than three decades, Clarke created a series of images that, taken as a whole, portray a narrative of changes in people, times, and places.

Perhaps the first people he met and learned about when he began taking his photographs were "jobbers," individuals whose work regularly took them from communities to logging camps and back again. Clarke needed their knowledge of where the logging camps were and their goodwill to move freely between camps. Jobbers, usually among the more educated and enterprising of the logging population, contracted with lumber barons to harvest, usually season by season, specified timber tracts owned by the barons. Jobbers generally acted as their camps' day-to-day woods bosses or were in constant contact with the individuals they hired to manage their logging operations. Jobbers had overall responsibility for hiring; housing the crew, feeding them, and managing their work assignments; keeping records; and dealing with myriad other tasks associated with work, as well as nonwork, concerns.

"Cruisers" also worked for the lumber barons and for large-scale timber owners. Their job was to estimate the number of trees to be harvested and the total board feet of lumber on a given site—prior to cutting. Owners armed with that knowledge could then specify quotas to be produced by the jobbers who contracted to log the sites, or, if clear-cutting was specified, the owners had an idea of the probable size of the yield. If a quota had been specified in a jobber's contract, the jobber invariably passed the requirement along to the loggers he hired or those who subcontracted with him (see plate 22). If they didn't make the quota, their pay was reduced or denied. The jobber's contract was filled with "what ifs" and was inherently risky, and much the same can be said for the loggers' returns.

To provide their employees with board and room, the jobbers hired carpenters to construct primitive temporary logging camp buildings, consisting of, at a minimum, a kitchen and dining area, lounge, bunkhouse, and barn, with larger camps also having a springhouse and root cellar. Carpenters also made "rough hemlock furniture: tables, benches, sinks, water trough, and barber chair," according to memoirist Leo Bailey, who said he couldn't recall "a single piece of factory-made furniture in camp."[2] Jobbers hired logging crews seasonally. The combined logging and bark-peeling season usually ran from autumn freeze-up to Independence Day. Camps were spread throughout north-central Pennsylvania in an ever-shifting pattern. The larger camps sometimes contained as many as one hundred men, who lived deep within the forests. Loggers performed a number of different, but highly interrelated, jobs. Some men fulfilled several different roles during the course of the season. Most, if not all, workers were not paid until the jobber's contract was completed at the end of the season. Meanwhile, their immediate needs were met by the jobbers, who provided housing, food, and small stores, selling or advancing against future wage payments such needs and wants as tobacco, pipes, matches, gloves, clothing, toiletries, suspenders, boots, liniment, and sweets.

Jobbers also were connected to workers whose business was railroad or road related. For example, logging in the machine age depended upon "swampers," men who cleared spaces through the forests to set up and maintain railroad spurs, as well as rudimentary roads used by horses and men that permitted movement of logs and bark from the forests to skidways located near railroad spurs or to waterways. After the spurs were built, swampers might find a place in the logging camp. New roads needed to be built as the logging season progressed and trees were reached, cut, and removed. In addition, existing roads required constant repair or icing.

Logging camps also were populated by self-described "wood hicks" (choppers, sawyers, trimmers, buckers), who notched, sawed, limbed, and cut the trees into uniform moveable lengths (see plates 23–25).

The wood hicks' primary tools were a double-bitted ax, a two-man crosscut saw, a felling wedge, and something with which to pound the wedge to keep it tightly driven into the tree cut before the tree fell.[3] Felling a tree began with the chopper's creating a sizeable notch in the direction in which the tree was to fall. Then, sawyers sawed from the other side of the trunk, angling slightly downward toward the notch. Sawyers used felling wedges, driven into the cut behind the saw as sawing progressed, to keep the tree from settling onto the blade and pinching it.

In the woods, sawyers had to watch for "widow makers" during all phases of felling a tree. Branches or treetops, either dead or weakened by disease, could come down while sawing was in progress. The moment the tree toppled after sawing was most dangerous. Traditionally, the sawyers shouted "TIMBER!" to warn others of a falling tree. However, they could not accurately predict whether the base of a tree trunk would kick back or twist, perhaps hitting their heads or their bodies, or fall toward and on them.[4] Even worse, they faced particularly dangerous situations when they tried to free partially fallen trees that had become hung up on standing trees. That gave them two or more trees to worry about simultaneously. Once a tree was down, trimmers cut the branches off with an ax. Men who lopped off branches from the huge trees often stood on or straddled tree trunks while chopping with their razor-sharp axes. A slip or miscue could mean a nasty sliced leg or severed foot and death from bleeding.[5]

Another large group of workers, known as "bark peelers," or (playfully) "bark savages," cut or chopped around the circumferences of downed and limbed hemlock tree trunks with an ax or saw, scoring through the bark at four-foot intervals in preparation for peeling the bark from the tree with a tool known as a "spud." The spud was made of iron, with a thin blade on one end, becoming thicker as it culminated in a handle. The spud's blade was rounded at the end and ground at an angle on one side to make a sharp edge. Bark peelers pried intact bark sections from the trunks (see plate 26).

The bark was then flattened and stacked to dry on top of an undergirding of poles. Dried loads were transported down the mountains to railroads and thence to leather tanneries (see plate 27). Bark slabs also could be used to construct temporary woods huts when needed (see plate 28).[6] Once the logs were peeled, "buckers" sawed them into uniform movable lengths (see plate 29). Wood hicks and bark peelers were constantly on the move, leaving behind peeled tree trunks cut into uniform sizes and piles of bark to dry as they progressed toward uncut forest trees.

Whenever feasible, logs subsequently were moved on "slides," using gravity to convey them down the mountains (see plate 30).[7] The slides were extremely dangerous when a speeding log escaped the trough. Woe to anyone or anything in the log's path! Leo Bailey recalled that as logs moved down the long slides they reached speeds too great for the slides to handle and that escaped logs sometimes shot hundreds of feet through the air like missiles and sometimes killed men or horses in harm's way.[8] To dampen such speeds, spikes (also known as "sprags") were driven into the slides.[9]

If slides could not be used, horses and teamsters pulled loaded sleds over roadways that often consisted of parallel small trenches the width of the sled's runners. A related method of moving logs, known as "snaking," "twitching," or "trailing," had horses pulling logs in a trench, used on flat ground or where hill grades were not steep. In winter, water was sprinkled into the trenches to freeze. If there wasn't a sprinkler sled, an alternative method to freeze the trench surfaces was to use water barrels with a hole on either side, filled and pulled inside the trenches, keeping them glassy. Horse teams, harnessed so they did not have to walk in the trenches, could move sleds or long strings of logs along the iced ditches with much less effort than over bare ground (see plate 31).

If the method of moving log loads was by sled, the loads were assembled by "skidders." The loads were then pulled by horse teams driven by "teamsters" to the railroad or to streams, rivers, or splash dams where they'd eventually be delivered to sawmills. Horses were well cared for, since they were a vital link to success (see plate 32). Blacksmiths shod them and the horses had adequate food and water and were quartered in barns. Horses' health was of major concern.

Sometimes the teamsters were jokingly accused of preferring their horses to their fellow workers. They woke before the others to feed their horses and get them harnessed for work. When they came back to camp, they removed harnesses for the night. After the teamsters ate their evening meal, they went back to the barn to water the horses and feed them oats and hay. The horses needed to be curried and washed. They were examined carefully for injuries or illnesses, however small or insignificant, that they might have incurred during the day's work. The teamster had to be an on-the-scene veterinarian if he and his horses were to work for long. And, of course, the barn had to be cleaned regularly. For their efforts and skills, teamsters were paid from $1.50 to $1.60 a day.[10]

Skidders who moved logs used peavey tools with a long wooden handle capped with a sharp pointed metal end that usually was stuck into the log's midsection near the top. A hinged gripper caught the log underneath. When the logger pushed the peavey's long handle, the peavey acted as a lever to lift and roll the log away (see plates 33 and 34). The gripping mechanism was disengaged by lowering the handle, releasing the lever pressure, and pulling the spiked end out of the log. Sometimes logs were worked into orderly piles (usually at the edge of a river or stream, or below a splash dam, for quick movement into rushing springtime runoff water for transportation by flotation). Horses were used to pile logs whenever feasible, but if men were the muscle, as many as three men working as a team might be needed to lever a medium-sized log up an incline of two or more logs to the top of the pile. The largest logs stayed on the ground at the bottom of the pile. In the early railroad lumbering era, logs were moved directly from rough-and-tumble chaos onto rail cars by horse and man with the help of log ramps and ropes or chains. Eventually, steam-driven mechanical log loaders simplified the rough-and-tumble loading process.

Most peeled logs from virgin stands were huge. They were round and slippery when moist. They were heavy. They had to be lifted and moved. They mostly rested on sloping ground that could be icy, snowy, wet, or spongy. In short, the logs were an ingredient in a potential recipe for danger and injury to men and horses. Men who worked in the woods took precautions. For example, their boots were "corked" with sharp iron pins driven through the soles for grip and footing when they walked over uneven terrain and on top of precariously piled logs.[11] Log piles could shift unexpectedly and break a man's foot, knee, leg, hip, spine, rib, arm, shoulder, neck, or skull in the blink of an eye. Loaders had to be wary and agile, ready to spring in the right direction at the slightest unexpected movement of a log.

The men did not have ready medical assistance when needed. This era predated the concept of health insurance. If they were injured and couldn't work, they might be moved to the nearest community to be cared for by medical personnel at a rudimentary clinic or hospital. If the loggers couldn't work, they didn't get paid: "As was usual with the average lumberjack, he had saved no money. [However, in] the usual procedure in the woods when someone got hurt or was sick for a long while, a paper was passed for a collection [to help] pay some of his immediate doctor and board bills." Often, each person in a camp crew would donate a dollar.[12] The poorhouse was the likely last stop if a man was permanently maimed or seriously ill. Woods work was a young man's job, but then the predicted average life span nationally at the turn of the century was only slightly more than forty-eight years for a baby born during that period and a little more than sixty-two years for a man aged twenty.[13] E. Willard Miller, reviewing Pennsylvania timber history, said with chagrin, "In spite of the legend and lore of the mighty woodsman with his axe, life was cheap and expendable. Indeed, the only thing cheaper than human life was forest land itself."[14]

The cruisers, jobbers, swampers, carpenters, wood hicks, bark peelers, skidders, and teamsters were outdoorsmen, constantly out in the elements and on the move. Some of the railroaders, who arrived en masse in north-central Pennsylvania at about the same time as Clarke, also were in the woods to load logs and bark onto rail cars, or they were transporting their loads to sawmills and tanneries. Once empty, the cars were hauled back to the skidways for new loads.

William T. Clarke photographed the woods workers in their workplaces. Surviving images depict them also with fellow workers in camps at the beginning or end of their workdays or on the rare day off (usually Sunday). Clarke crowded his jobber camp images with many people, if possible. Economically, this maximized his opportunities to sell more prints and minimized the need to transport his heavy equipment (camera, tripod, glass plate negatives, and so on) to places where the natural light might not be as favorable as in open areas around the logging camps. There also were fewer men congregated in any one location at the work sites.

Indirectly, Clarke was dependent upon the jobbers he met, who were economic gamblers. They dealt with many unknowns, including sometimes stubborn, unreliable, and independent-minded employees and a hostile natural environment. If employees or weather or both didn't cooperate, jobbers had a bad season and could lose their shirt. For a jobber's crew, the impact was repeated. For Clarke, this also was a calamity, since there would be fewer people to buy his photographs during or after a poor or disastrous season. The prevailing economic laws of the woods for financial deals between "outsiders" and camp personnel was cash on the barrelhead for inexpensive items or a cash deposit for more costly items and the balance when the goods were delivered. If jobbers and their crews failed to complete logging a tract and to provide the promised peeled trees and bark in the specified location at the proper time, no one got paid, or people were paid less than they'd anticipated and hoped when the season began. Reduced cash flow would be felt by outsiders as well. A variety of workers had to work together in order to have a successful season. In addition to those who worked in the woods, there were others who stayed at the camp. At a minimum, there were clerks and storekeepers who handled inventories and kept accounts, blacksmiths, carpenters, saw sharpeners, cooks and cookees (cooks' helpers), and lobby hogs. Together with wood hicks who wielded axes, manned saws, and limbed the trees; skidders who rolled the logs with their peaveys; bark peelers and stackers; swampers; teamsters and their horses who transported logs and bark; cruisers and scalers; skidders; railroad log loaders; and men who ran the locomotives and trains, they had to form a cooperative unit, supervised by the jobber (see plate 35).[15]

William T. Clarke, as an itinerant photographer who traveled from camp to camp, survived, and his photography thrived, so it's fair to say he adapted well to a remote country existence after leaving city life behind. The visual legacy he left demonstrates his skill in photographic processes, but it also bears witness to his organizational abilities and his aesthetic sense, as he staged humans in a landscape that often was unfriendly to the photographer and his bulky equipment. That he survived economically is testament to the faith his photographic subjects had in him and his work. When men had to work as hard as they did for rather modest and uncertain pay, it must have required some persuasion to convince them to take time to pose and to pay him for the results. If the occasion involved making images on behalf of the Goodyear operation, there was even less incentive for workers to stop work for the photographer. C. E. McGuire (see plate 23) posed after notching a tree and appears somewhat impatient, as if he is saying, "C'mon, Clarke, take the picture and get it over with! I've got work to do."[16] The wood hicks and others in his photographs have the look of strong, hardened, gritty workers who were up to handling the physical demands of the strenuous and dangerous job of logging. They no doubt wanted realistic photographs that documented life in the raw as they experienced it.

The camps and the men who lived in them were not idyllic or romantic. The buildings they lived in were temporary, utilitarian, unsanitary, uncomfortable, and mostly ugly. The men who inhabited them were not well groomed or sophisticated. Clarke, who spent a lot of time in the camps, couldn't help but be well aware of the shortcomings of the camp and of the down-to-earth personalities of the men who dwelled in them. He made photographs that perfectly caught and preserved their lives in those surroundings.

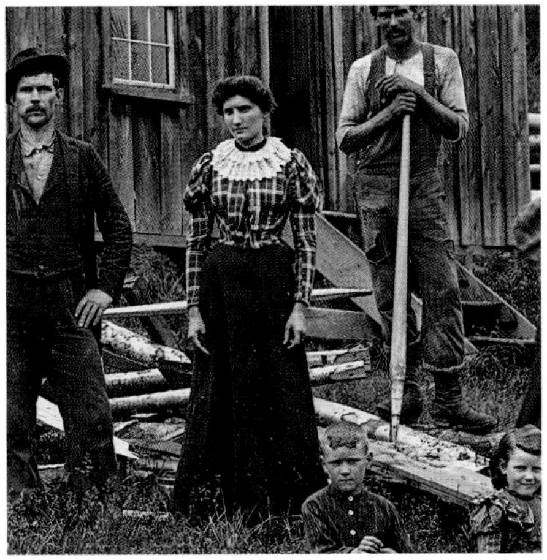

CHAPTER 4

CAMP LIFE

The north-central Pennsylvania woods logging camp of the late nineteenth and early twentieth century had no electricity, no running water, no indoor toilets, no showers, no washbasins, no washing machines, no innerspring mattresses, no La-Z-Boy recliners, no dishwashers, no microwaves, no refrigerators, no telephones, no regular newspapers or magazines, no routine mail service, no radios, no TVs, no Internet, no automatic furnaces, no recycling, no garbage pickup, and no automobiles, pickups, or mechanized vehicles other than the loaders and locomotives that pulled logs and bark out of the backwoods. There was, however, plenty of isolation.

Many inhabitants were first- or second-generation immigrants. The Swedes, Finns, English, and Dutch, on the East Coast, and French from Canada were Pennsylvania's earliest European settlers, followed by groups from Austria, the former Czechoslovakia, Denmark, Finland, Germany, Greece, Hungary, Ireland, Italy, Norway,

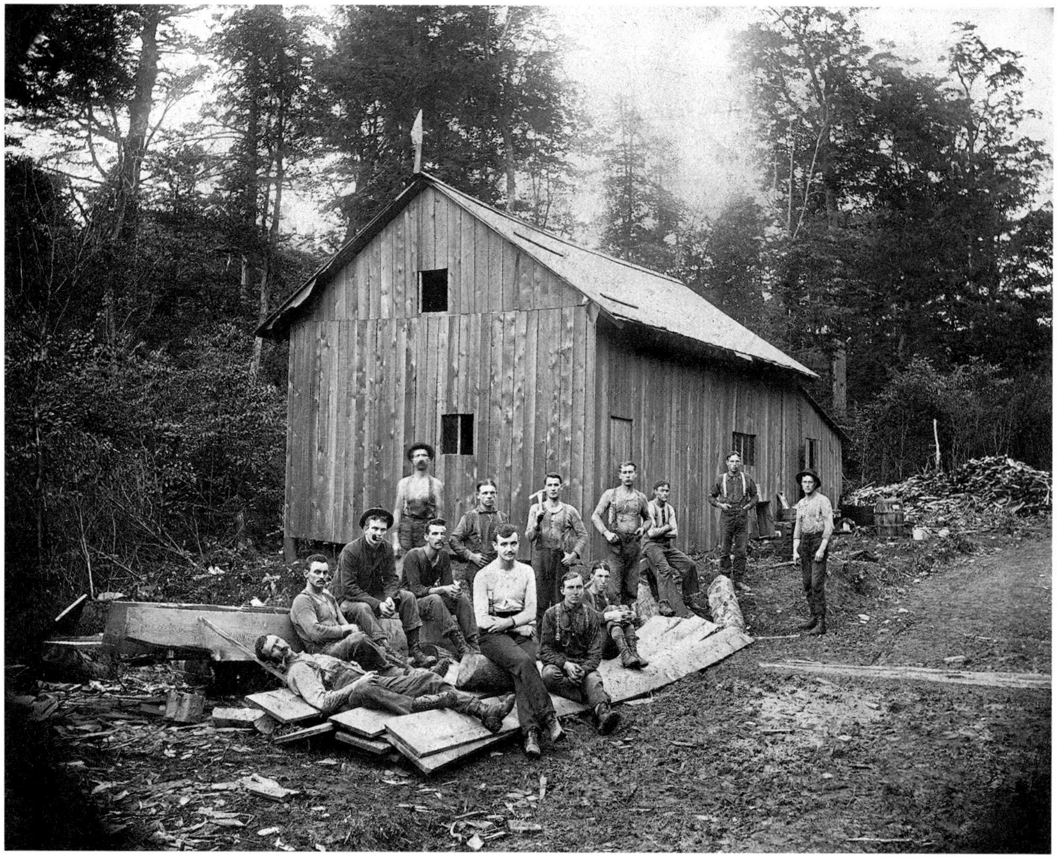

FIGURE 14
Men pause from working in the woods and on a building project

Poland, Russia, Scotland, Slovakia, and Switzerland. Before and during the 1890s, immigrants came from countries where people had survived primarily through traditional, self-sufficient, manual agriculture. Many did not speak English and the jobs in their new country often were industrial, complex, and technologically sophisticated, requiring a remarkable degree of adaptation.[1] By contrast, early woods workers logged for the most part as generations had before them. Immigrants unwilling or unable to adapt to industry tended to gravitate to places like north-central Pennsylvania. Such individuals tended to prefer life in the rough, with physical labor in challenging jobs; had few personal ties; and often were on the move. Even after the machines invaded the garden, figuratively speaking, there remained many jobs that were unchanged from those of earlier decades and generations (see plate 36).

Woods workers were not lazy. Wood hicks, bark peelers, and woodsmen worked from daybreak to sunset, up to fifteen hours a day, six days a week. Logging during the late 1800s and into the following century was rugged work. William T. Clarke's photographs of loggers show very few men who carried extra pounds on their bodies—they were lean and muscular. The number of calories they burned daily is legendary. Concomitantly, the amount of food they consumed was impressive. During the day, while men were laboring in the forests, the camps were not deserted: the cooks were at work.

The camp cook was one of the first hired after railroad spurs were constructed and a camp was ready for the full complement of workers (for larger camps, more than one cook would be needed). After the jobber, the cook was the second-most-important person whom loggers considered when they decided which camp would be their home during the logging season. A good cook and plenty of promised good grub helped the jobber assemble and keep a happy logging crew. Wood hicks known as "short-stakers" would not remain in a camp with a cook who produced poor food; they derisively called such a cook a "stomach robber," "boiler," "sizzler," "scorcher," or "can opener." The men who stayed for a season (part of the "long route") would work harder for a camp with a good cook. Good "hash flingers" were so in demand that if they moved from one jobber's camp to another jobber's camp from one year to the next, a contingent of men who'd happily eaten their fare would likely follow. Cooks received top dollar for their skills and services—in the neighborhood of $3.50 a day.[2] Jobbers were willing to pay such high wages, because they knew a bad cook could spell doom (see plates 37 and 38).

The cook was assisted by a helper known as a cookee. The cookee began his career by learning the art of cooking as sous-chef, doing the unglamorous chores of basic food preparation and service.[3] As a helper, he fed wood into the cook stove, stoked it, kept the fire going at the right temperature, hauled ashes away, and set the tables. The cookee also was responsible for calling the men who were working or waiting near the dining room for the meal. This was done by ringing a large dinner bell or by striking a piece of railroad iron hung on a nearby tree and after the clang shouting, "Take the bait!" or "Gobble the hash!" Once the men were settled at the table, the cookee served food and drink. Once they'd eaten and left for the lounge, the cookee cleared and wiped the tables and afterward washed and dried the pots, pans, and dishes. The mess hall, or cookhouse, was the cleanest area in camp. Woods workers often could not spare time to return to camp for midday meals, in which case the cookee packed take-along meals before they headed to their work sites, or delivered lunch baskets (sometimes called "nose bags," a term usually associated with feeding horses) with warm food to forest workers.[4]

Once the men assembled for camp meals, the cook, often typed as temperamental, ruled the kitchen and dining area. Loggers generally held to a rule of very little table talk, except to ask for an item of food or seasoning to be passed. Cooks and cookees worked seven days a week.

Eating was one of the greatest pleasures available in the camps. Meals were not fancy, but variety was expected and provided. For example, a breakfast (often served at 5:30 a.m.) consisted of the cook's choice of an assortment drawn from a typical menu of hot baking powder biscuits with blackstrap molasses on the side, oatmeal or cornmeal mush with butter and honey or sugar, molasses and buckwheat cakes, pancakes and syrup, doughnuts, salt pork, prunes or dried apples or other dried fruit, and coffee or tea. Dinner staples were wheat and cornbread, sauerkraut, rice, corned beef, fried potatoes, smoked ham, and salt cod. Eggs and milk were used in cooking and usually were not served fresh. Some large camps kept and butchered their own pigs and cattle. There was no refrigeration other than the springhouse, a small building over a nearby flowing spring, cool enough to keep meat fresh for a week or so, depending upon the weather. A root cellar that stored food underground was another alternative to refrigeration. Some camps near settlements relied on regular fresh meat deliveries from town butchers. In a pinch, salt pork, smoked ham, or cured bacon could be stirred into a stew or vegetable dish (see plate 39).

Cooks were expected to bake bread and simple pastries and to prepare a dessert, such as pie or pudding, for dinner. Baked beans tended to be on hand at all times. Vegetables that kept well, such as carrots, potatoes, dried corn, turnips, and cabbage, were purchased from local farmers during the harvest season and stored in rudimentary root cellars. A cook who relied too heavily upon canned food was suspect. Fresh fruit other than local berries in season was a rarity, but as canned fruit became available it was welcomed. If the camp was small, the dining area also was a relaxation area for conversation or a game of cards or checkers after the cookee cleared and wiped the tables.

The "lobby hog" held another full-time-job role. The lobby hog usually was a former logger who was too old, injured, sick, or feeble in mind or body or was no longer sufficiently alert (hearing and sight were critical to survival) to work in the woods.[5] It was not a charity job, however. The lobby hog usually sawed and split firewood and stockpiled it for the camp's kitchen, bunkroom stove, and blacksmith's shop. He did rudimentary cleaning of living and sleeping areas. He helped haul water to the camp from a nearby spring or stream. In short, he was the camp's odd job man. Not all disabled woods workers were relegated to the lowly role of lobby hog. An accidental injury, such as the loss of an arm, suffered by a young man did not necessarily disqualify him from working in the woods. Such a young man might choose to partner with another worker on a two-man saw.

Successful logging also required saw sharpeners (also known as "filers"), ax grinders, blacksmiths and, as camps mechanized, men who were handy with mechanical maintenance and repair. A saw sharpener and his filing workstation were located in the camp (see plates 40 and 41). Likewise, blacksmiths were camp dwellers because they required buildings, fires, forges, bellows, heavy iron and tools (such as anvils, hammers, pry bars, pinchers, and tongs) and bulky specialized clothing (such as heavy leather aprons) (see plate 42).

The "scaler" was an important person who lived in or periodically visited the logging camp and log skidways. He typically was hired by the timber owner and was watched carefully by the jobber and the woods workers. The scaler used a marked stick or rule to calculate board feet. Loggers often called this the "thieving stick." They were apt to assume scalers were dishonest and prone to underestimating the camp's yield in favor of the timber owner. As a result, loggers initially were inclined to ostracize a scaler. It took a wise scaler with a pleasant personality and scrupulous, demonstrated honesty to gain the loggers' confidence.[6] Occasionally, a scaler might even be generous toward their cause. In that case, everyone in camp treated him very well. Of course, that scaler would be putting his job at risk, since the timber owner was his ultimate boss (see plate 43). If the contract called for clear-cutting, the scaler would inspect the work areas to see how the job was progressing, keeping track of how many logs were cut, comparing the rough totals to what was expected based on the cruiser's estimate. His job was to count and measure trees cut by the loggers, keep production records, and report progress to the timber owner and the jobber.

Depending upon the man's skill and job, he could expect to earn from $1 to $3.50 a day, with the better pay going to saw sharpeners, blacksmiths, and cooks, who did not go into the woods, but remained in the camps.[7] Saw sharpeners needed stationary saw-filing stations. Blacksmiths required fires, bellows, forges, iron, hammers and ancillary tools, and anvils. Cooks and cookees worked with knives, pots and pans, stoves, dishes, and flatware. Bookkeepers and storekeepers, with specialized talents and responsibilities in stores, accounting, inventories, and record keeping, required desks, filing systems, account books, and dry storage space. Those who stayed in camp supported the woodsmen with serviceable tools, equipment, suitable housing, food, records, and personal needs.

Jobbers sometimes hired foremen (a foreman was called the "bull," the "push," or the "woods boss"), but in small camps jobbers often acted as their own foremen. Whether the jobber or a hired foreman was in charge, that person had to be a tough and capable authority figure in order to win the workers' respect (see plate 44). Most jobbers and their foremen did not allow card playing for large sums of money or the drinking of alcohol, predictable causes of turmoil, in their camps. Any activities in the "lobby" (lounge) that could lead to arguments or fights were monitored closely and squelched, if possible. Because the men essentially were confined to remote logging camps for long periods of time, it was important that they got along in order to be effective. Undependable or troublesome workers were fired summarily and turned out of camp.

While it varied from camp to camp, Sunday was a day off most of the time. The men could mend clothes, wash their clothes ("boiling up"), repair their gear, get a haircut, read, play cards, write letters, and so forth. If there was another camp within reasonable traveling distance, the men might go for a visit. However, jobbers were reluctant to see their employees leave camp for visits to nearby camps or communities, knowing that many workers on leave were notoriously undependable, prone to drinking, carousing, fighting, and missing work days. Of course, if there was an itinerant "sky pilot" (preacher) in camp eager to save a few souls or if a camp was located within walking distance of a settlement, a few men might attend Bible studies or worship services (see plate 45).[8] Toward the end of the season, particularly if the camp was behind the log and bark quota through bad weather or other reasons, Sunday might find the men in the woods trying to catch up ("highballing").

Photographer Clarke was probably welcome in the logging camps. There, he'd usually find overworked inhabitants who were terribly isolated, starved for information about the outside world, and eager for the novelty of a visitor with whom to talk. Too, the diversion of having their picture made gave the men something to think about and discuss among themselves afterward. In that sense,

Clarke was in the class of occasional and unscheduled visitors, like the pack peddlers who walked from one lumber camp to the next with two packs, one of clothing and another of notions (such as collar buttons, shoestrings, suspenders, pipes, razors, pocket knives, and the like). And according to George Korson, even itinerant "dentists" and "physicians" (with only rudimentary dental and medical training and instruments) were welcome alternatives to the usual self-administered camp "yarb" remedies mostly made from herb ingredients: goose grease, pennyroyal, mandrake root, deadly nightshade, burdock, bear fat, boneset, elder bark, star root, fawn's tail, rhubarb, smartweed, rattlesnake oil, camphor, vinegar, hartshorn, pokeroot and many others.[9] Thus, wanderers, peddlers, and transient travelers of all types usually were assured a good meal and a night's lodging in the logging camps. Other than merchants, farmers, and suppliers who made occasional day deliveries of camp goods and supplies, or the railroad men a woods worker might occasionally meet as logs and bark were hauled away from landings, he was unlikely to see anyone other than a fellow camp dweller for weeks or even months on end.

Any entertainment was of the camp's own making. Clog, jig, and reel dancing; reading; good-natured kidding; and practical jokes abounded. Singers were popular. They most likely would croon a heroic ballad, a lovesick song, or sentimental tune (such as "The Jam on Gerry's Rocks," "The Face on the Barroom Floor," "When Irish Eyes Are Smiling," "Marching Through Georgia," and "Grandfather's Clock"), perhaps accompanied by a musical instrument. Men who could play the fiddle, banjo, mandolin, harmonica, Jew's harp, drums, or rattling bones were in popular demand. Other amusing pastimes included magic and card tricks, board and checkers games, cribbage, matching or pitching coins, carving or whittling, or a game of pitching horseshoes.

Storytellers who could spin a yarn and tell a tall tale were much appreciated. Leo Bailey recalled "Old Beecher" as a man who told "tales of his life in southern and eastern Pennsylvania. He spoke of attending the Philadelphia Centennial Exhibition in 1876 and even showed a part of a jig that he said his nephews had performed there. 'One-Armed Ed English' once said that if all of Old Beecher's stories were true he was nine hundred eighty-four years old. But his stories were always good and, if he lied, it was only to entertain people."[10]

Reading a newspaper or magazine was a popular activity when those media were available. Bailey's memoir mentioned a "camp library that consisted of a few well worn 'dime novels,' the Pennsylvania *Grit* and a copy of the *World's Almanac*."[11] He explained, "the *Grit* was an excellent source of news and entertainment as it had a number of features, including a comic section and a weekly magazine with short stories and always a serial story of high quality. Jimmy, Pop, and some of the others had many historical political arguments that consisted of much loud talk and the presentation of a few facts. The *World Almanac* was the final word in any question involving facts."[12] The "library" might be passed from man to man if all were literate. Or there might be an especially good reader who could read aloud a popular story, a novel, or an autobiography one chapter at a time, thus passing the early evening around a barrel stove and giving the men something to look forward to the following evening.[13] The stories and topics read the evening previous no doubt would be further pondered during the next day, with some brave listeners advancing theories about what was going to happen next, to whom, and so on.

There always seemed to be at least one camp clown, who was filled with surprises and the unexpected (see plate 46). And men being men, and tough men at that, there was plenty of good-natured random roughhousing within the camp communities. Sometimes feats of strength and boasting led to challenges, resulting in wrestling and boxing.

Part of any camp's routine was the evening's review of who had that day's bragging rights. Mrs. John F. Kuhns put it this way: "These woodsmen [had] a monotonous life, where there was great competition as to who could do the greatest amount of work."[14] Who cut the most trees? Who peeled the most bark? That day? That week? So far that season? Those discussions could get contentious. No doubt every camp had peacemakers and methods of alleviating troubles, conflicts, and potential animosities.

In another aspect of camp life, however, the men could show their softer, kinder dispositions. Camp life often consisted of more than human-to-human interactions. The teamsters, of course, had their horses. But there also were cats or dogs to befriend in many camps. Animal pets are shown prominently in a number of Clarke's camp life photographs. Cats were useful for controlling rodent populations. Dogs could put the run on or at least alert the men to larger unwanted or dangerous animals that may have been lurking in the vicinity. Beyond those roles, cats and dogs made good companions for groups of men. Ear scratching and petting probably were so common that the behaviors were almost automatic (see plates 47 and 48).

Clothes might be washed by hand on Sundays. Often, such tasks were done cooperatively. Washing clothes, for example, could easily involve five men: one man to start and keep a fire going, as well as to haul and boil water, another to soap and soak and stir clothes in a kettle or cauldron and to scrub them on a washboard, a third to thoroughly rinse, and two to wring and to spread the washed articles to dry on top of a bark pile, or in trees, or over logs (see plate 49). And, as long as they were at it, perhaps a man would take a bath then, too. But washing and bathing were not mandatory and were nearly impossible in cold weather. In fact, some men owned only the clothing they wore and believed the medieval superstition that washing oneself caused disease. Another complication came during the winter, when washed heavy clothing (often wool) hung by a wood stove could not be counted upon to be dry in time for Monday morning's labors (see plates 50 and 51). Younger men, either in the camps or working for the lumber railroads or road crews, were likely to own a change of "good" clothes and a fancy hat that would be donned for going to town on special occasions (see plate 52).

Writing letters was another "day off" activity, even though it might be some time before they could be mailed. However, not all the wood hicks and bark savages were literate. Other Sunday favorites were sleeping in, gossiping, whittling, and perhaps shaving or getting a ten-cent haircut from someone handy with scissors. When the bicycling craze hit the United States in the 1890s, some loggers caught the bug, but the rough conditions of the few roads available must have made their treks laborious and hazardous (see plates 53 and 54).

While the activities of the day off were unstructured and spontaneous, the workers' normal routine had them awake at 5:00 a.m. Each day's entertainment, after the evening meal, was brief. In truth, the men usually were exhausted after long hours, dawn to dusk, from working in the woods or in camp. "Douse the glimmer," the call for lights out (extinguishing lard oil or kerosene lamps and candles) came early, around nine o'clock. Lamp chimneys, globes made of clear glass, were bulbous above the end fitting over the lamp's wick and then narrowed to form a round, smaller-diameter chimney top at six to ten inches above the wick. The chimney's inside often became black because of untrimmed wicks, or wicks that were turned up too high, which made the fire burn brighter initially, but eventually created more black sooty smoke. The lobby hog's odd jobs included cleaning lamp chimneys (see plate 55).

The buildings, hurriedly constructed, often from recycled lumber from earlier buildings on other sites, were not insulated. Buildings often had open cracks between the wallboards and a thin tarpaper layer nailed to the roof boards to cover gaps, a poor design for any season. In 1975, longtime jobber Clyde Lynch recalled a camp on Skinners Creek near Port Allegany: "The bedroom had double beds on each side. Two men slept in a bed. The beds were wood framed with metal slats upon which the mattresses lay. The camp was hotter than blazes in the summer. It had only two windows, one at each end. In the winter the only heat was from the kitchen and lobby stove on the first floor."[15]

When the men went to bed, outer clothes were hung near a stove to dry, if necessary (*if* there was a stove in the sleeping quarters). A man usually slept in his "union suit," a cotton body smock that buttoned in the front from groin to neck, with a back flap over the buttocks held up by two buttons. The men collapsed dead tired into rough bunks in cramped quarters. The fires died soon afterward. Meanwhile, the temperature outside often was subzero. Wool blankets, but no sheets, were standard bedclothes over straw- or cotton-filled mattresses. Under these conditions, as Clyde Lynch noted, it was not uncommon for two men to sleep together in order to share body heat. No doubt many men objected to the practice, acquiescing only because of the extreme cold. Often, the small windows in the buildings' gable ends were covered with boards for the winter. Despite the cold, one can imagine that soon after retiring, the fatigued men were sawing symbolic logs with their snores.

When the weather warmed, the buildings did a complete temperature flip-flop and became like ovens. In camps where men slept above the mess hall, kitchen, and cook's room, the wood cooking stoves below sent up unwanted summer heat. Ventilation always seemed problematic, regardless of the season. In cold weather, ventilation seemed drafty. In the stifling hot weather of spring and summer days and nights, ventilation seemed negligible, even though the winter's boards that covered the open windows were removed in the spring, summer, and early autumn. There were no summer screens to prevent ready access for mosquitoes, gnats, flies, and other insect pests (called "punkies"). However, the little ventilation that resulted was deemed preferable to a few bug bites. The workers were philosophical, accepting punkies, bedbugs, and lice as part of their lot in life: "They can bite and be damned!"

Whoever they were, wherever they came from, the men in lumber camps were tough and wiry. They lived in primitive shacks constructed hurriedly and crudely. They enjoyed few amenities. Camp conditions and habits often were rudimentary, with men using the same towel, drinking from springs or bunkhouse water buckets with the same nearby drinking cup or water dipper, or guzzling from the same water jug while in the woods. Garbage and dirty dishwater were dumped anywhere and everywhere. Sometimes, there may not have been a designated privy. At least, no buildings that look like traditional outhouses have been identified in camp photographs. Under those circumstances, we can speculate that when nature called, the men tended not to stray too far from camp. They put up with a lot of dirt, primitive washing and bathing facilities, and unsanitary conditions.

Occasionally, there would be a halfhearted attempt to rid the sleeping area of the pests. The lobby hog would spray beds and sleeping quarters with a fine mist of kerosene oil mixed half and half with turpentine to keep the vermin at bay. But in an environment of wood or coal fires for warmth, kerosene and turpentine were used with caution. The lobby hog also made up the bunks and kept the floors in a semblance of cleanliness, hampered by the fact that many of the men chewed (and often swallowed) plug tobacco, used snuff or "snoose," or smoked pipes and weren't fastidious about where they spit or dumped their ashes and burned matchsticks.

Jobbers had "but one goal: it was production. Logs! The best crew was the one that put the most logs on the landings. It was as simple as that."[16] They dealt with many unknowns, including rough, tough, and hard-to-bluff employees and a hostile natural environment. Around the Fourth of July, the bark tightened on the logs and the bark peelers couldn't continue even if they had wanted to. Besides, some of the woods workers had been in the boonies since the previous autumnal freeze-up and were anxious for a change of scenery and some excitement. Independence Day was a major holiday and everyone was ready for a big party (known by the lumbering men as a "convention"). Wood hicks, bark peelers, and other woods workers from nearby camps gathered in local communities for a few days of boasting, competing, drinking, gambling, fighting, whoring, and letting off steam. Each camp had a champion tough guy. When the camps broke up for the Fourth of July celebrations, contests sometimes were arranged when camp strongmen were pitted against one another. However, spontaneous barroom brawls were more frequent when men of various camps congregated and imbibed.

There was usually a lot of end-of-the-season money in their pockets at convention time, and the loggers often weren't interested in saving or investing their season's pay. It was much more common for the men to blow their money on a good time, consisting of wine, whiskey, loose women, gambling, and other vices.[17] Some men were "rolled" (robbed) when they were drunk and passed out. Cautious men sometimes asked someone else (often a madam, bartender, pool room operator, or hotel keeper) to hold part of their pay, which they intended to use for necessities or previously incurred debts. However, they ran the risk of being bilked out of part or all of the money they'd set aside. Unscrupulous "bankers" might shortchange them, overcharge them, or claim they'd come for their money while drunk and it was gone.

Men with a connection to family, friends, or a community were more likely to gravitate to celebrations where activities and companions were more civilized, social, and sedate (see plate 56). Some had farms or family members with farms and returned to them after Independence Day to take up farming, at least until harvest, when there was a temptation to "go a-loggin' agin" that might lure them away. Jobbers were hiring crews for the upcoming season.

For the wood hicks and other logging laborers who had spent their money and enjoyed the good times of idleness in pursuit of various sins, all too soon the conventions were over, the languorous, hazy, crazy days of midsummer were gone, and the cool winds and nights of autumn were upon them. Then, they could be seen with their few belongings in a rucksack, walking the railroad tracks, in search of their camp and berth for the season. The big fun was over and no doubt more than one of the men had but a befuddled memory of what had transpired. A few probably vowed to shape up, live right, and reject their wayward habits. The woods beckoned, with visions of good, regular food, honest work, good companions, a bed for the night, and a structured, purposeful life once again.

Clarke's images of logging camps and lumbermen would not win blue ribbons in fine pictorial art shows. Nor did they hang in New York City's prestigious 291 Gallery, which displayed Photo-Secession images by his contemporaries, prestigious photographers such as Alfred Stieglitz (1864–1946) and Edward Steichen (1879–1973).[18] However, when Clarke made portraits for clients living in the communities that grew up during north-central Pennsylvania's logging era, he created photographs that required a different approach and a sensitive touch. Whether they were made in his studio, inside his client's homes, or out-of-doors, he captured compelling images of men, women, and children.

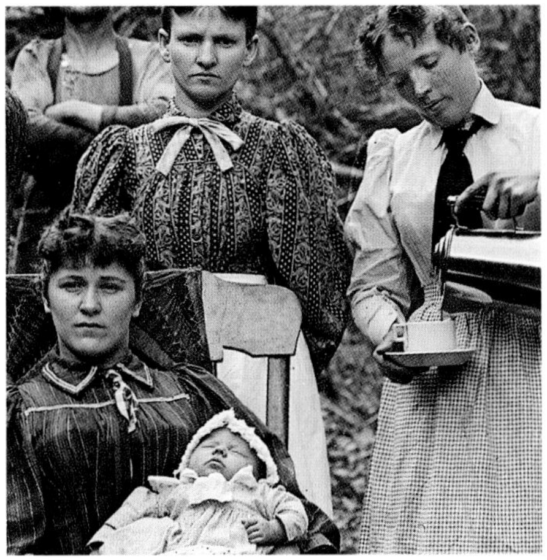

CHAPTER 5

COMMUNITY LIFE

As time went on, the small communities the woodsmen left behind as they trudged back to the lumber camps each autumn were becoming more and more civilized. Gender roles were changing as the result of the increased pace of inventions, manufacturing, national transportation, marketing, goods and services consumption, and work. Some men were turning from self-employment, subsistence and cash crop agriculture, and small businesses to factory work and service jobs. While women's domestic burdens were relieved somewhat by the advent of tools for laundering, cleaning, ironing, and cooking, a majority continued to work at home and to care for children. However, conditions were changing, slowly and inexorably. More and more people aspired to and became middle class. Families increasingly lived in houses featuring amenities such as curtained double-hung windows, indoor electricity, electric or gas lighting, telephones, parlors, clapboard siding, shingled roofs, porches, and brick chimneys.

The evolving communities suggested a permanence not found in the earliest wilderness communities that had grown up as extensions of expanded logging activities. As communities grew, flowers, gardens, lawns, and fences decorated many plots of land where homes were situated. Citizens sent and received mail from established post offices and participated in community social activities that frequently revolved around schools and places of worship. Belongings and store goods increasingly were standardized, but more choices were available for purchase. Mail service and railroad transportation led to the growth of mail order catalogs.[1] Local merchants, who had once had a monopoly on goods with inflated prices, were forced to become more competitive and to offer consumers increased choices and alternatives in order to compete with what was available through mail order and rail delivery. Merchants began to provide more local services, such as free delivery, warranties, on-site repairs, credit, and flexible installment payment options. The proliferation of consumer goods led to increased niche markets, targeted advertising, brand names, and corporate identities.[2]

Clarke's images of communities show family homes that were built more solidly and therefore suggested more permanency, compared with the temporary logging camp buildings occupied primarily by males who intended to leave as soon as the timber had been harvested. The photographs show dwellings that displayed homey touches and that were located in neighborhoods. Business and industrial sites were integrated into communities, and permanency was suggested by sidewalks, roads, railroad tracks, electricity, and so on (see plates 57–60).

His photographs record north-central Pennsylvania communities that had a symbiotic relationship with the logging industry. For example, a typical small town in timber country might include a general store, butcher shop, pool hall, barbershop, livery, schoolhouse, church, post office, sawmill, lumber-dependent manufacturing industry, railroad connection, and "pig's ear," with its steady clientele trying to quench their thirst for liquor (see plate 61).[3] Logging stimulated employment for many working in ancillary occupations in those towns and villages. Post office workers received and distributed mail. Railroads employed stationmasters, baggage handlers, section crews, roundhouse mechanics, conductors, brakemen, firemen, and engineers (see plate 62). Storekeepers and suppliers provided workers with processed food and supplies. Butchers slaughtered animals and dressed meat. Liveries housed and tended to stabled horses. Barbers cut hair, provided shaves, and shared town gossip. Laundries and bathhouses allowed people to clean up—quite a novelty for some. Nearby farmers tended animals and grew vegetables and sold them along with dairy products and eggs. Some farmers processed specialized products such as maple syrup and honey.

Distillers and bootleggers brewed and bottled potent batches of liquor, wine, and beer. Marie Nuschke provides an intimate glimpse into the alcohol, gambling, and prostitution industries in "Industry Comes to Freeman Run."[4] Proprietors of pig's ears where men drank alcohol, gambled, and employed "fancy ladies" became bold enough to run businesses on main streets, businesses that often posed as respectable soft drink establishments, pool halls, restaurants, or drugstores, in a commonly recognized fiction. Madams and prostitutes sometimes took over the second floor of such buildings, or provided their own facilities for visits and services, usually on back streets or outside the town boundaries. Since local townspeople also frequented the fronts, town constables and county sheriffs tended to give them free rein, unless fights broke out, at which time the pig's ears' bouncers tossed offenders out the door to finish their brawls in the street. The audacity of the proprietors came to an end in Potter County in 1899 with strict enforcement of the Brooks High License Law. Many pig's ears thereafter retreated from communities and conducted more secret underground businesses in isolated locales.

As communities grew, physicians, dentists, lawyers, bankers, teachers, religious leaders, carpenters, journalists, innkeepers, tailors, police officers, undertakers, and others were ready when needed. Nearby timber-processing industries—sawmills, chemical plants, tanneries, and related businesses—converted logs and bark into consumer and industrial products. The local and regional north-central Pennsylvania economy of the late nineteenth and early twentieth century often revolved almost exclusively around logging and timber by-products.

A few residents became quite wealthy through lumbering (see plate 63). Ladies from well-to-do families kept abreast of trends in interior decoration by subscribing to national women's magazines and wore the fancy dresses and flowered hats that were in fashion elsewhere.[5] They traveled on graded roads, their buggies and fancy carriages pulled by sleek horses. They lived in large, well-built, painted, two-story houses firmly situated on stone foundations. Their nicely furnished houses suggested rooted permanence, with broad porches and walks. Plants and flowers were displayed in house windows, in window boxes outside homes, on the porches, and in cultivated garden plots. Solid sheds and sturdy barns were nearby.

Clarke adapted to the growth of such communities and became more than an itinerant photographer working for the Goodyear Lumber Company camps. He established a studio sometime between 1900 and 1912 in Conrad ("Hull's Settlement"), Potter County, and between 1912 and 1917 in Betula, McKean County.[6] This developing business sideline probably offered a welcome financial supplement to his primary work for the Goodyears. In time, as an established community and regional photographer, Clarke became trusted and respected by people from all walks of life living in the growing number of north-central Pennsylvania communities. His images had a wide range of subject matter, depicting, for example, men in primitive logging camps, a jobber and his family outdoors (see plate 44), and a well-to-do older woman inside her Victorian house (see plate 64). Although Clarke clearly had mastered outdoor photography, he needed a place to settle where he could develop prints from his negatives, made in camps when he was on the road, and to store his heavy and fragile glass plate negatives. His surviving studio and home interior images demonstrate that he had honed his lighting and compositional techniques. Increasingly, Victorian-era clients wanted to show off their sitting rooms or "first parlors" so as to suggest affluence, leisure, and comfort. Photographic images helped them satisfy that desire.[7]

Clarke took advantage of natural lighting from skylights in his studio. He relied on bedsheets to diffuse and reflect light as needed. His studio featured a faux backdrop that depicted a portion of a home interior, painted to suggest three-dimensional surfaces. Clarke posed his subjects carefully (see plate 65). The studio was small. Clarke's ingenuity permitted some changes of configuration; he had to be versatile when he worked with large groups in such a limited space (plate 66). Another studio image features a beautiful child, old enough to sit but young enough that a woman, barely visible to the right, appears ill at ease (see plate 67). The three studio images do not show his customers in everyday dress. More likely, the clothing represents the subjects' finest, because getting a portrait made was often for a special occasion.

Clarke made group photographs at schools, one of the most important institutions in the community. School life probably was like that in most small towns and villages throughout the nation (see plate 68). Schoolhouses almost always were built by community residents and consisted of fairly large one- or two-room buildings, heated by a wood- or coal-burning barrel stove in winter. School was held from a minimum of seven to a maximum of nine months (September to May) a year. Children walked to school and back home, often trudging two or more miles in each direction, in all types of weather. School usually started at nine in the morning and ended at four in the afternoon. The day would start with a short opening exercise consisting of a prayer, the Pledge of Allegiance, and perhaps a song or two. Students sometimes recited poems or read stories at morning assembly. Eight grades were taught, often by a lone teacher, very often an unmarried woman who boarded with a local family. The older children were expected to tutor the younger ones while the teacher was busy teaching students of a particular grade. The boys and girls of the first grade learned to "say" the alphabet and to read and to write a little. All grades had frequent lessons in spelling and grammar. Fancy cursive penmanship in the Palmer method was respected in the community and considerable time was spent developing this skill.[8] Arithmetic stressed solutions to practical business problems and bookkeeping. Any subjects beyond the 3 Rs (readin,' 'ritin' and 'rithmetic) were considered frills. Students welcomed the midmorning fifteen-minute recess, followed by a lunch break and another recess, at midafternoon. Well-behaved pupils were awarded the coveted jobs of passing out pencils at the beginning of the day and collecting them at the end. After school, erasing and washing the blackboard also was a favorite reward. Older children were expected to help with chores, such as carrying wood or coal, emptying ashes, sweeping and mopping the floor, and carrying water.

Clarke also photographed churches. Even in the most remote and hardscrabble areas of the timbered wilderness, acceptance of spiritual beliefs and established religious dogma was an important part of many lives. One reason, perhaps, was that churches proclaimed an eternal afterlife with one of several possible existences. The most attractive of the everlasting life options, heaven, promised Christian admittance, a very attractive sequel for many, because their earthly lives often were filled with work, hardship, disappointment, despair, and drudgery. The alternative perpetual destination, hell, was vividly described from many pulpits and those contrasting visions made a life of piety attractive (see plate 69). With communities splintering into different ethnicities and religious denominations, even small communities usually had several places of worship. Many Clarke church images are stereographic, consisting of double images, each slightly different from the other, made by a special camera. The resulting stereoview cards gave viewers the illusion of a single three-dimensional image when viewed through a properly focused stereoscope. Clarke made a number of stereo church pictures, suggesting that he was able to tap into a ready market in church congregations and was willing to photograph what would readily sell.

Clarke's images often depicted town residents who interacted with railroad employees at depots and with merchants in well-stocked stores. Their wagons, buggies, and carriages rolled over roads that didn't jar their bones with continuous bumps. But the drift toward gentility didn't happen rapidly. The communities also attracted rowdy nonresidents from time to time. For example, hamlets such as Cross Fork in Potter County were places of woodsmen's drunken debauchery, scandalous behavior, and even violence that erupted after men had spent months in the wilds of the forest. The railroad ran right through the town of Cross Fork, making the town a crossroads (plate 70). This "lusty, brawling town" was known for hosting the seamy side of life with its many pool halls and pig's ears. Once away from the camps with their pockets filled with pay, wood hicks exhibited behavior that was often erratic and vicious. Sooner or later, there would be no-holds-barred fights in which men were mauled and injured. The brawls usually started as a result of too much alcohol.

Marie Nuschke, who came from a long line of French wood hicks, witnessed such fights as a child and described the wood hicks who got into fights as "full of liquor, especially of the squirrel or red-eye variety. . . . The offender started by swinging with a right and a left until his opponent's eyes were black, a few teeth were out, and he was as groggy as a swing under an apple tree. Then came the finishing process. Every hick wore 'corked' (calked) boots. As soon as the opponent was down, the victor jumped him, often corking him in the head and face into unconsciousness. Worse yet, at one time, it became popular to chew one ear off the vanquished man on the ground so he would remember his complete defeat when he came to." As in the woods, physicians were summoned or a man brought to the doctor's office only if bones were broken, the bleeding was profuse and wouldn't stop, or a man seemed incapacitated and near death. Old-timers remarked that they could always spot fighting men: "They usually had an eye gouged out, an ear torn off, or most of their front teeth gone." The marks left on a man's face by the steel spikes in corked boots were known as "logger's smallpox."[9]

Cross Fork provides an illustration of a boom-to-bubble-to-bust community during the late nineteenth and early twentieth century. The town's growth was phenomenal. Two months after the opening of the state's most modern sawmill on May 3, 1894, the *Tribune* on July 3 reported that the "splendid specimen of the younger class of enterprising American towns" had a population of fifteen hundred persons and "a fire department, with a first class equipment," as well as plentiful housing built by the Lackawanna Lumber Company for rent at four dollars a month. There also was a company store where provisions and household effects were sold "at astonishingly low prices." Two months later, the newspaper reported that "Cross Fork, the new town on Kettle Creek, Potter County, has two drug stores, three meat markets, five grocery stores, three millinery shops, one hardware, one boot and shoe store, one cigar factory, one furniture store, five billiard rooms, four barber shops, three blacksmiths, four hotels, eight boarding houses, two liveries, one shingle mill, one clothing store, one opera house, two secret societies, one jewelry store, one bakery, one confectionary store, twenty-five building contractors, 120 dwelling houses, seventy-five carpenters, and 800 laboring men there and in the vicinity." Cross Fork had a YMCA, a coronet band, an Independent Order of Odd Fellows, a Knights of the Maccabees, and its own hose company with sixty members pledged to parades and fire protection.

The raw and seamy side of Cross Fork was tempered somewhat as prosperity continued after the early boom days. Dynamos from the newly rebuilt sawmill provided electricity. Four churches were established. The Baptists and the Methodists-Episcopalians supported regular pastors. The Lutherans met every other Sunday, sharing a pastor headquartered in Port Allegany, McKean County. And the Roman Catholics worshipped with a rector who traveled from Austin once a month.

Newspapers throughout the United States tended to be boosters of their hometowns during the late nineteenth and early twentieth century, carrying glowingly positive news and editorials that were notoriously optimistic and filled with puffery.[10] The boasting and bragging helped inflate a north-central Pennsylvania bubble. While the boom communities grew, the woods were swarming with men from logging camps. However, the bubble was to burst.

As time went on, hillsides for miles around looked the same—shorn of trees. Hemlock and other species were consumed voraciously by sawmills, secondary wood facilities, chemical plants, and tanneries. Trees are not an annual crop, and once the big hemlock stands came down, it would take decades to replace them. The timber supply was exhausted. The logging activities of Cross Fork Creek, the Kettle Creek valley, the Big and Little Kettle Creeks, the Hevner Run, the Trout Run, the Carter Camp, and hundreds of others were abandoned. With the demise of the forests, loggers and prosperity vaporized. Cross Fork deteriorated rapidly over the course of a few years. The kindling wood factory burned in July 1907 and was not rebuilt. The voracious sawmill was gone by 1908. Forest fires on January 5 and April 13 in 1909 destroyed significant portions of the town's buildings. They were not rebuilt. On March 18, 1911, the company store, the Kirk Hotel, the YMCA, and several other buildings were lost to fire. The last vestige of industry, the Pennsylvania Stave Company, called it quits in 1912. Buffalo and Susquehanna and Oleona and Germania Railroad railroading ceased May 31, 1913. After that, there were only two general stores, a cigar factory, and two hundred residents in Stewardson Township. As businesses closed, workers and their families moved away. One of the few work activities left in Cross Fork was the melancholy dismemberment of homes and buildings that once housed hopeful families and businesses that had prospered and then faded or failed. Salvaged building materials were recycled elsewhere.[11] And then Cross Fork became very, very quiet.

Surviving communities struggled on. Businesses often relied on immigrants and other laborers who did not have sufficient money to build their own houses. Most potential laborers probably knew their jobs would be transitory and they'd be moving along after awhile anyway. As an inducement to attract a dependable labor force and to make money at the same time, companies often built and rented boardinghouses for single workers and company houses for families (see plate 71). Construction was quick and cheap. Houses were identical. The company store, which handled almost anything the population needed, was a similar monopoly, extending credit to workers and their families and then deducting hefty purchase prices from paychecks. By 1916, electricity was available in many of the company towns, supplied to consumers (again, at a profit) by dynamos installed primarily to power the plants' machinery.

Harry W. Edenborn and Dessie Severson estimated that more than seventy wood chemical plants were in operation on the Allegheny High Plateau of north-central Pennsylvania from 1880 to the mid-twentieth century. The chemical factories used destructive distillation methods to manufacture charcoal, creosote (wood tar), turpentine, methanol (wood alcohol), calcium acetate (acetate of lime), and acetic acid (ethanol acid) for industrial uses, using mostly smaller hardwoods at the rate of ten to twenty cords of wood daily, with the largest of the factories consuming an average of eighty and a half cords daily.

During the region's timber heyday, thousands of workers were involved in the processing, packaging, transportation, marketing, and sale of wood and related products. When the timber was gone and the lumber-related industries left or died, the communities sometimes died as well or became dramatically smaller and poorer as residents faced the future and the need to reinvent their economic base. A substantial portion of the population simply chose to move to places where economic conditions were better. In keeping with the fate of many of those whom he photographed, William T. Clarke was one of those who packed up and left.

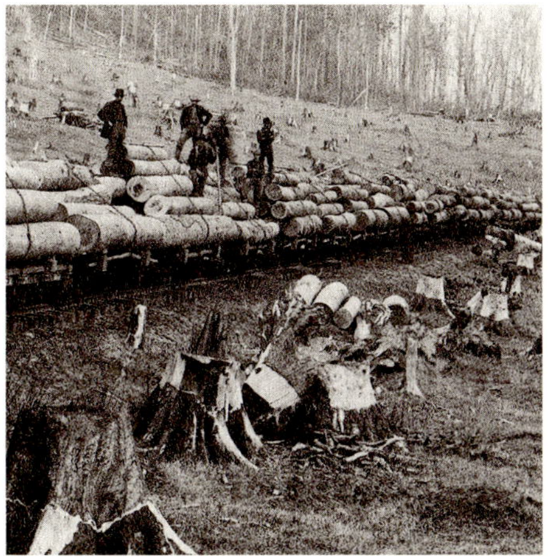

THE PENNSYLVANIA DESERT

CHAPTER 6

When Europeans discovered the Western Hemisphere, the region now known as the United States of America was a vast forest extending from the Atlantic Ocean westward—limited only and finally by lack of moisture—encompassing some 900 million forested acres, or 47.4 percent of the land surface.[1] Conventional wisdom holds that aged forests, which had been left alone for centuries and subjected only to disruptions such as floods and fires, would have continued to thrive for centuries more. As the original forest grew and renewed itself, the land became increasingly fertile. In 1847, Henry Wadsworth Longfellow celebrated a "forest primeval," an unspoiled virgin forest untouched by humankind.[2] Longfellow's romantic outlook increasingly was challenged and the idea of an unspoiled virgin forest existing for centuries untouched by humans has been declared mythical by an emerging scientific point of view that holds that forests suffered "flagrant destruction" by humans for agriculture and commerce, leading to waste and land abandonment.[3]

A major challenge to the idea that the nation's forests and other natural resources were unlimited and inexhaustible and could be harvested by humans without restraint did not come until the late nineteenth century, however. Recent research on the American forest since 1600 presents evidence of continuous changes to U.S. forest ecology by natural forces and pre-European populations. Fire (from lightning strikes or set by Native Americans), wind, climatic change, insects, fungi, animal activity, and other factors have had constant impacts on the forest and led to changes.[4] This science-inspired perspective was late in coming to Pennsylvania even though Gifford Pinchot, foremost promoter of the modern conservation movement, was a Pennsylvanian.[5]

The original forest of William Penn's domain of 1681 was estimated to cover 90–95 percent of the land.[6] By the early 1840s, it had shrunk to two-thirds of the Commonwealth.[7] Part of that shrinkage can be attributed to the "Age of Wood." In the years prior to 1840 almost everything humans needed and wanted could be made of wood. The transition to the "Age of Metal" was evolutionary. Between 1840 and the beginning of the twentieth century, wood retained its basic importance and was a primary fuel in the production of iron, the most basic metal. Nationally there was insatiable demand for Pennsylvanian wood when William T. Clarke photographed the logging industry.[8]

As shown in an 1814–15 survey in Warren and McKean Counties and studies of the 1930 "Heart's Content" old-growth forest in the same part of Pennsylvania, the natural forest of nineteenth-century north-central Pennsylvania consisted mostly of hemlock, beech, and maple with smaller stands of chestnut and white oak.[9] As loggers moved from camp to camp, they left behind clear-cut hills and valleys where once the seemingly endless Black Forest had stood. Succession growth without planned cutting, reforestation, and selective fire control leads to forests of different tree species. Logged white pine and hemlock on upland soils are replaced by even-aged hardwood trees.[10] Much succeeding growth in Pennsylvania's devastated forests was not planted or planned. It took a long time for the landscape, clear-cut and often burned, to look like it held anything but shorn and forlorn scrub brush. The scarred region came to be known derisively as the "Pennsylvania desert," a name popularized by Joseph Trimble Rothrock, the first commissioner of the Division of Forestry under the Pennsylvania Department of Agriculture, a position he held from 1895 to 1904 (see plate 72). Local self-deprecating descriptions of the area included "desert of five million acres."[11]

In 1903, Robert Laird McCormick, president of the Mississippi Valley Lumberman's Association and secretary of the Weyerhaeuser Timber Company, spoke with chagrin of the "vandalism" the lumber industry had caused: "The necessity for care is not seen until the harm has been done and the results of it are bitterly felt. We are often unable to supply the enormous demand for lumber which we have fostered. We must look to the production of a second crop upon lumbered lands or prepare to stop lumbering when the first crop is gone. In the old days it was easy to lumber one forest and then turn to another. To-day, however, we cannot count upon new fields to turn to. We must make the best of what we have, and wherever it pays we must lumber conservatively."[12] Timber titans did not follow their fellow harvester's advice to replant trees. When the tall timber was exhausted, they abandoned Pennsylvania (and elsewhere) with their companies. Their former land was forfeited for unpaid taxes or they sold it for a pittance. As the companies moved, so did many workers. This was nothing new. Timber industry and logger migration from state to state was routine. Many men who logged Pennsylvania came from the depleted forests of Maine and New York, and when they were done in Pennsylvania, many migrated to the South, the Upper Midwest, or the Pacific Northwest.[13]

Industries built by timber titans were dependent on a slow-growing natural resource, and their levels of greed could not be sustained. In north-central Pennsylvania, the Goodyear brothers ran two of the largest sawmills in the state, one in Galeton (producing 92 million board feet annually) and the other in Austin (72 million board feet anually). The boom-to-bubble-to-bust community of Cross Fork was home to the Lackawanna Lumber Company, the Commonwealth's third-largest sawmill operation at that time (69 million board feet annually). It was apparent from the beginning of their sawmill operations that production could not be sustained. Clarke, in an August 1912 letter to Henry Shoemaker, lamented the voracious speed of a Goodyear mill, possibly the one in Norwich: "What a hurry for the fastest mill ever run in this country is now eating up the trees at a rate of 275,000 to 300,000 per 24 hours." Their timber holdings feeding the Cross Fork sawmill consisted of forty thousand acres, mostly hemlock, in the mountainous lands of the Stewardson and Abbott townships in Potter County. There, the Department of Agriculture in 1897 estimated that forests would supply fifteen years of logs (an estimate that proved right on the money). The Lackawanna enterprise helped Potter County reach a peak population of 30,621 in 1900. By 1906, there were fifty-seven large mills in Pennsylvania. Hundreds of smaller mills were also operating.[14]

As the timber industry declined, the wood consumption industries continued to produce sawdust and to belch smoke as they consumed logs that had to be hauled from increasingly distant forests. Industries dependent upon wood adapted to conditions of scarcity. For example, the paper mill in Austin, built in 1900, along with the nearby Emporium Lumber Company's hardwood mill, used smaller and smaller remaining trees as time passed. Finally, the loggers and mills conducted a mopping-up operation, using natural resources bypassed during the gravy days of lush, dense forests (see plate 74).[15]

Once major logging operations ceased, fires became common in north-central Pennsylvania. Wood waste had accumulated and was tinder dry. Massive piles of sawdust were abandoned near mills. Industries engaged in wasteful manufacturing processes, simply dumping unused timber nearby. There also were vast waste branch and foliage slashings left behind in the former forests after wholesale cutting and removal of bark and tree trunks. Such conditions rendered that part of Pennsylvania susceptible to massive destruction. Fires, ignited by lightning or started by people, often burned rapidly and out of control, leaving dead animals, smoldering stumps, and barren hillsides.[16] The huge ancient forest of old-growth trees was but a sad memory (see plate 75). Eventually all that was left in many places was a bleak, burned, black wasteland. Governor Pinchot, who held many public service roles in conservation and forestry, said in 1923 that 5 million richly forested acres had been decimated, reduced to desert. He wrote, "The one word which properly describes the facts surrounding the forest and lumber situation in Pennsylvania is 'bare.'"[17]

Communities suffered as a result of the way the land was vacated and the exodus of many former residents. After the fires passed, the scrub trees and brush that sprouted naturally on the cutover and burned land were not the same species that had been decimated. Sarcastic critics noted there often was more wood in the area's railroad ties than on the hillsides along the rail lines (see plate 76). Dealing with what remained, a district forester had three major responsibilities after the demise of the timbering era: to suppress fires; to organize and conduct reforestation; and to respond to the growing public demand for recreation, hunting, and fishing.

Environmental problems in addition to fire became apparent. Erosion was common, and many communities, built by rivers and streams, suffered from flooding and pollution (see plate 77). Waste products from a number of timber-related chemical industries were toxic. As a result of these problems, many communities continued to be abandoned, becoming ghost towns.[18]

From the "Pennsylvania desert," loggers headed south to Louisiana and other states of the gulf or to upper midwestern states like Michigan, Wisconsin, and Minnesota, where the same scenario would be repeated.[19] Some migrant loggers, encouraged by the railroad companies, ended up in the Pacific Northwest, where the mighty ancient redwood, giant sequoia, ponderosa, lodgepole pine, and Douglas fir would once again inspire unrealistic dreams of inexhaustible timber resources.[20]

Farming, a traditional alternative method of earning a living, was only modestly successful (see plate 78). Farming also was susceptible to depletion of natural resources or was inappropriate because of environmental conditions (see plates 79–80).

Promising alternative hopes for north-central Pennsylvania focused on the fledgling tourist and outdoor-sports industries (see plate 81). However, tourism depended upon an infrastructure that had not been fully developed or that had become unprofitable in north-central Pennsylvania. For example, transportation was hindered by the railroads' abandonment of lines, removal of rails, and movement of equipment elsewhere. Hunting and fishing were popular outdoor sports for some tourists and former residents who lived outside the region (see plate 82). However, wildlife had become extinct or exceedingly scarce through the destruction of fish and other animal habitats and through the increase in fishing, fur trapping, and hunting for food.

With the discovery of oil at Titusville, Crawford County, in 1859, the population experienced a new exhilaration. Crude oil, once refined, had many uses. Oil, geologists knew, extended from western Pennsylvania, at the New York border, to the southwestern corner of the Commonwealth.[21] McKean County, which had been a lumber producer, contained a large reservoir, the Bradford Field, producing thirty-one million barrels of oil, or 83 percent of the nation's output.[22] Potter County, east of McKean County, was not a big oil producer, and it's likely that a well that Clarke photographed in the East Fork Valley between Hull and Jamison turned out to be mostly dry (see plate 83).[23]

Both soft and hard coal (bituminous and anthracite, respectively) were abundant in Pennsylvania. Soft coal was found mostly in western and north-central areas, while hard coal was located in east-central and northeastern regions of Pennsylvania.[24] In the 1870s, soft coal fueled the growing railroad system. When the hills were shorn, workers in some north-central Pennsylvania counties (for example, Cameron County) were able to turn to coal mining as an alternative source of livelihood (see plate 84).

Another Pennsylvania industry in Clarke's era was glass making (see plates 85, 86). A strong demand existed for windowpanes, lampshades, tableware, photographic plates, and bottles and containers of all sorts, and there were available local resources, including glass sand, silica, potash, and bituminous coal or natural gas to fire the furnaces. Glass plants began in north-central Pennsylvania after the discovery of gas around 1900. In Potter County, manufacturing plants were located at Coudersport, Roulette, and Shinglehouse.[25]

Clarke's photography business sputtered to a halt. His last known Pennsylvania address was Betula. He left the Commonwealth for good sometime soon after November 1917, when he completed images of the once mighty Goodyear mill at Norwich. Norwich would become a ghost town by 1920, having existed for ten years. By the time he returned to Rochester, New York, where his extended family lived, most of the land of the former Black Forest that he had left behind was a sorry sight. Many workers and communities were without much hope for the future. Before leaving Pennsylvania, Clarke filled a sturdy wooden crate with 131 glass plate negatives that included many of his earliest efforts (the cache Lois Barden discovered in a shed many years later). Clarke took the crate and apparently did not leave a forwarding address or even tell anyone where he was going. Pennsylvania came perilously close to losing all his work, now one of the finest visual troves of the lumbering era.

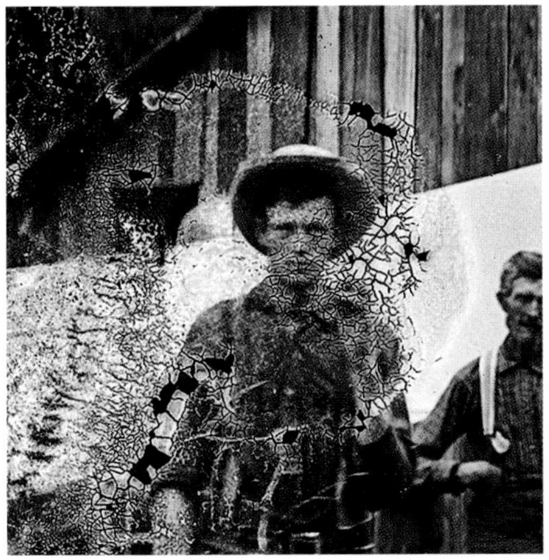

A MIGHTY TRANSFORMATION

CHAPTER 7

At the end of the nineteenth century, visionaries and futurists looking at north-central Pennsylvania saw economic trouble ahead. They saw that an economy dominated or heavily dependent upon a single natural resource would not be sustainable. As a resource, trees were renewable but at a very slow rate, not enough to maintain the peak levels of employment and production that the logging, lumber, and wood industrial processes had achieved for decades.[1] In short, north-central Pennsylvania was running out of the trees that were preferred for commercial logging and manufacturing (see plates 87–89). Moreover, the long-term effects of clear-cutting the forests were negative (see plate 90). Meanwhile, the population, dependent upon the harvesting and processing of forests, had settled in (see plates 91–97). There were well-established communities, a developed rail transportation system, a developing roadway network, retail and service businesses, government offices, banks, electricity, schools, religious and fraternal groups, and so forth (see plates 98–100). All this translated into a firm sense of pride of place. Alternatives would need to be found in order to continue the area's development and to reward individuals and families who wished to remain. Individuals more flexible in their plans might choose to move. This included people who had accrued wealth (see plate 101). They also would have to find new means of generating capital, or leave. While the forests were viable for economic purposes into the first decade and a half of the twentieth century, the day was coming when big decisions and workable new ideas would be needed.

A disaster highlighted one of the unintended, but important, consequences of indiscriminate logging. In 1909 the Bayless Pulp and Paper Mill constructed a dam 50 feet high and 534 feet long to supply a dependable source of water and power to the mill. The dam burst. The Austin flood of September 30, 1911, that destroyed the communities downstream resulted in part from tremendously high water levels after rains and the subsequent pressure that caused the dam to be breached. Raging floodwater knocked down nearly everything in its path as it surged eight miles through the valley into the Austin and Costello communities (see plates 102, 103).[2] There were seventy-eight confirmed deaths; thirty-eight people were reported missing and presumed dead.[3] Although the dam had recently been "repaired," it later was found to have structural defects and the company was held liable for negligence. However, the fact remains that excessive water behind the dam as a result of erosion was a major contributing factor to its failure. The erosion was attributable to tree removal and the loss of vegetative understory that formerly absorbed rainwater, preventing excessive runoff.

The challenges were great, but key thinkers, activists, public servants, community-minded businesses, and industries met the heavy demands of the era. Some alternative means of making a living, such as agriculture, immediately came to mind as people cast about for a way of making a living. Agriculture preceded and had coexisted with the logging era, but with some exceptions, the land was not well suited to farming (see plate 104). The task of transforming the way of life of north-central Pennsylvania was enormous. Diversification of economic enterprise was fraught with risks. Some ideas and schemes did not work. Fortunately, enough were successful to characterize the changes wrought in north-central Pennsylvania over the twentieth century as "a mighty transformation."

One of the best alternative ideas for economic growth grew out of the leisure activities of logging and farming families and tourists, who had enjoyed the natural recreation and aesthetic benefits of north-central Pennsylvania before the wholesale harvesting of timber (see plates 105–109). The land and areas of water were developed to promote tourism and recreational activities. Those activities remain very popular as more and more nonresidents discover the delights of contemporary north-central Pennsylvania. Today, residents and tourists alike experience life in a thriving, scenic environment featuring the natural beauty of parklike mountains, canyons, plateaus, forests, and sparkling waters.

Moving from place to place over long distances is facilitated within the comfort of personal vehicles and bus travel on smooth, safe highways. Cities, towns, and villages offer quality public and private institutions and businesses. Citizens and visitors enjoy safe and comfortable accommodations, services, and conveniences. They have many places where they can shop, dine, and enjoy arts and culture. There are institutions of higher education where people can inform themselves and enjoy high-quality enlightening entertainment. There are plenty of opportunities to pursue a wide variety of leisure activities: to sightsee, hike, jog, cycle, raft, tube, swim, boat, canoe, kayak, camp, picnic, golf, horseback ride, watch birds and animals, take nature photographs, hunt and fish, drive an ATV, take a historic railroad excursion, and participate in diverse individual and team sports. During the winter, cross-country and downhill skiing, snowshoeing, tobogganing, skating, and snowmobiling are popular. Today, the forests and waters are shared by logging and other industries, by residents, and by tourists. The many state parks and the Lumber Museum are among the appealing venues in north-central Pennsylvania that people look forward to with anticipation.

Most possibilities in today's north-central Pennsylvania weren't even dreamed of during the thirty-odd years during which William T. Clarke lived in the state, from the late nineteenth through the early twentieth century. As a young man, he'd witnessed the majesty of the old-growth Black Forest. In adulthood he'd seen years of single-minded, unrelenting clear-cutting of those ancient forests, a harvest that yielded great economic wealth for some and modest prosperity for others. And by the time he packed up to leave Pennsylvania, things looked grim. As he traveled to Rochester, New York, where he'd live until his death in 1930, Clarke must have been dismayed and disgusted. The timber titans and many who worked in their industry had devastated and abandoned north-central Pennsylvania, looking for new greentop gold. Clarke had stayed long enough to photograph the ruined condition of the land and the early searches for alternative incomes. But his livelihood, like that of many others, had diminished. He must have left the state reluctantly. Photographers like Nelson Caulkins and George Clark would fill his void.

In writing of slashings (parts of trees and vegetative debris left behind after logging), Clarke referenced the inefficiencies and wasteful practices of timber processing during the late nineteenth and early twentieth century. Pennsylvania had witnessed a timber boom that ceased after the bubble burst. The major question everyone asked was, what now? Residents had to immediately deal with survival. There were two broad alternatives: join the exodus and leave for real or imagined opportunities elsewhere or stay and figure out alternative ways to make a living. Most decisions about staying or leaving were made on a personal level by individuals, families, and small groups.

One who definitely stayed was Joseph Trimble Rothrock, who was active in the state's forestry program until 1914.[4] As forestry commissioner, Rothrock inherited the problem of a north-central Pennsylvania left in ruins, a circumstance that troubled him enormously. Considered Pennsylvania's "father of forestry," Rothrock decried the fact that the forests of white pine and hemlock were gone. He told audiences during his speeches that some sixty-four hundred square miles containing more than 9 million acres of the state were cut and that fire was an ever-present danger. According to Eleanor A. Maass, his biographer, he wrote his wife that much of the land he saw on a train trip along the Clarion River (once heavily forested) was a "worn out, fire wasted region," with "mile after mile. . . simplylaid waste. I could not get 'Abomination and Desolation' out of my mind. . . . No words can depict it."[5] Commissioner Rothrock achieved two major accomplishments. First, he initiated an ambitious state land acquisition program by purchasing land from the lumber companies. Second, in 1903, he urged Governor Samuel W. Pennypacker (1843–1916) to create the Pennsylvania State Forestry Academy (one of the first in the nation, now known as Penn State's Mont Alto) to train foresters for service in Pennsylvania.[6] Rothrock, moreover, was an American environmentalist who constantly promoted conservation and the creation of state parks and state forests. He encouraged the people of Pennsylvania to keep three goals in mind: preserve the land and natural resources, protect them, and propagate the forest. He was a strong advocate of reforestation to prevent erosion of the soil.[7]

Happily, in addition to Rothrock, there were other Pennsylvania visionaries and leaders, often removed from the immediate area (such as Harrisburg), who took a systemic and institutional view of the problem. They weren't simply dreamers. They were activists who worked through a series of interlocked steps necessary to achieve long-range goals, as manifested by north-central Pennsylvania as it exists today. They saw and defined big-picture problems. They considered alternative actions to create solutions, both ideal and practical. They established goals for their vision and mission, set sustainable priorities, and mobilized support. They decided what activities had to be done and in what sequence. They wrote and passed enforceable rules and laws concerning behaviors that would not be allowed or that would be encouraged and rewarded. They created new organizations to implement their plans. They debated ideas and arrived at workable compromises, after which they worked together to coordinate activities. They measured successes and failures and adapted their activities if needed. They stressed communication within their institutions and to the public. Small successes were celebrated and setbacks were analyzed and new methods designed and implemented to prevent similar failures. Their work was cumulative and long range in orientation as they worked through successive phases to achieve their dreams. New, unanticipated problems were not ignored but were studied and, to the extent possible, nipped in the bud. As the imagined new north-central Pennsylvania began to be realized, they thoughtfully promoted activities that would enhance quality of life and economic prosperity for residents and that would provide unique, pleasurable experiences for visitors. Education, research, articulation, and advocacy of positive environmental and social values became part of the modus operandi. Powerful allies were recruited to sustain and enhance gains and to prevent possible losses. Increasingly, nonprofit and special interest groups found common ground with what had been primarily a publicly funded state enterprise. The balance of this chapter will note a few of the important individuals and institutions responsible for the magnificent and mighty transformation that is north-central Pennsylvania today, as compared with what it was when William T. Clarke left in the early twentieth century. Each individual is discussed separately, but in reality, they all worked together, either by plan or by happenstance, knowingly or unknowingly, and in real time and at different times to press forward their visions for the future.[8]

Mira Lloyd Dock (1853–1945) was a prominent Pennsylvanian from Harrisburg, educated at the University of Michigan in botany, geology, and chemistry, and a good friend of Joseph Trimble Rothrock.[9] She was an author, speaker, teacher, informal lobbyist, and reformer. As a member of the American Forestry Association, she participated in the Progressive Era forest-rebirth movement through education and conservation advocacy.[10] In 1900, with William Stanley Ray, Dock published a bibliography on nature study for public school education.[11] From 1901 to 1913, she served as a member of the Pennsylvania State Forestry Reservation Commission, which was charged with purchasing land for the state.[12] She often visited forest sites that were for sale and reported back to the commission, thus becoming a highly credible source of information and suggestion.[13]

Similarly, Dock dedicated her active volunteer life to spreading the word about the need to support the emerging conservation movement in Pennsylvania. She gave testimony at state hearings and lectured widely on a variety of topics, including the need to protect and preserve forest lands. She advocated for forestry education, relevant laws, parks, and playgrounds. She made a short bibliography on forest topics available to her audiences so they could follow up her talks with additional information.[14] An avid photographer, she took pictures of trees and used glass lantern slides in lectures in Mont Alto forestry classes.[15] And as an active advocate for women in careers, in 1910 she published a chapter, "The Profession of Forestry," advancing forestry as a suitable vocation for trained women.[16] As chair of the General Federation of Women's Clubs, Forestry Committee, Dock published a bulletin titled *State Forestry Today*.[17]

Clearly, the conservation movement had to be promoted outside north-central Pennsylvania. Government action, civic change, and social improvements in Harrisburg also were high on Dock's priority list. She advocated for beautification projects, such as parks and gardens. As a college-educated woman, she knew the importance of basic sanitation and worked hard to see that Harrisburg had the healthy benefits of a sewage system, paved streets, and clean water. She did not receive many accolades during her lifetime. However, her courageous and outspoken stand for the causes she believed in and the successes achieved for her fundamental concerns made her a respected community citizen. The many improvements that were achieved largely because of her persistence, writings, and speeches led to her widespread recognition after her death in 1945. Her example has inspired many young girls and women, as well as boys and men, to follow her activist lead for the betterment of their communities.

With pioneer conservationists paving the way for him in Pennsylvania, Gifford Pinchot (1865–1946) was able to make a great positive impact on forestry following the Pennsylvania desert years.[18] His father, James W. Pinchot, came from a family distinguished for great business success. His mother also came from a wealthy family. His father had suffered from "hemorrhages of the lungs" as a young man and spent a long time recuperating during a horseback tour through a number of southern states, where he came to love nature.[19] Later, James urged his son to study forestry at Yale University, in one of the first forestry programs in the United States and a program James had helped found.[20] Gifford Pinchot followed his father's advice. After completing his undergraduate program, he spent an additional year at the French Forest School in Nancy.[21]

Thus began Gifford Pinchot's lifelong commitment to the progressive conservation movement.[22] He advocated forester professionalization and promoted scientific management of forest harvests.[23] He emphasized an efficient market economy and was an effective communicator for practical forest planning, cutting, elimination of waste, and renewal.[24] Unlike many charismatic figures of the era, he did not believe in preserving the forest for its beauty and recreation alone.[25] He wanted a commercial "working forest" for "working people." Pinchot advocated government planning and management assistance (for a fee), with government regulation and oversight of logging.[26] His model promoted modest family logging operations by private owners and discouraged trends toward huge logging syndicates and timber monopolies that he feared would continue to create "absolute devastation" of forests through clear-cutting.[27]

In 1898, Gifford Pinchot was chief of the Division of Forestry under President William McKinley. He began a national dialogue devoted to a shift in national emphasis from harvesting entire forests to modern selective and sustainable forest management. After McKinley was assassinated in 1901, Pinchot joined forces with his friend President Theodore Roosevelt to make conservation a premier issue of American politics, serving in that role until 1910.[28] As part of Roosevelt's administration, Pinchot gained more resources and credibility for the cause of conservation and for his newly established United States Forest Service.

Pinchot was a "big picture" planner. As first chief of the Forest Service and chair of the National Conservation Commission, he oversaw a three-volume report that for the first time in U.S. history inventoried national forest, water, soil, mineral and other natural resources.[29] The report demonstrated that "only about three-eighths of a tree was finally put to use" and decried the colossal waste that had taken place.[30] Pinchot used the monumental reference as a resource in talking and arguing for a new interpretation of people and their environment, for pushing a new national conservation ethic, and for fostering other developments associated with an optimistic American natural resources program advocated by Roosevelt.

In his first annual message to Congress in December 1901, Roosevelt, the great conservationist who thought that forests and water problems were the most vital domestic issues the country faced in the new century, put action regarding irrigation and forestry very high on the nation's to-do list. He spoke more specifically on the nation's forestry problems to members of the Society of American Foresters in late March 1903. He elevated the problem of forest depletion from a provincial business concern, seeing it instead as meriting national alarm, warning that the nation's prosperity had to be preserved and increased: "The United States is exhausting its forest supplies far more rapidly than they are being produced. The situation is grave, and there is only one remedy . . . the introduction of practical forestry on a large scale."[31] Not everyone agreed with Roosevelt, however. Natural succession growth in the clear-cut forests seemed to ensure an eventual second cutting, but this easy approach had an unforeseen flaw—the natural source of tree seeds often was scarce to nonexistent.[32]

Pinchot may have provided the first drafts of Roosevelt's earliest conservation speeches. At any rate, the two men certainly were in tune and "on message," as demonstrated when Gifford spoke to the National Lumber Manufacturers Association in Washington, D.C., on April 20, 1903. Gifford offered help to the timber users in order to meet Roosevelt's goals, saying, "It is perfectly obvious that the supplies of certain kinds of timber are rapidly disappearing, that the lumber trade is falling back year by year on poorer material and longer hauls. . . . The question of continuance is already demanding an answer. . . . Should you cut off your timber and let the land go back for taxes?"[33]

Despite Gifford Pinchot's vision of scientific forestry and his offer of practical advice and instruction on how to secure a profitable second crop from their land, leaders within the Pennsylvania lumber industry largely disregarded him and the president and abandoned much of their land after clear-cutting. They simply headed for greentop gold elsewhere to repeat the cycle all over again. Indeed, the expertise and wealth of lumber barons like Frank and Charles Goodyear, Frederick Weyerhaeuser, John DuBois, William E. Dodge, and Harry McCormick were invaluable in starting large-scale logging operations elsewhere.

After Joseph Trimble Rothrock stepped down as commissioner of the Division of Forestry in the Pennsylvania Department of Agriculture in 1904, subsequent government leaders seeking to restore the forests and environment in north-central Pennsylvania realized that a photographic history of the boom-to-bust timber cycle would be an invaluable aid in efforts to inform and warn the population of problems and to inspire new efforts to regain what had been lost. Clarke's photographs, taken over a time span, permitted critical interpretations of the effects of machines in the garden. The government leaders were in agreement with Clarke's personal texts and captions written on the backs of some photographic prints indicating his disillusionment with the changes machines had wrought.

When he was governor of Pennsylvania for the first time (1923–27), Gifford Pinchot brought an impressive set of national credentials and accomplishments to his state service. Because state law prohibited governors from serving successive terms, Pinchot had to wait for his reelection, to a second term served in 1931–35. During his time in the state governor's office, Pinchot continued to preach and promote a "conservation ethic" (a term he preferred to "natural resources"). He also advocated for better roads and a quasi-public, government-sponsored, federally regulated giant electrification project. After finishing his terms as governor, he went on to write influential books and continued to work for conservation causes.[34]

Like Governor Pinchot, Joseph S. Illick exerted great influence through his publications. In addition to his Pennsylvania Department of Forests and Waters technical research writings, aimed at scientists, Illick made forest information and photographs available to naturalists, working foresters, tourists, students, and youth through numerous educational forest field guides and assessment and evaluation publications.[35] The popularity of his publications is attested by their frequent revisions and reprinting.

His popular writing and subsequent work with influential opinion leaders and activists had beneficial effects on the public agenda. For example, Illick knew and respected another Pennsylvania environmental pioneer of the early twentieth century, industrialist Jacob Nolde (1859–1916). Nolde, a self-made man, remembered the majestic European forests he saw as a child. Born in Germany, he moved to Pennsylvania in 1880 and, thanks to a tireless work ethic and his business acumen, soon owned a large manufacturing plant based in Reading, Pennsylvania. Demand for the hosiery he manufactured made him wealthy. His lasting contribution to Pennsylvania, according to legend, was inspired by a single white pine growing in a meadow. With his eye on that majestic tree, Nolde discovered the second passion of his life. He hoped to eventually "convert a great number of acres of farmland into the most beautiful pine forest in Pennsylvania."[36] As he worked to fulfill his dream, his interests in forests broadened and he helped found the Berks County Conservation Association and became active in other environmentally oriented groups, such as the American Forestry Association.

Nolde took many cues from what the Pennsylvania government was doing to reforest the logged and burned-over sections that once held the north-central Pennsylvania forest. Like federal and state leaders, Nolde saw that professional foresters were needed. In 1912, he hired William G. Kohout, an Austrian educated in modern forestry management. Kohout knew about the advances in forestry that Pennsylvania was making and how the state was going about making progress, having worked with Rothrock at the Division of Forestry for several years. With Kohout's encouragement, Nolde hired workers to construct an extensive network of roads to interlace his forest (which eventually grew to some 665 acres south of Reading) for firefighting purposes. They imported trees from Europe and built local nurseries to grow seedlings. The Nolde-Kohout planting program featured white, yellow, Norway, and Douglas pine, supplemented with red elder and Japanese larch. Nolde's dream eventually was realized. The plantation became the Nolde Forest Environmental Education Center under the aegis of the Pennsylvania Department of Conservation and Natural Resources in 1970 and continues to serve as a demonstration of the powerful importance reforestation has for contemporary life.[37] Encouraging private reforestation and setting up state-sponsored environmental education centers, such as the Nolde Forest Environmental Center, were a great success.

Another key facet of Pennsylvania's leadership in the environmental movement and in fostering positive conservation attitudes and behaviors is demonstrated by Maurice K. Goddard's career in education and government service. Goddard (1912–1995) had undergraduate and graduate degrees in forestry, a subject he taught at the Penn State Mont Alto School of Forestry in 1935–37 before moving into administration as director of the school. Mont Alto was followed by a directorship of the Pennsylvania State University Forestry School. His major impact on Pennsylvania's mighty forest restoration came when he turned his attention to state government. Cabinet officer for six Pennsylvania governors, Maurice Goddard was a driving force behind the creation of forty-five state parks, adding an additional 130,000 acres to today's 136 Pennsylvania state parks (not counting conservation areas). Striving toward a goal of a state park within twenty-five miles of every state resident, he and his staff nearly accomplished this feat, all the while staying in touch with ordinary people and urging them to become active in forestry.[38] He served as secretary of the Pennsylvania Department of Forests and Waters, later known as the Department of Conservation and Natural Resources, from 1955 to 1970. In 1970–79, he oversaw the creation and professionalization of the Department of Environmental Resources, serving as the department's secretary. The unit was formed by combining a number of existing state departments.

After he retired, he remained active as a board member of organizations, including the National Wildlife Federation, the Pennsylvania Environmental Council, and the Pennsylvania Forestry Association. Maurice K. Goddard State Park in Mercer County of northwestern Pennsylvania is named for him, celebrating him as a master planner and advocate for Pennsylvania parks and natural and wild areas with clean air, forests, and waters.[39]

How well did Maurice Goddard's management model work? For that matter, what are the cumulative legacies of Joseph Trimble Rothrock, Mira Lloyd Dock, Gifford Pinchot, Joseph S. Illick, Jacob Nolde, and the many other environmental and conservation activists who have dedicated their efforts to restoring and improving Pennsylvanian woodlands and natural resources?

Historical records of the Department of Conservation and Natural Resources and similar state agencies that are preserved at the Pennsylvania State Archives give testimony that great strides have been made in sustainable forest management. The state's first purchase of land for a state forest, orchestrated by Joseph Rothrock and Mira Dock, was in 1898 for the Young Woman's Creek area in Clinton County. Continued purchases of state land, often replanted with trees grown in state seedling nurseries, led to impressive increments in state-owned forest acreage: by 1912, there were 966,295 acres; by 1917, 1,017,773 acres; by 1975, 1,926,772 acres; by 1995–2007, 2.1 million acres. State forest lands are found in forty-eight of Pennsylvania's sixty-seven counties, representing 12 percent of the state's total forested land. Forests have regenerated. About 17 million forested acres cover 59 percent of the state. Whereas once, in the early part of the 1900s, fires burned out of control, they are now controlled (in a 97 percent reduction from the 350,000 acres that once burned annually). And the forests aren't locked up in untouchable fenced areas. Wood is part of a vibrant Pennsylvania economy. Forest products account for $16 billion in sales annually, with a total economic impact estimated at $27 billion, leading to more than eighty thousand jobs.[40] Dealing with problems and issues of environmental and natural resources at the public and private levels was a task that began slowly in the early twentieth century, accelerated, and continues today.[41]

William T. Clarke left Pennsylvania nearly a century ago, no doubt discouraged and downhearted after decades of photographing the people who lived and worked in the forests. He had witnessed the glory days of the Black Forest; these eventually became the sad and bitter days of the blackened, scorched, denuded earth of the Pennsylvania desert. Clarke's great achievement was to document visually a boom-to-bubble-to-bust era of timber harvesting, showing viewers today what great splendor existed in his time and what horrific tragedy ensued. He reminds us now that *we* must be aware of the consequences of what we do and how we do it in the name of making a living and consuming products and energy by harvesting natural resources.

Clarke probably wouldn't believe his eyes if he were to revisit the areas he once traversed. Happily, Pennsylvanians could show him today that with the active participation of the people and with a government, timber industry, and culture that values nature and the outdoors, it is possible to regain majesty that once existed and was then lost.

Time, effort, science, management, education, infrastructure, dedication, and investment are required. Resilient north-central Pennsylvania today is grand testimony to that will and the great accomplishments that have been achieved in the century since Clarke last walked the plateau, hills, and valleys. One big difference that he would observe today is that the forest is not now primarily white pine and eastern hemlock. Now Clarke would see and perhaps gladly photograph timber stands also consisting of hardwoods.

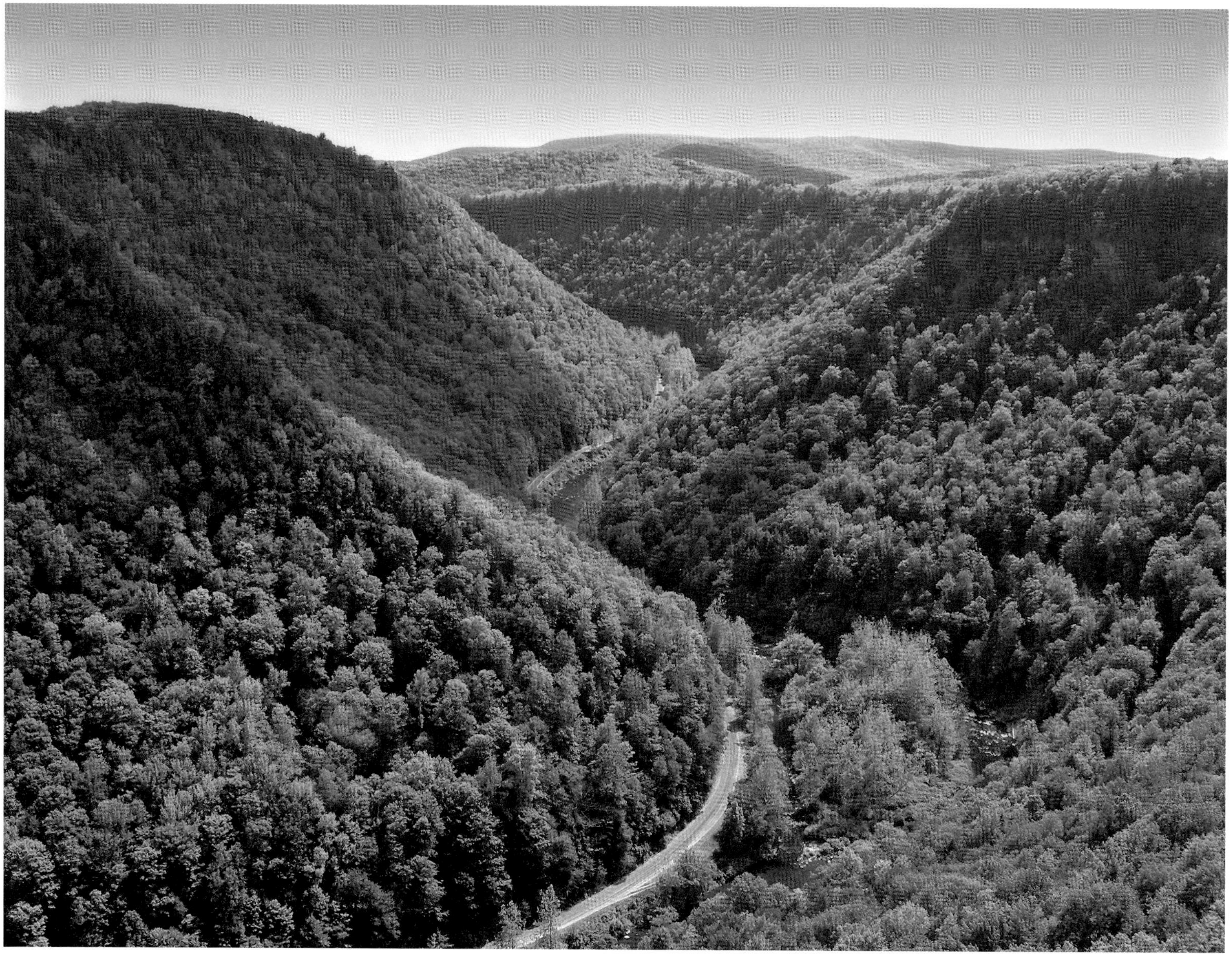

FIGURE 15

Contemporary image by Nicholas A. Tonelli, Pine Creek Gorge, Tioga County, the same area pictured in plate 3 (Grand Canyon of Pennsylvania)

It will be a long, long time, however, before future conservationists can make the claim made by William Dodge of Phelps, Dodge and Company, developer of the lumbering capital of Ansonia beginning in 1836. Dodge boasted that the company cut trees that measured 168 feet from butt to tip and soared ninety feet before the first limb was formed. Some of those logs were fifteen feet in circumference.[42] But perhaps the future may again permit bragging rights to such lofty statistics.

A museum near Galeton, in the Pennsylvania Wilds and the Lumber Heritage Region in the heart of the Susquehannock State Forest of north-central Pennsylvania, helps curious visitors put things into perspective. The Pennsylvania Lumber Museum, administered by the Pennsylvania Museum and Historical Commission, is dedicated to keeping the world of forest industry and technology that Clarke knew in the late nineteenth and early twentieth century alive and relevant. It is situated along Nine Mile Run, a place where Clarke climbed the Goodyear-owned hills to photograph loggers in action and where he froze the image of an impatient C. E. McGuire in time and place (see plate 23).

If possible, a visit should be planned to coincide with the Bark Peelers' Convention, an annual celebration held at the museum and coinciding with Independence Day. The Convention (as the old-time wood hicks and bark peelers simply referred to the gathering) marked the end of peeling hemlock bark from cut trees for the tanneries and chemical industries. By the Fourth of July, the season was over. Bark peelers had worked up to sixteen hours a day nearly nonstop, beginning in the spring when the frost was gone to the time the bark "tightened up" in late June or early July. They and their fellow wood hicks were ready to spend some of their money, blow off some steam, and have a good time, gathering in communities and villages to compete in activities such as greased-pole climbing, "birling" (two men stand on a log floating in a pond, each spinning it with their feet, trying to dunk the other), horseshoe pitching, clog dancing and singing, storytelling and tall tale boasting, competing in crosscut sawing and chopping speed, precision tree felling, ax throwing, bark peeling, wrestling, and so on. At the Lumber Museum, visitors can view a reconstructed logging camp; a restored Civilian Conservation Corps cabin; and displays of steam-powered locomotives, log loaders, and other mechanical equipment of the past.[43] On occasion, they can also see work demonstrations, such as of log skidding, blacksmithing, sawmilling, shingle milling, wood turning, and craft making. From time to time, the museum has demonstrated a functioning birch still, modern chainsaw carving, broom making, and musical performances. Of particular interest is a display integrating examples of Clarke's photographs in a newly redesigned museum building exhibit space.

Most of the forsaken land of one hundred years ago, photographed by Clarke, is now occupied by many state parks, state game lands, and the Susquehannock State Forest. The region is also once again blanketed by a green carpet of trees, thanks to the Commonwealth's massive and aggressive tree planting campaigns from 1910 to 1930 and continued later by the Civilian Conservation Corps. The streams have been cleaned and promote world-class trout fishing. Many plant and animal species absent from the "Pennsylvania desert" one hundred years ago have returned. These lands are now managed as part of the continuing rebirth of Pennsylvania's forest. Its survival is ensured for the foreseeable future. The lands, forests, and waters will be there for generations to come as part of the public trust to promote livelihoods, enjoyment, and nature appreciation. There are regional hiking trails over the remnants of the old graded beds left by the Buffalo and Susquehanna, the Goodyear Lumber Company railroads, and the many spur lines up the valleys. The town of Cherry Springs, where Clarke once photographed the grueling rail switchbacks over Hogback Mountain, is now part of Cherry Springs State Park; one can hike these switchbacks and walk on history. Now reclaimed by the forest, Cherry Springs State Park also is one of the best places in the eastern United States for stargazing, and it is the site of the annual Woodsmen Show sponsored by the Galeton Area Rotary Club.

Likewise, at Lyman Run State Park, where the Goodyear operations carved a logging rail spur into Lyman Run Valley between 1905 and 1908, visitors can hike the Spur Line Trail. In 1905, Frank and Charles Goodyear had a large camp and engine terminal built on the site that now serves as the park's day use area. Clarke photographed many activities in the Lyman Run Watershed, ranging from logging, to homes in the present ghost town of Lyman Run, to the Goodyear rail facilities, to the National Chemical Company plant (see plates 20, 24, and 26).[44] Some state park visitor centers in the region offer displays of area history and hold programs featuring Pennsylvania's conservation legacy. For example, there is a tribute to Rothrock, Dock, Pinchot, and Goddard in the Sinnemahoning State Park exhibit area.

It will take perhaps another century to wholly heal the land and return the trees to near their size when William Penn obtained his charter for Pennsylvania in 1681. But we have learned much about the wise use and management of the forests. Timber companies now recognize and respect the benefits of replanting after harvesting. State park, forest, and game lands are managed for all to enjoy and to reap the benefits of the outdoors. We are the stewards of Penn's woods. William T. Clarke would no doubt smile and approve.

PLATES

PLATE 1

Original forest along Indian Run, near Norwich, McKean County, 1910

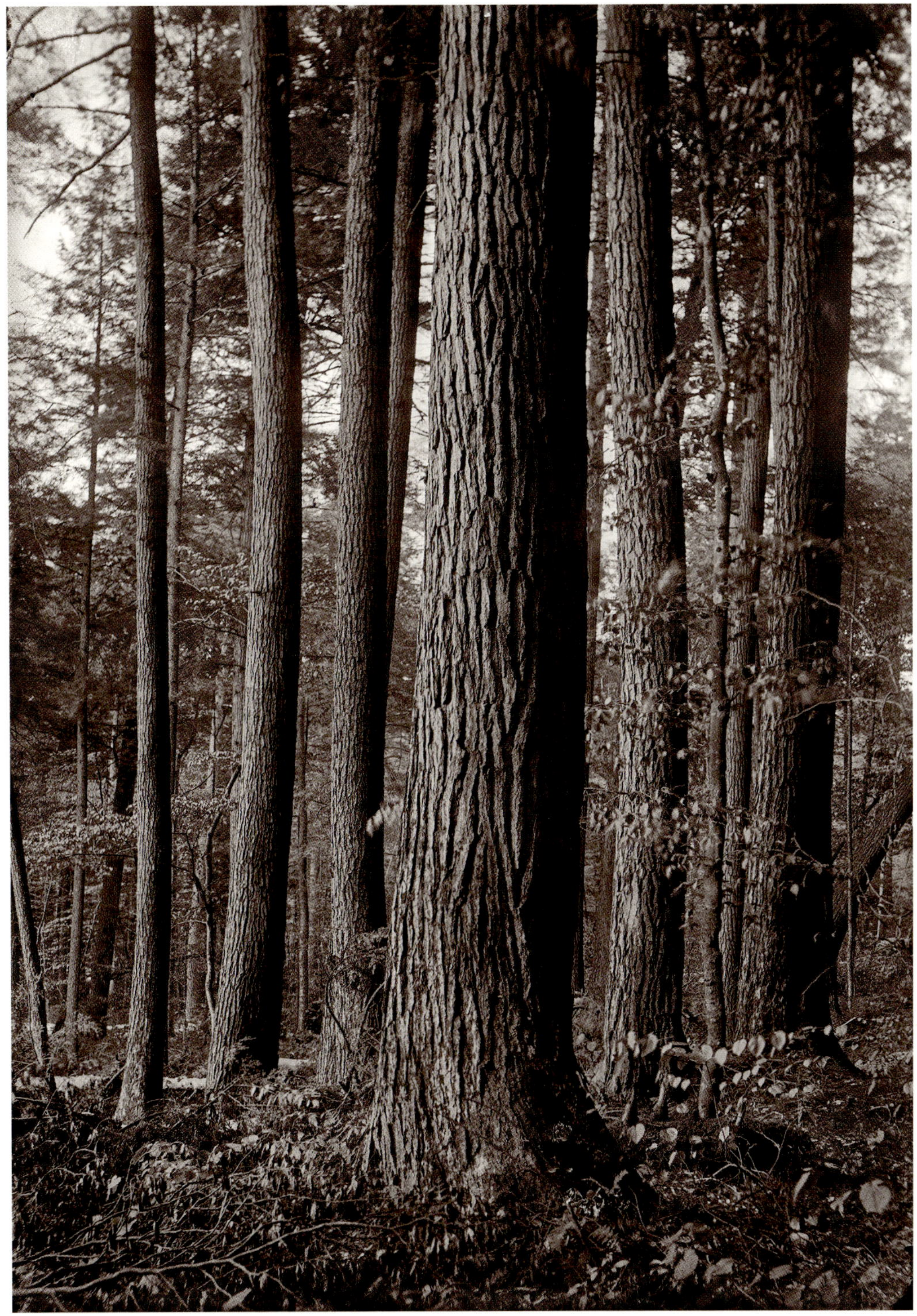

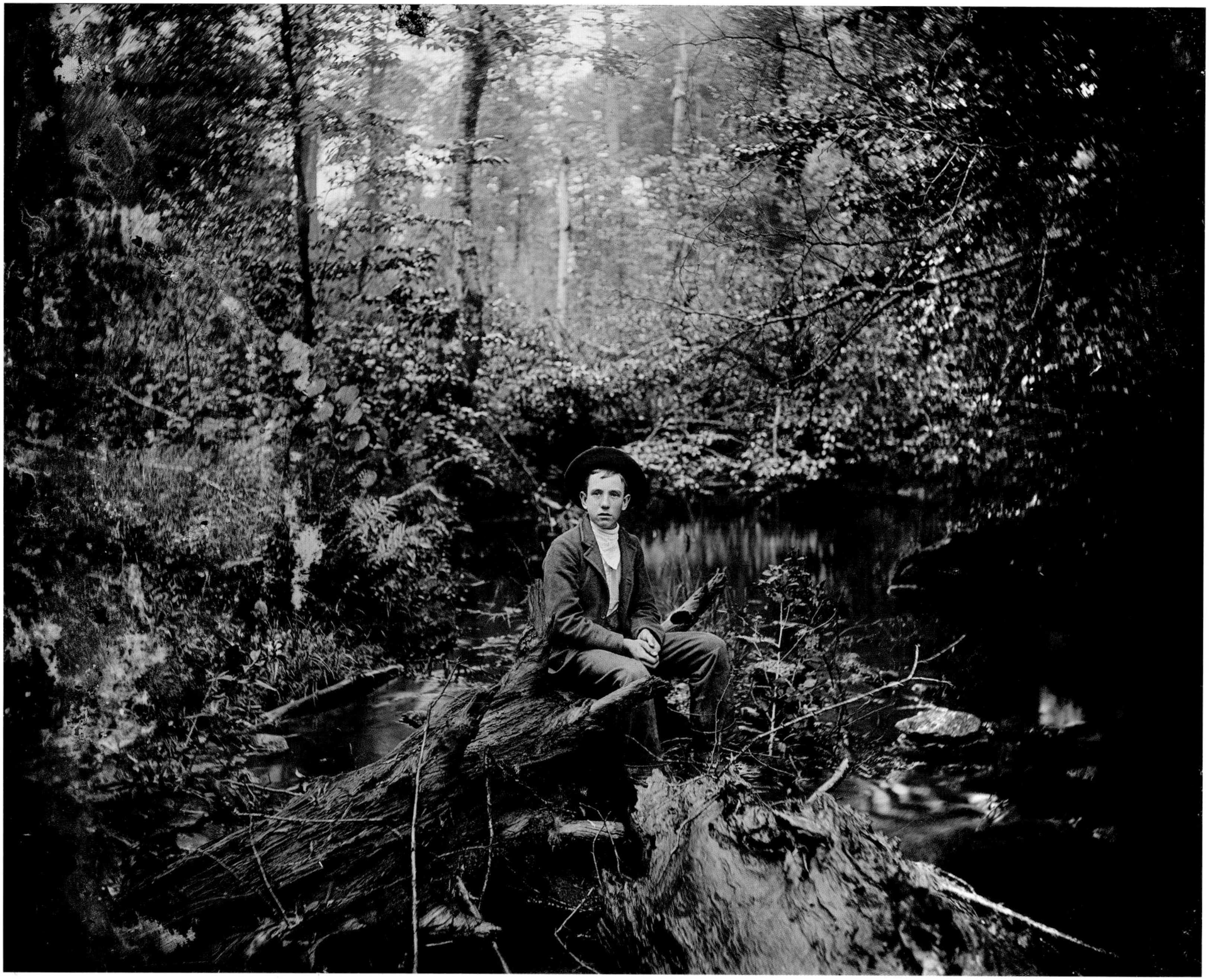

PLATE 2

Boy near a stream in the north-central Pennsylvania forest

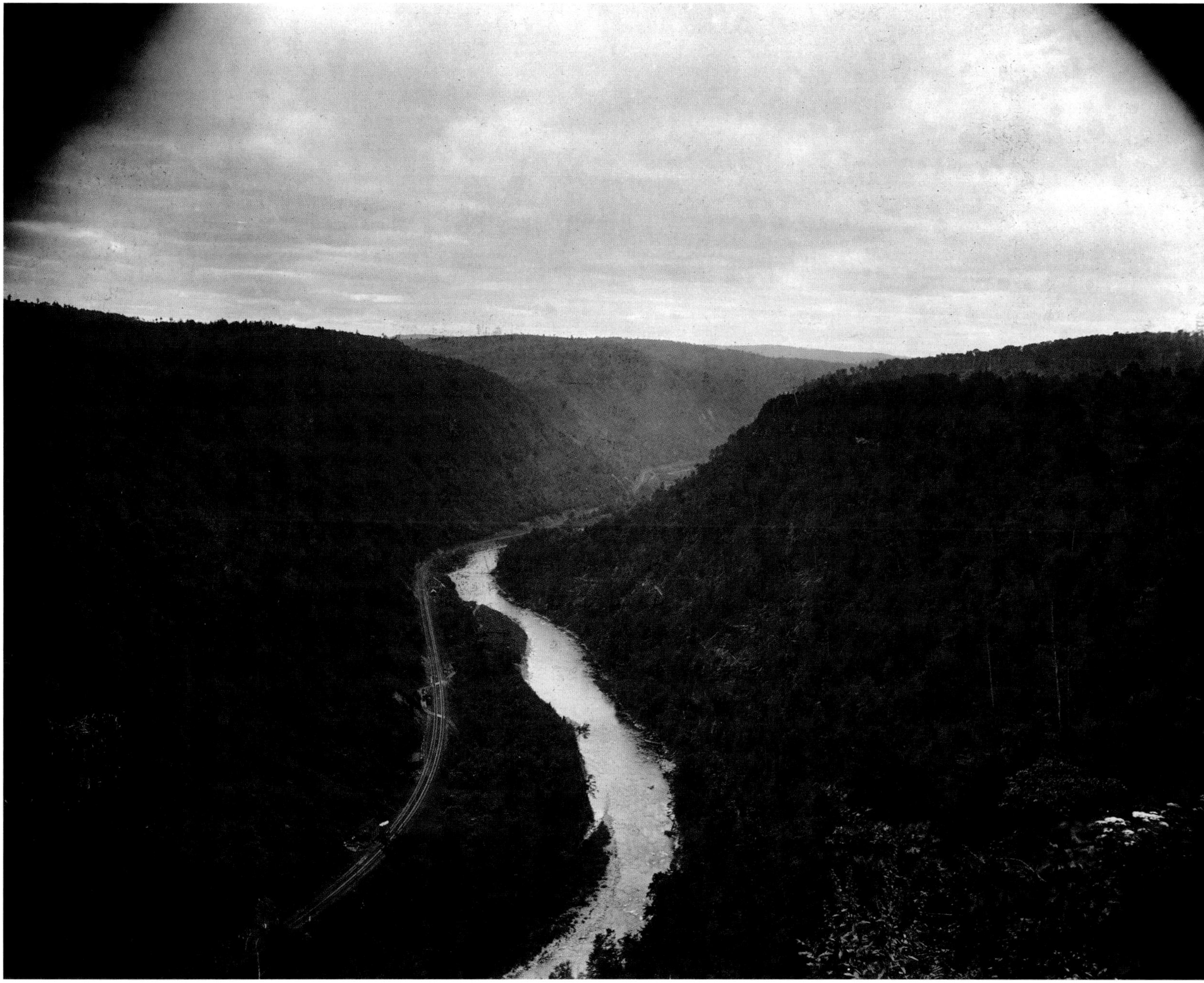

PLATE 3

Barber's, three miles below Ansonia, looking up Pine Creek toward Ansonia, Tioga County

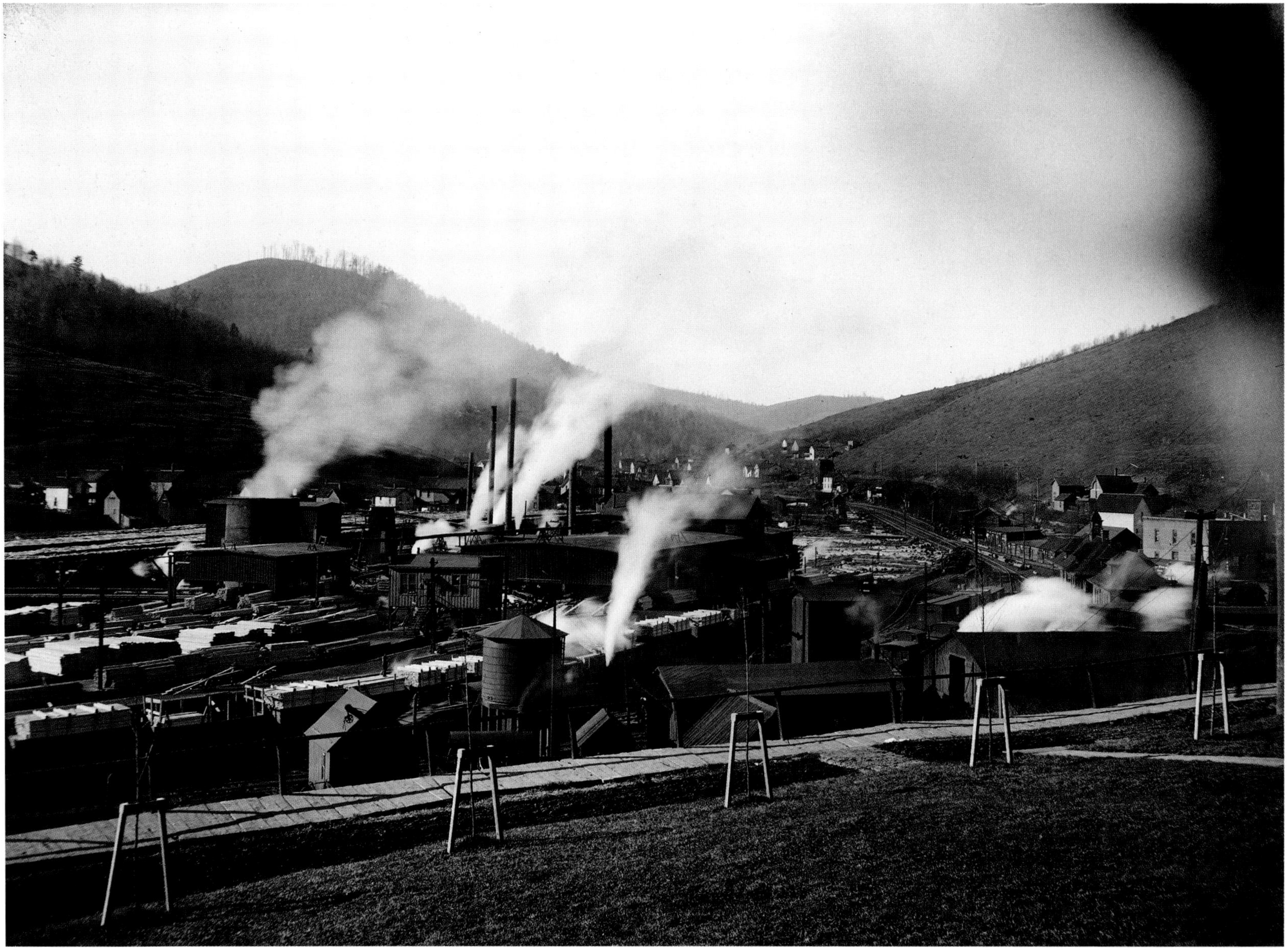

PLATE 4

Goodyear Company sawmill, Austin, Potter County

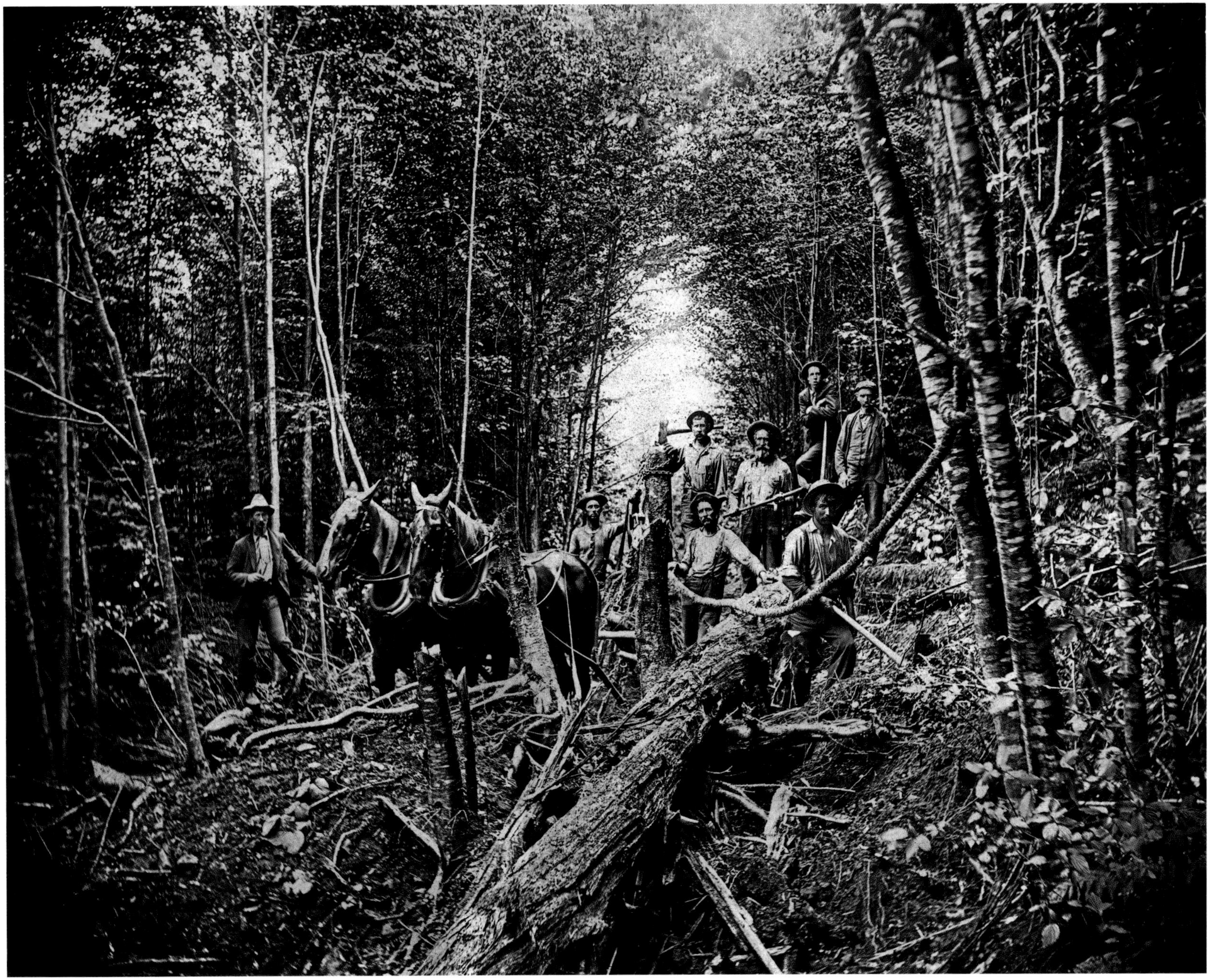

PLATE 5

Swampers carving out a mountain road

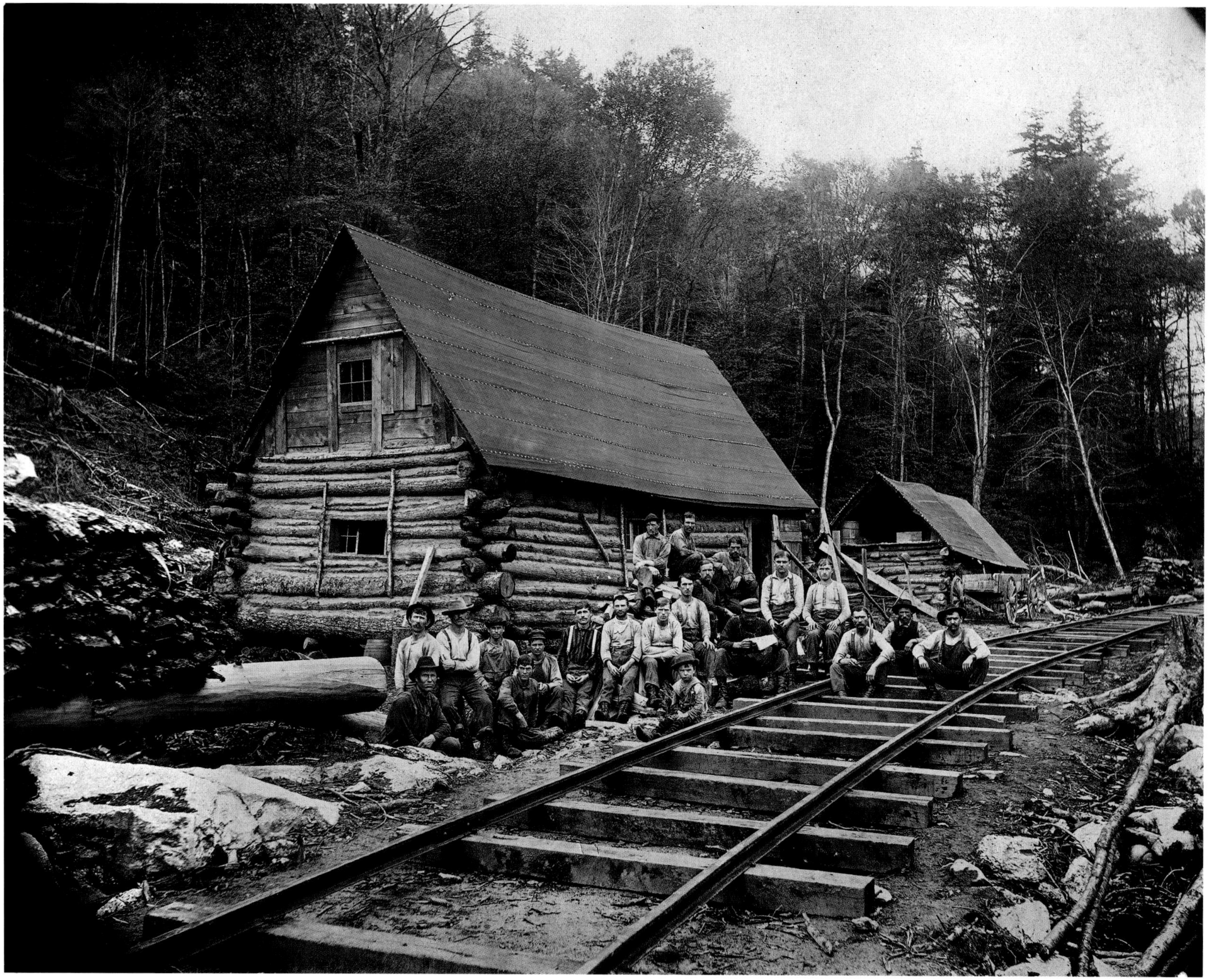

PLATE 6

Spur rails on ties with no underlying bed at a lumber camp

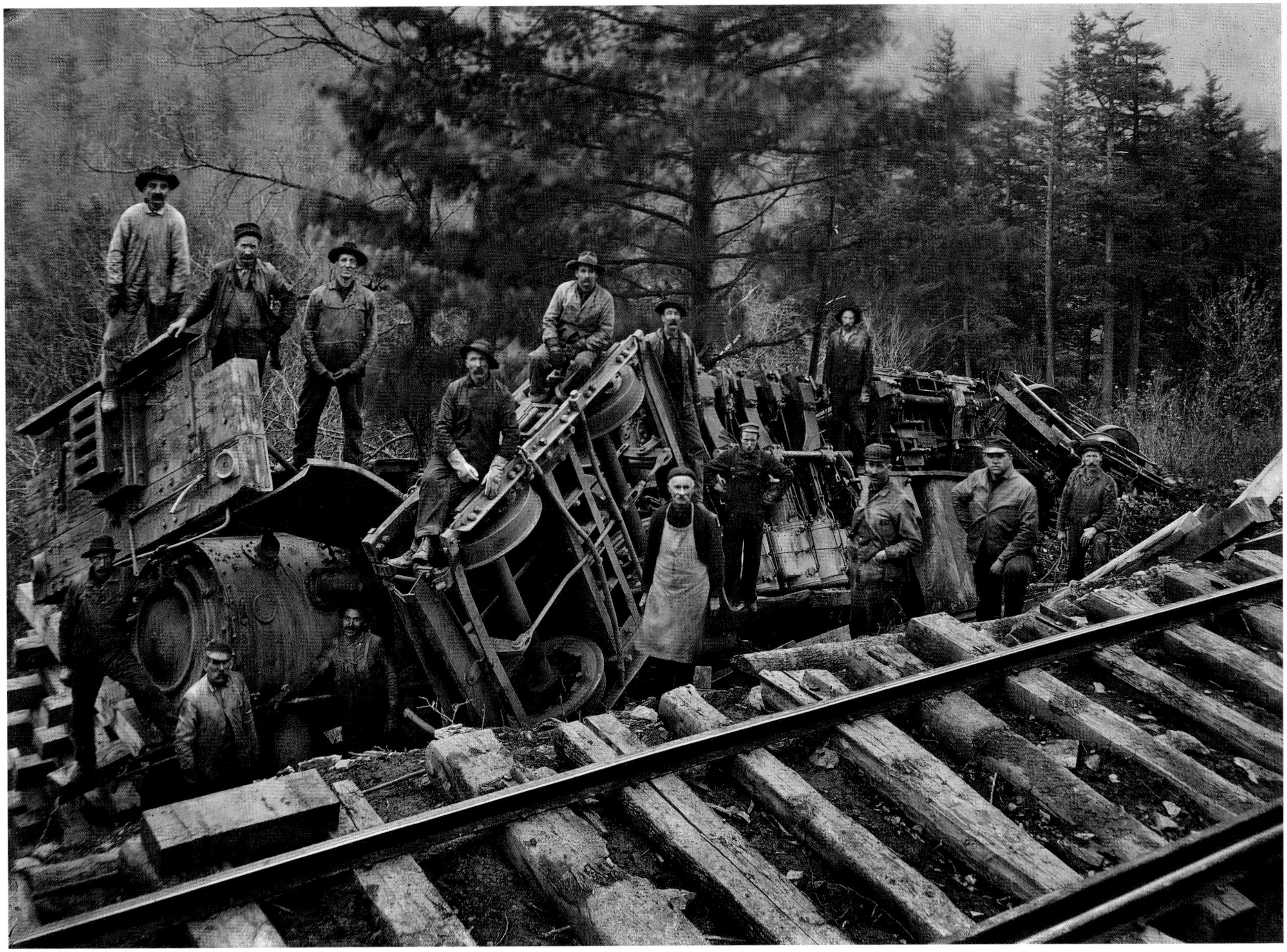

PLATE 7

Help wanted. Hammersley region, northwestern Clinton County.

Locomotive and cars derailment

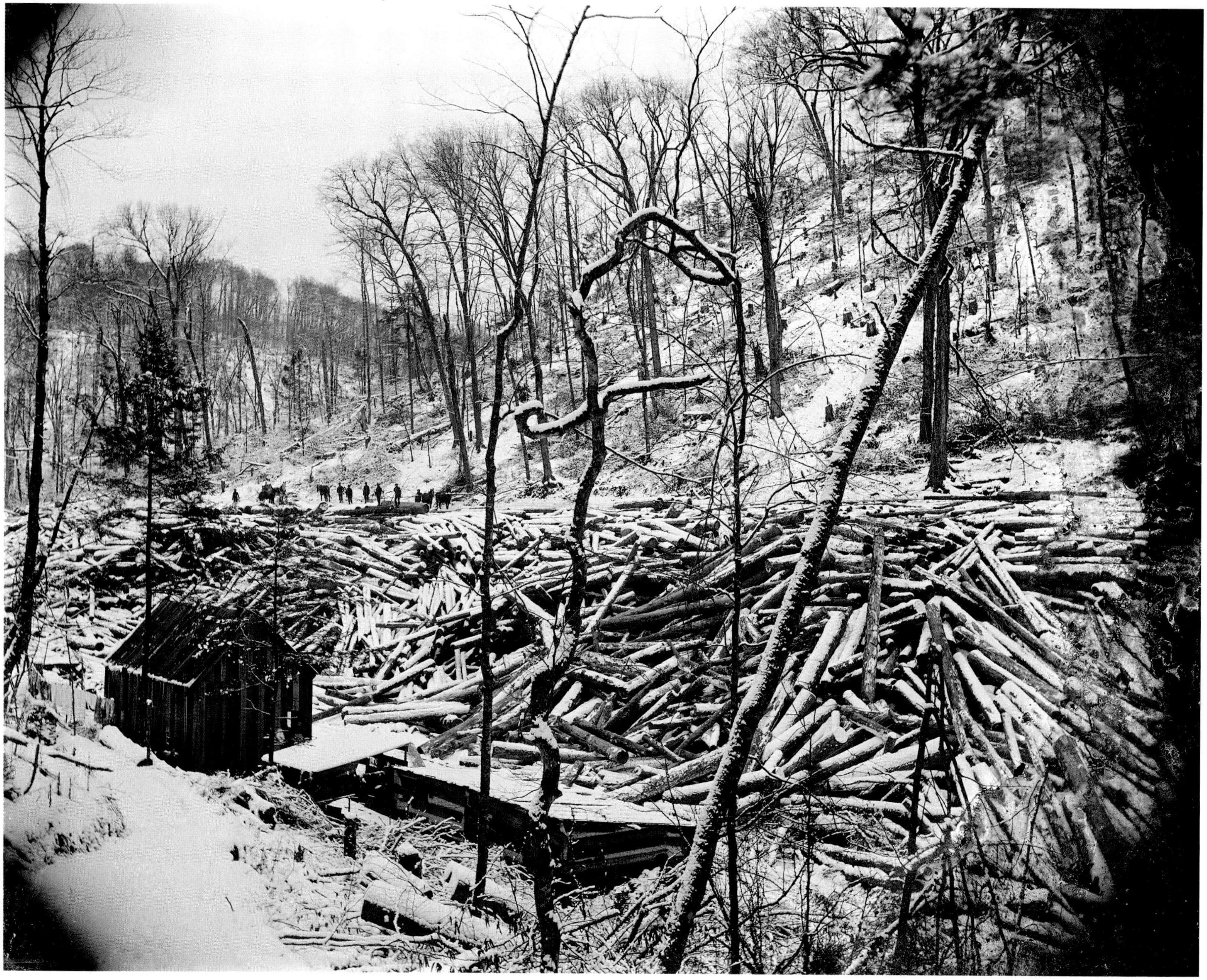

PLATE 8

Rough-and-tumble skidway logs await rail transport to the mill

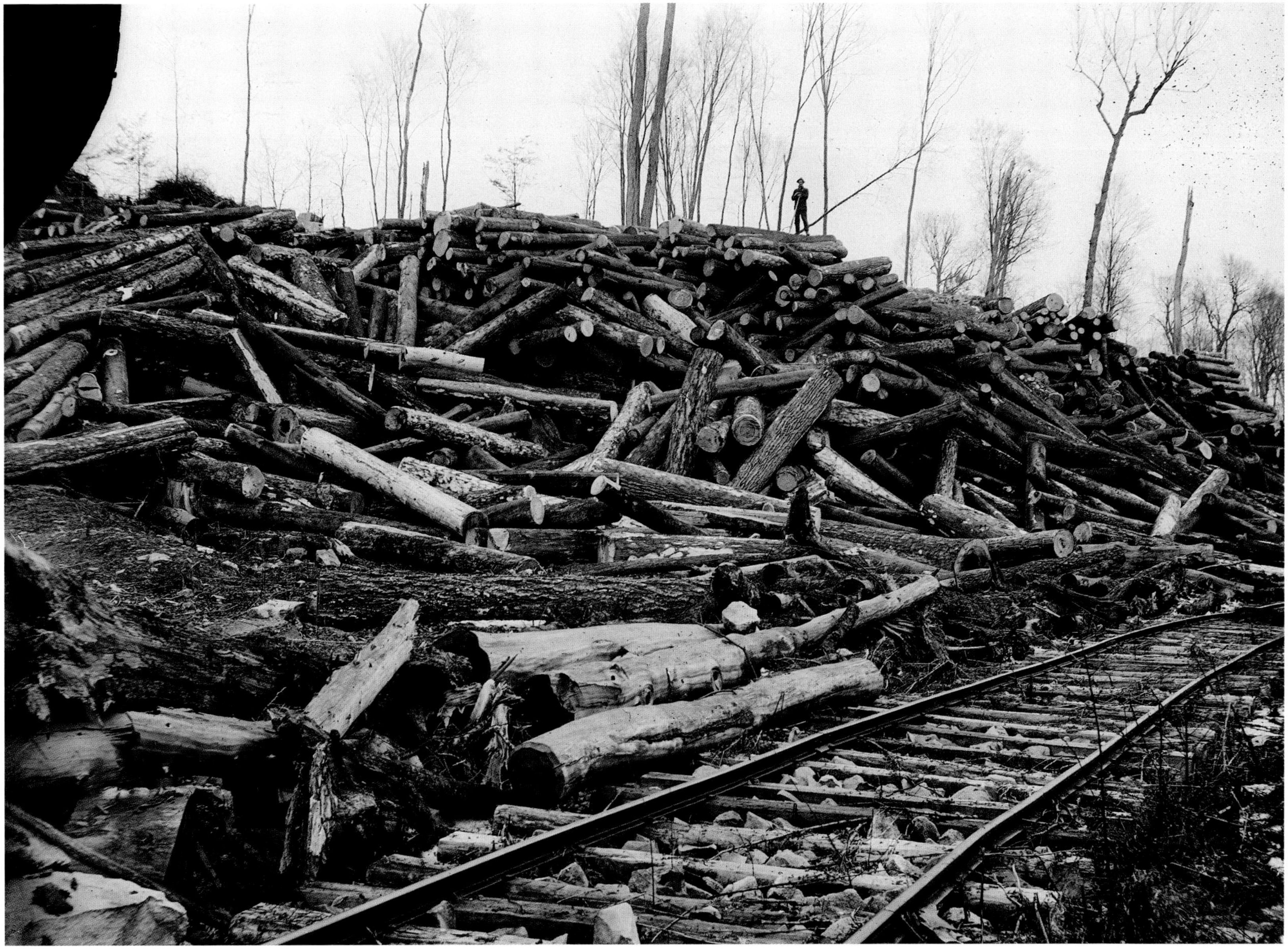

PLATE 9

Skidway logs by railroad near Austin, Potter County

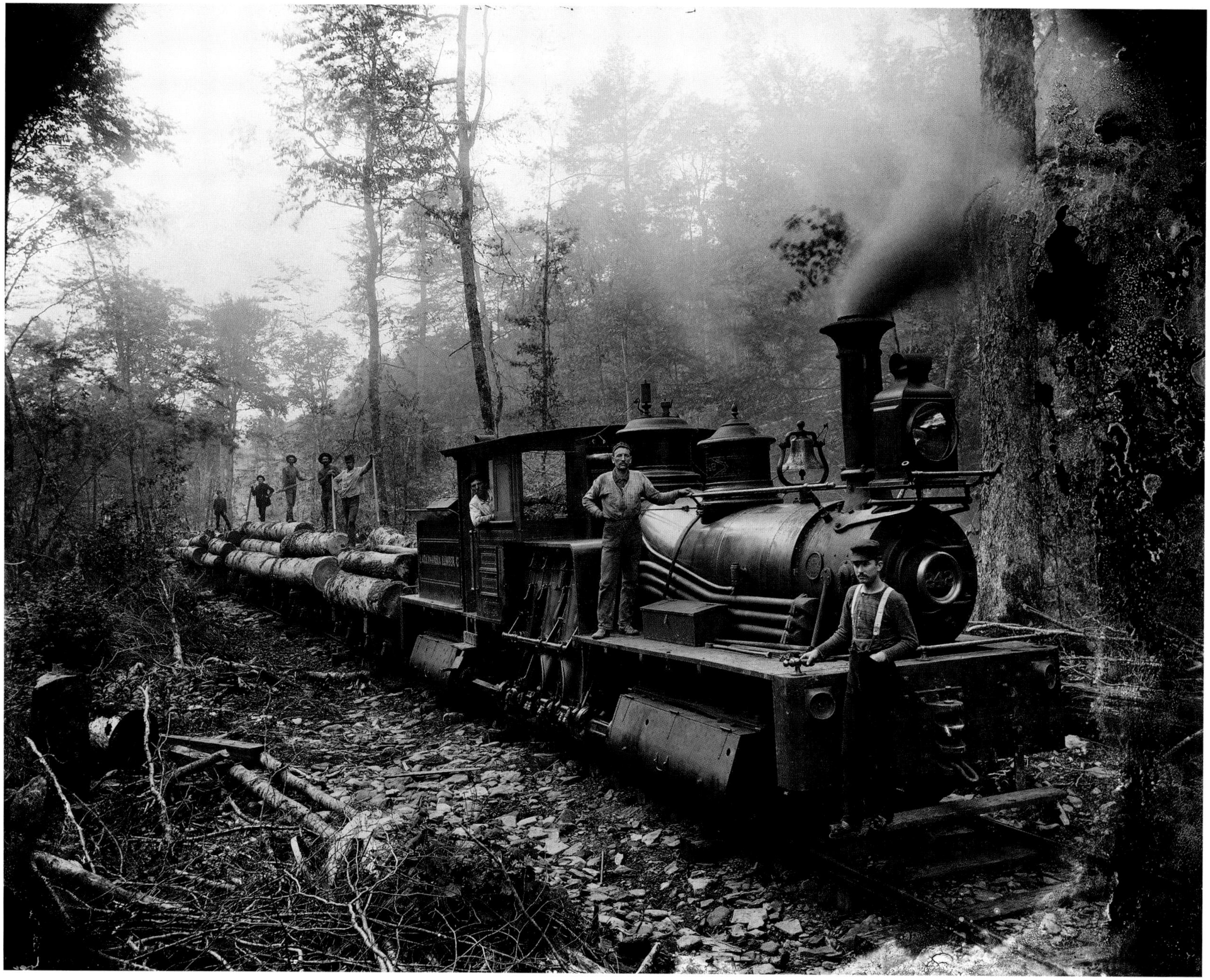

PLATE 10

Shay locomotive No. 242, Lackawanna Lumber Company

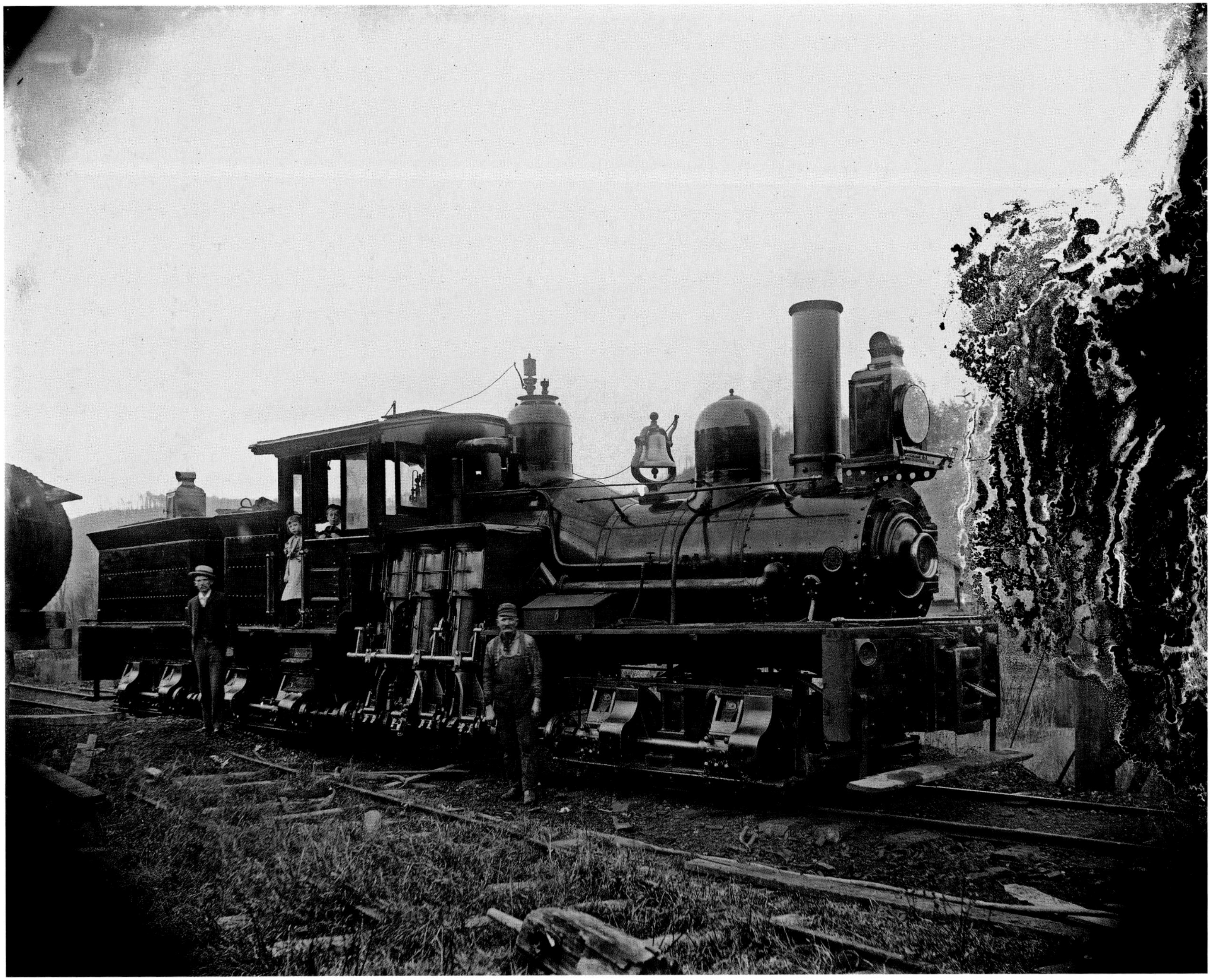

PLATE 11

Three-truck Shay locomotive, built in Lima, Ohio, 1892

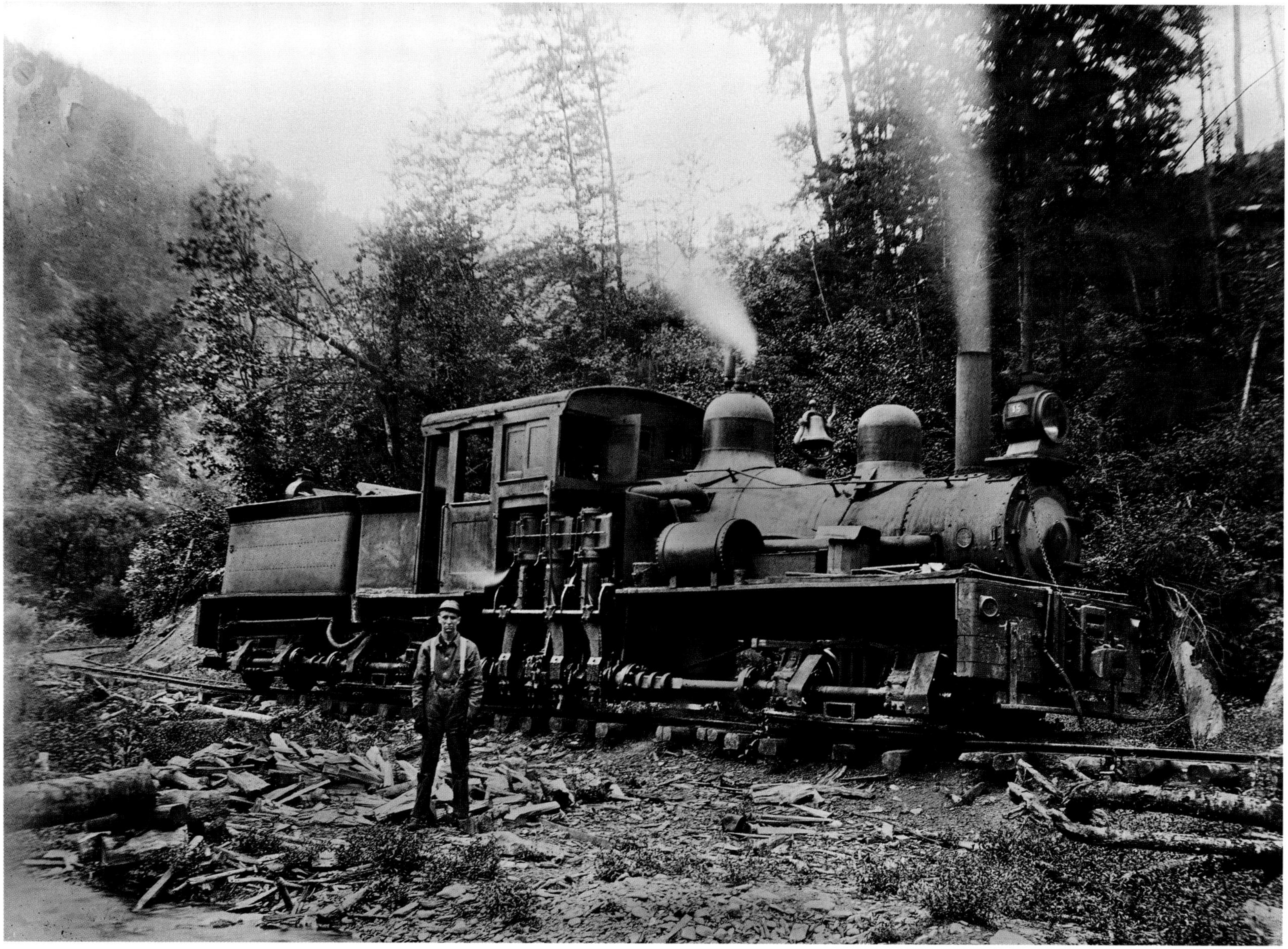

PLATE 12

Brothers, Hammersley region, Clinton County.

Goodyear Shay locomotive No. 15, built in 1901

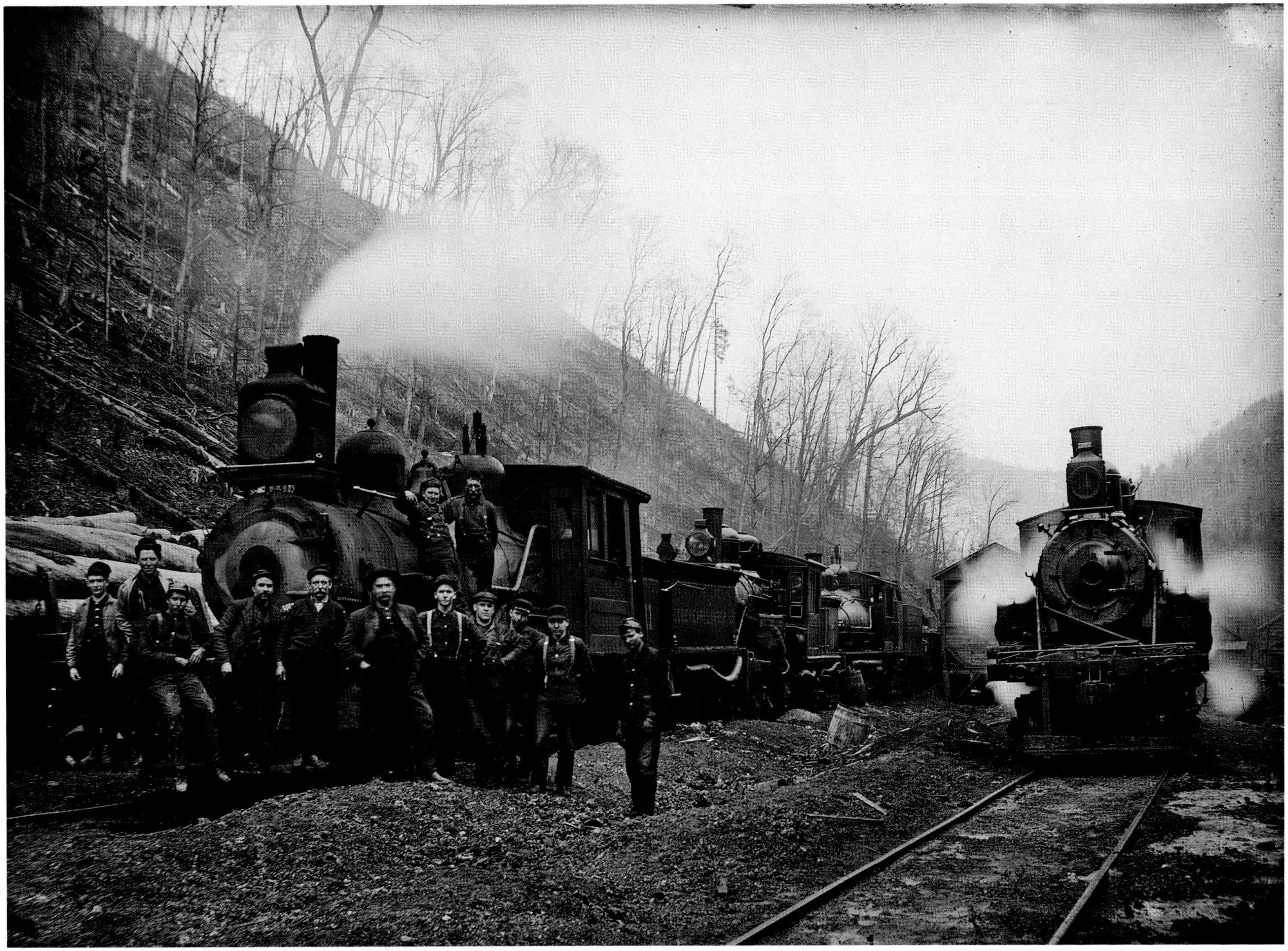

PLATE 13

Goodyear Lumber Company crew and engines,
Hammersley region, Clinton County

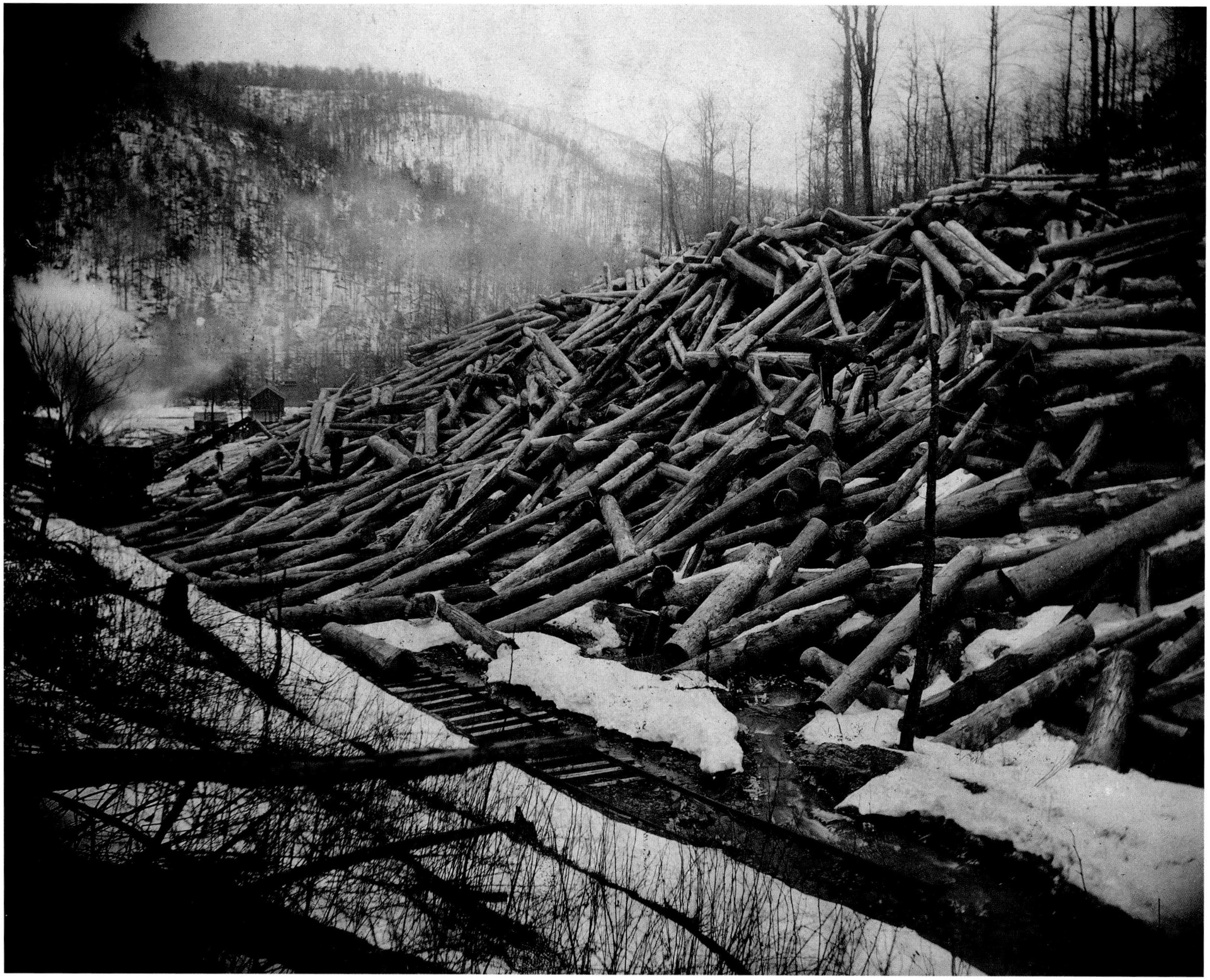

PLATE 14

Jake Swanson's camp near Hull, Potter County, 1902

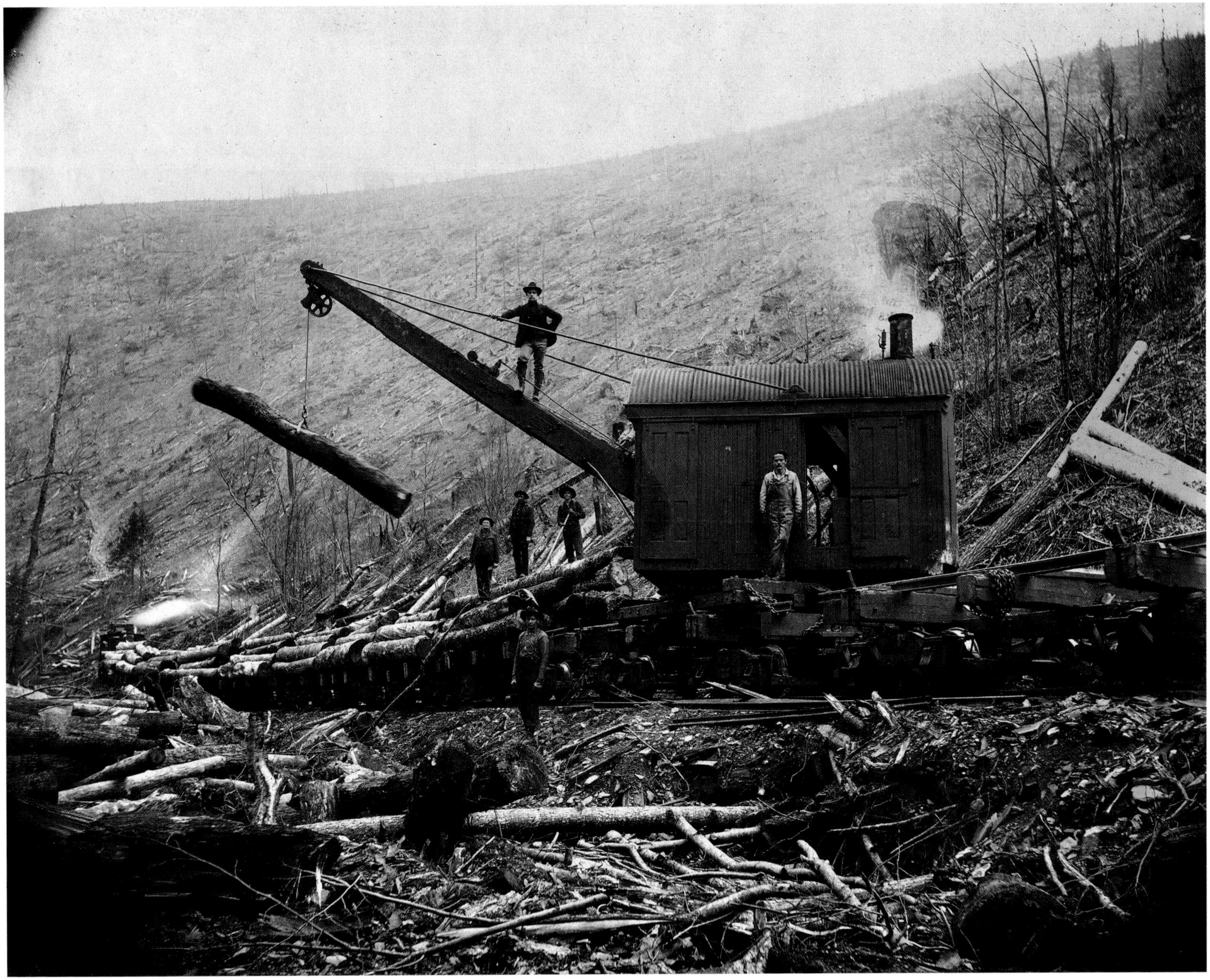

PLATE 15

Mechanical log loader, Freeman Run, Potter County

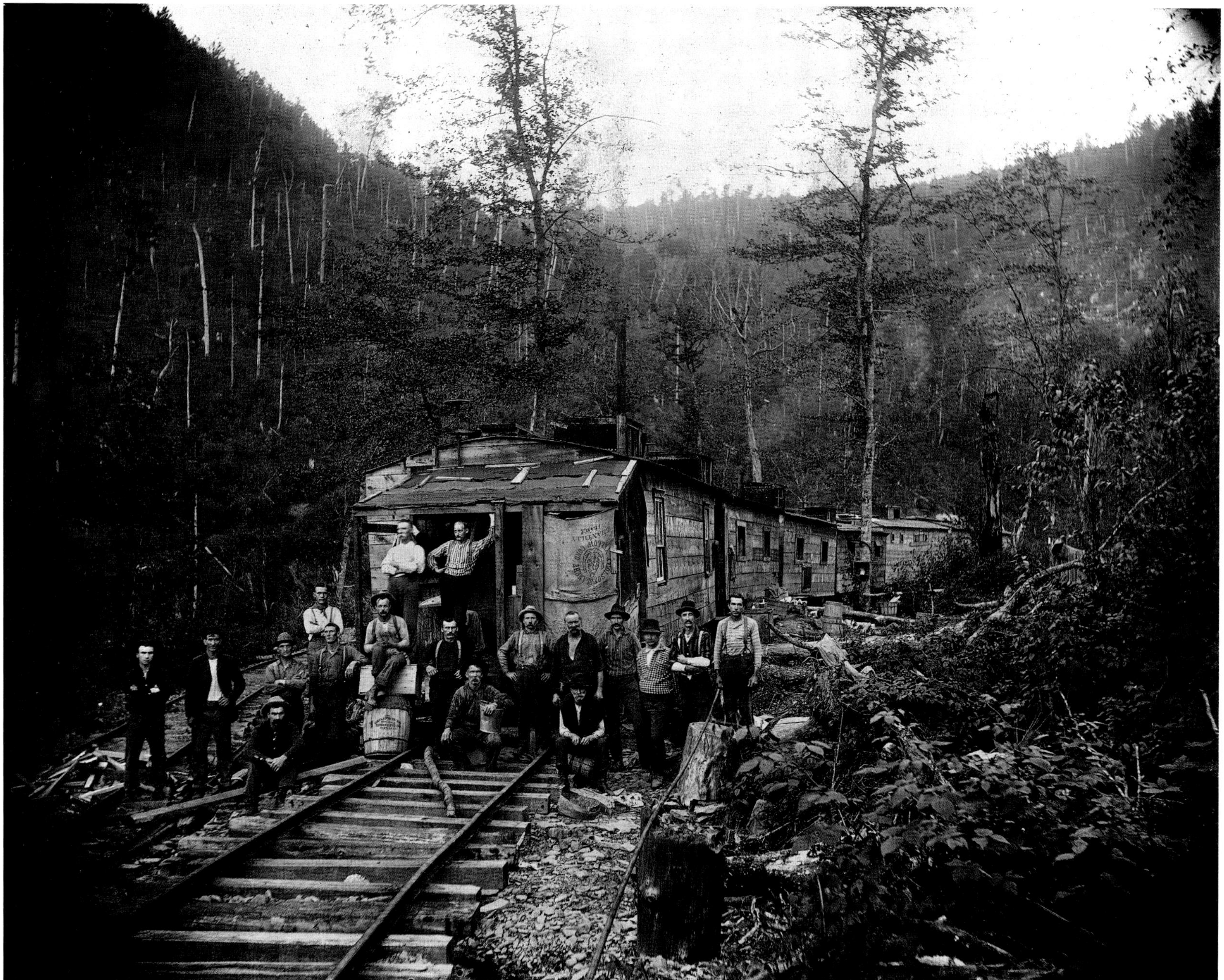

PLATE 16

"The Shacks," near Cross Fork, Potter County

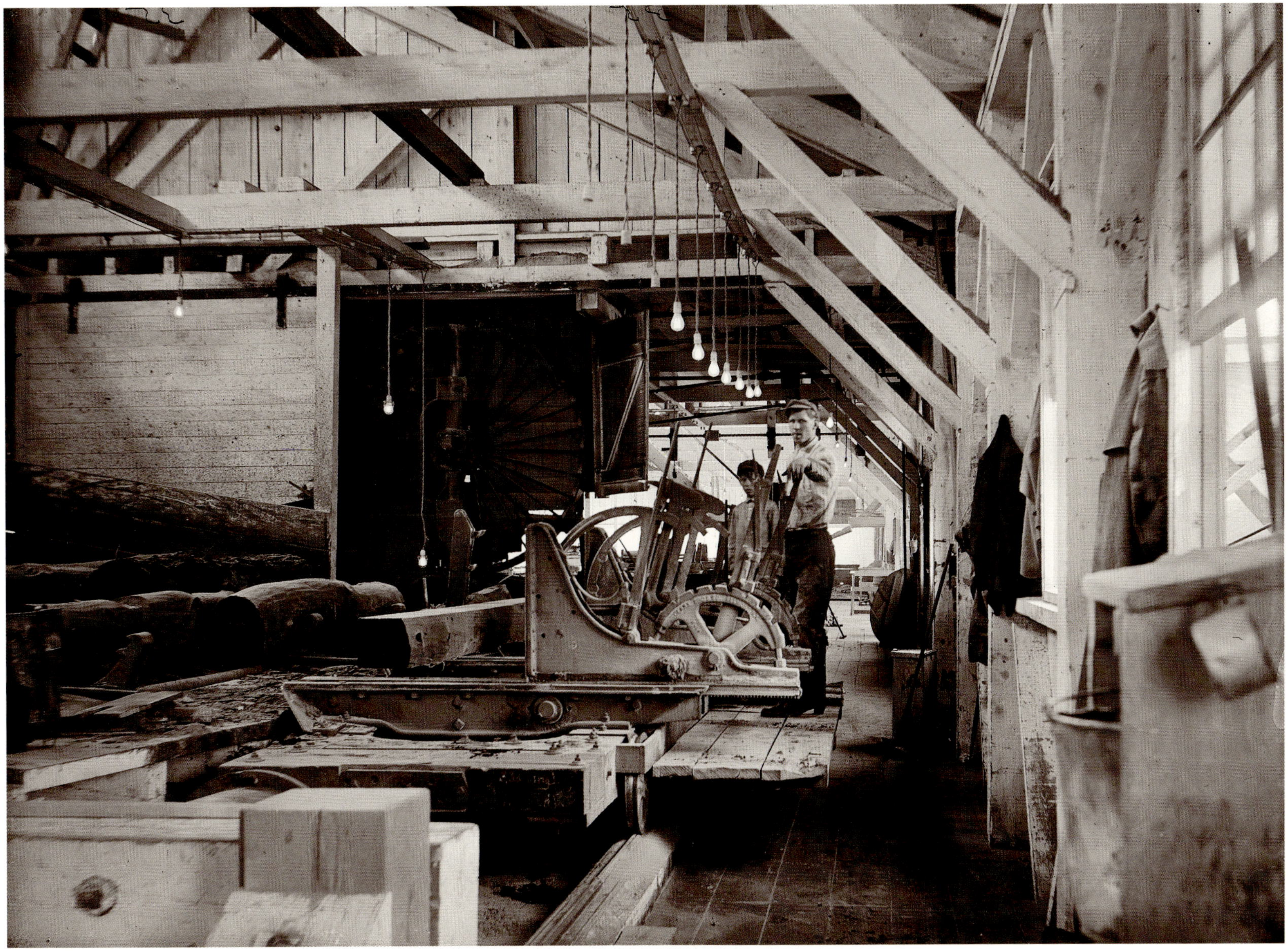

PLATE 17

Interior of sawmill, Norwich, McKean County, November 7, 1917

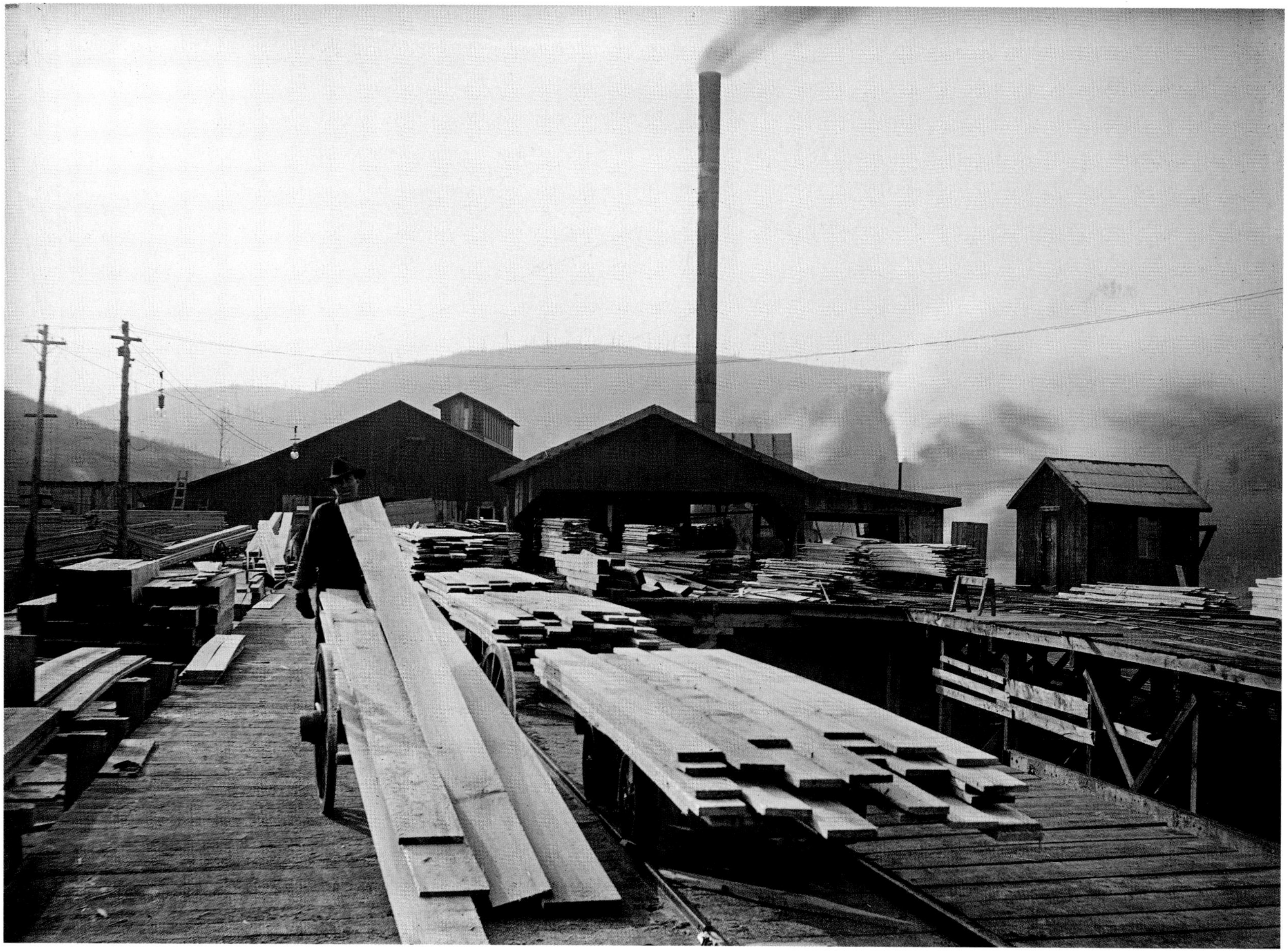

PLATE 18

Sawed lumber coming from the mill, Roulette, Potter County

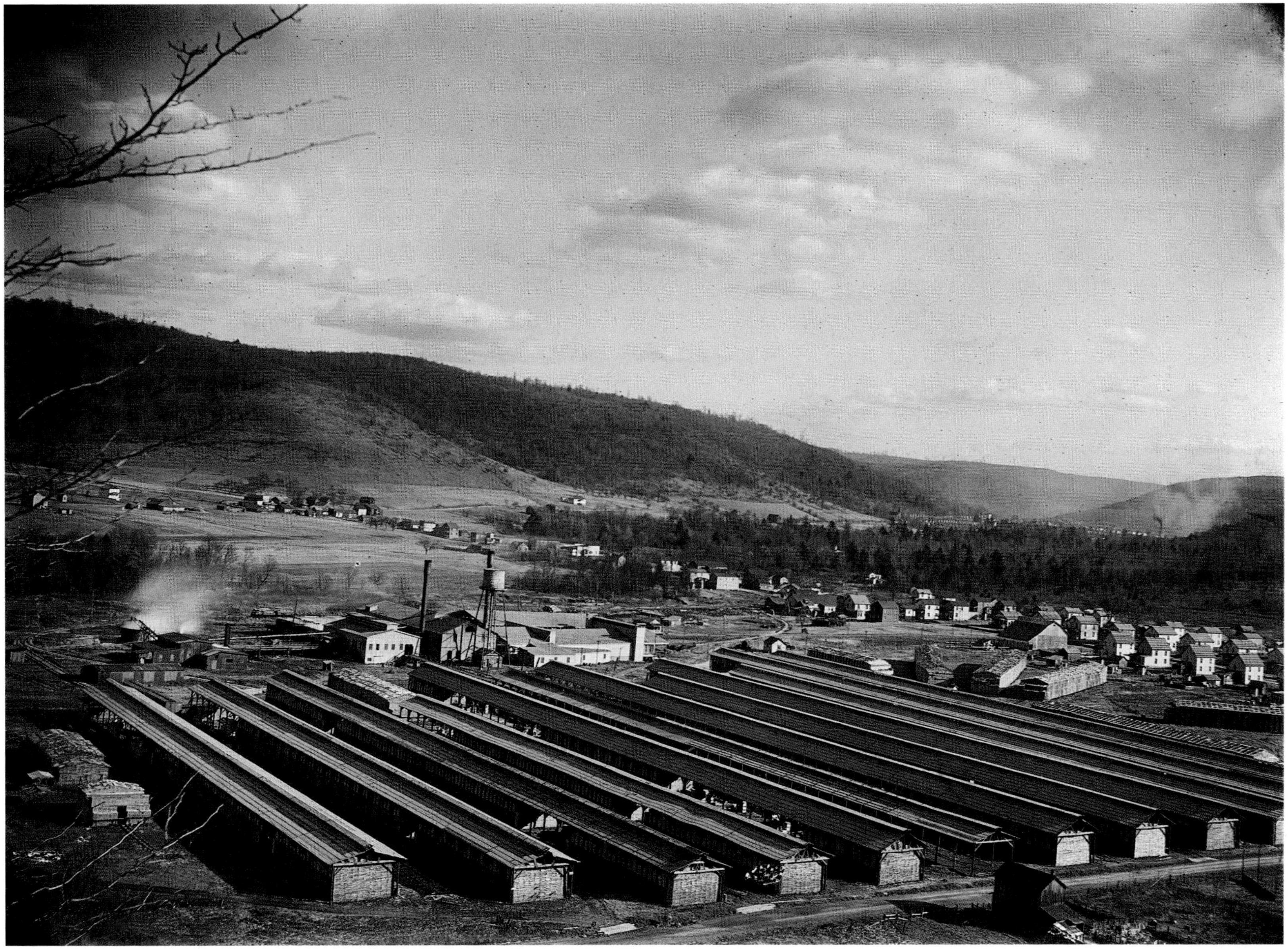

PLATE 19

Heading and stave plant, Betula, McKean County, November 3, 1916

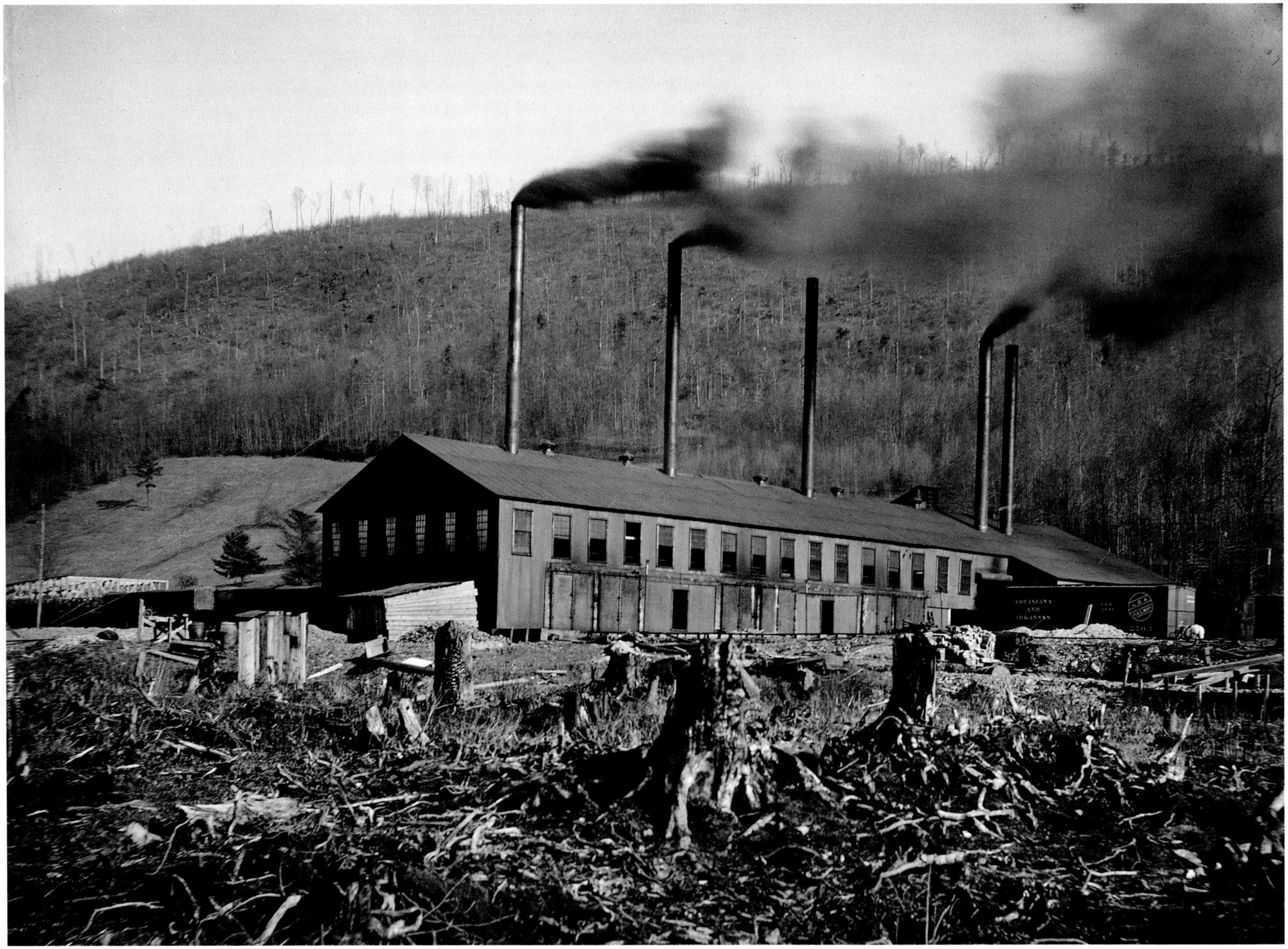

PLATE 20

National Chemical Company plant, Lyman Run, Potter County

PLATE 21

Stacks of lumber at the Goodyear Company sawmill, Austin, Potter County

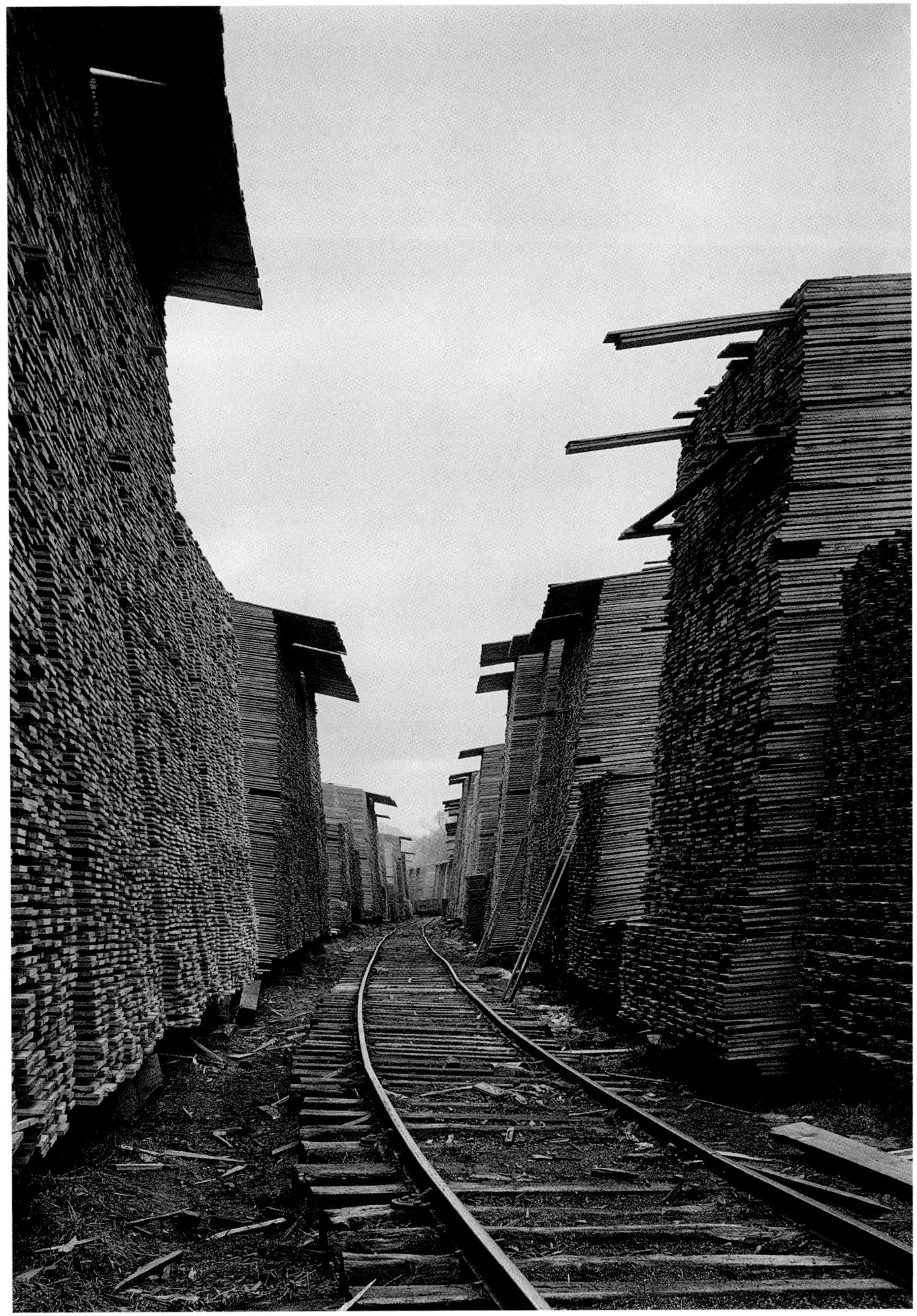

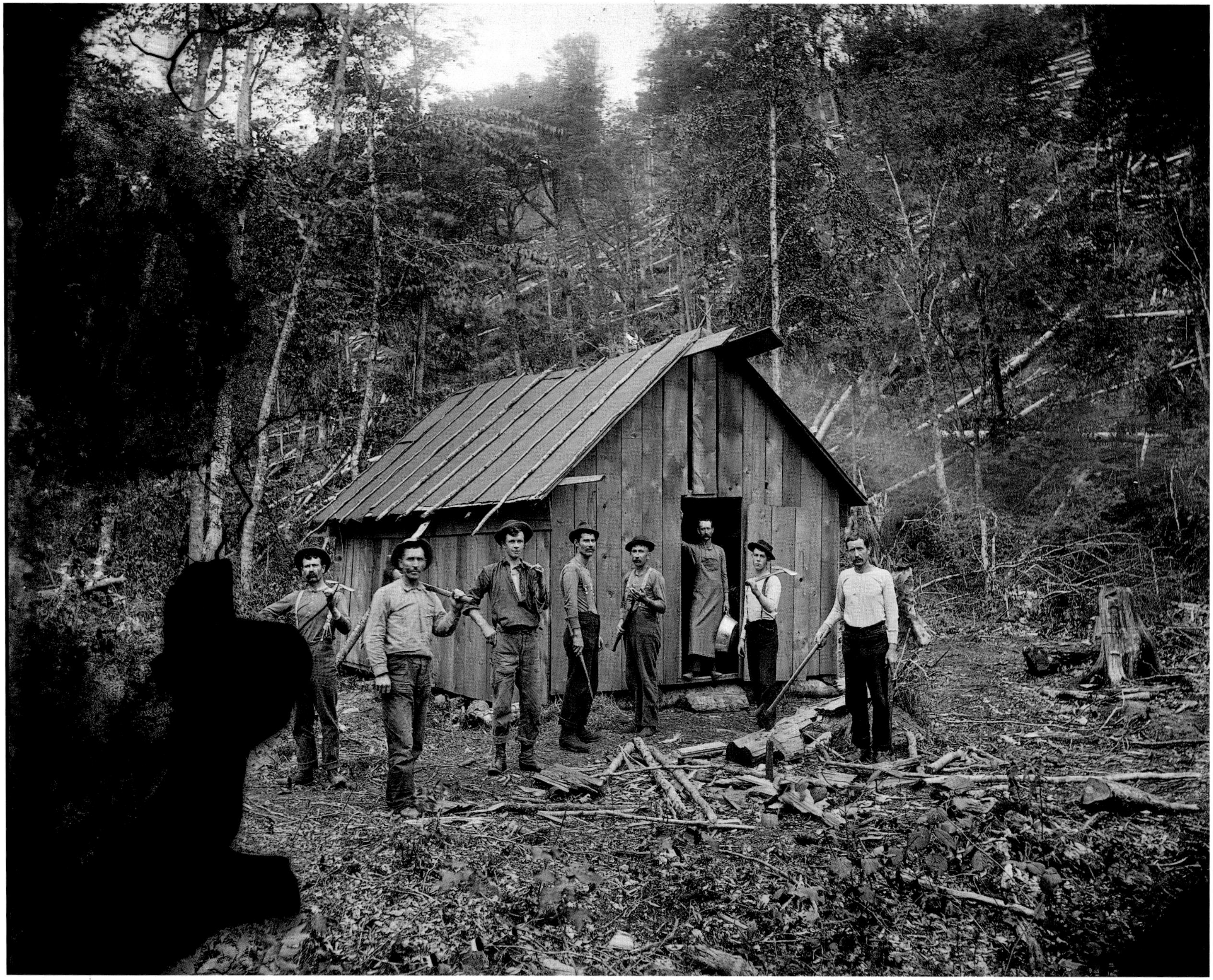

PLATE 22

Logging subcontractor's building and crew

PLATE 23

A notch, the first step in the felling of a tree, Nine Mile, Potter County

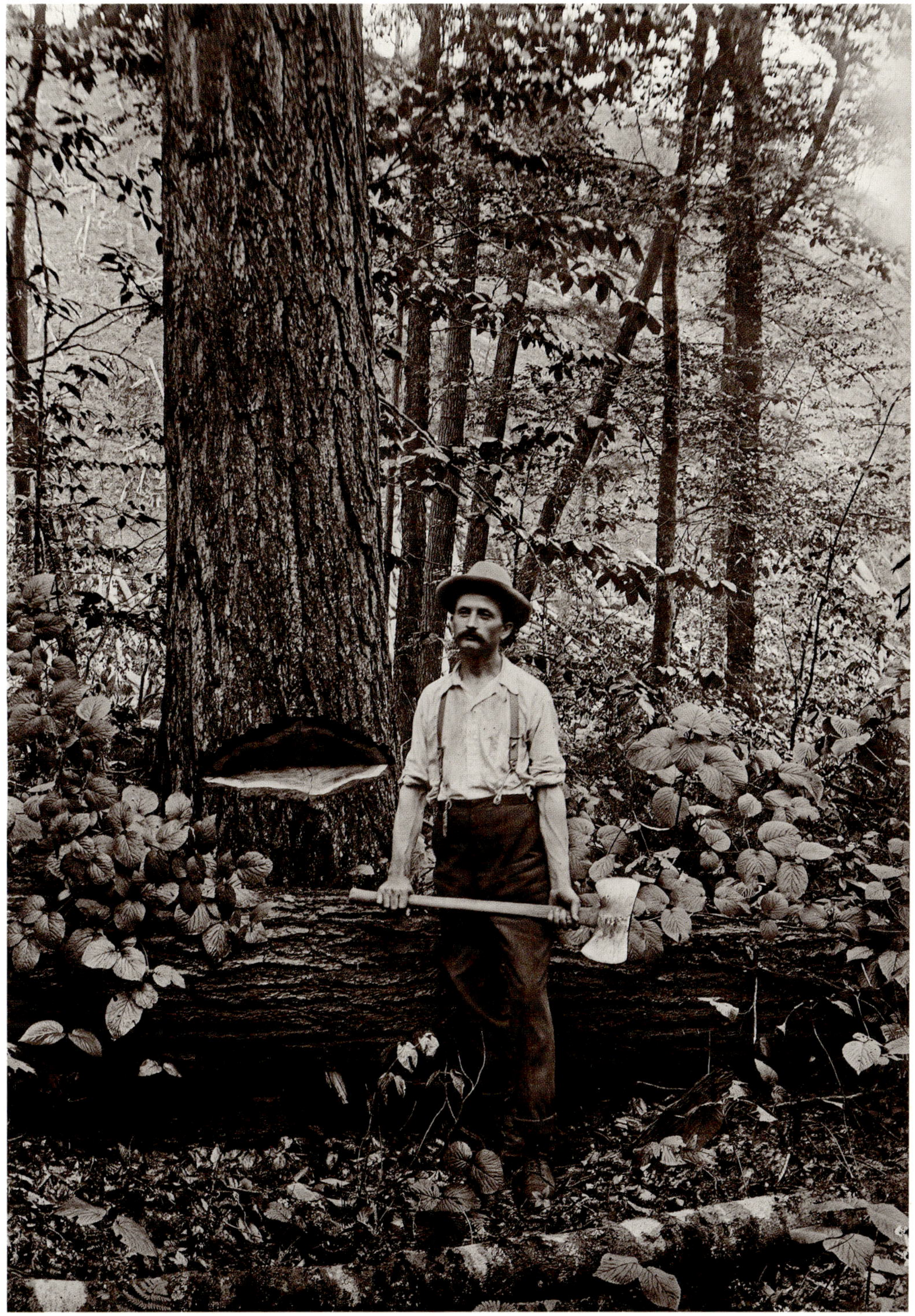

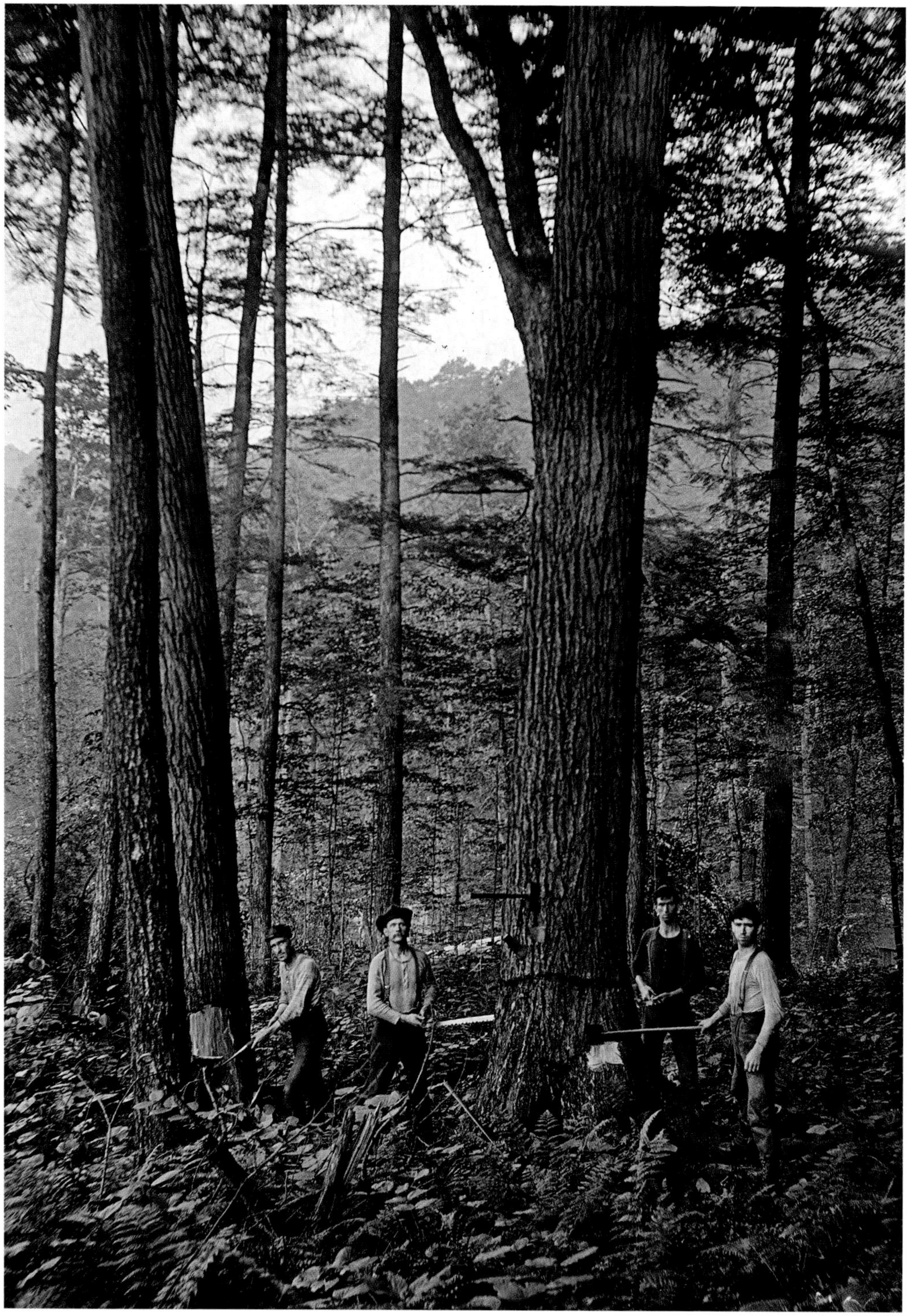

PLATE 24

Felling has begun, Lyman Run, Potter County

PLATE 25

Two-man saw, cutting notched log

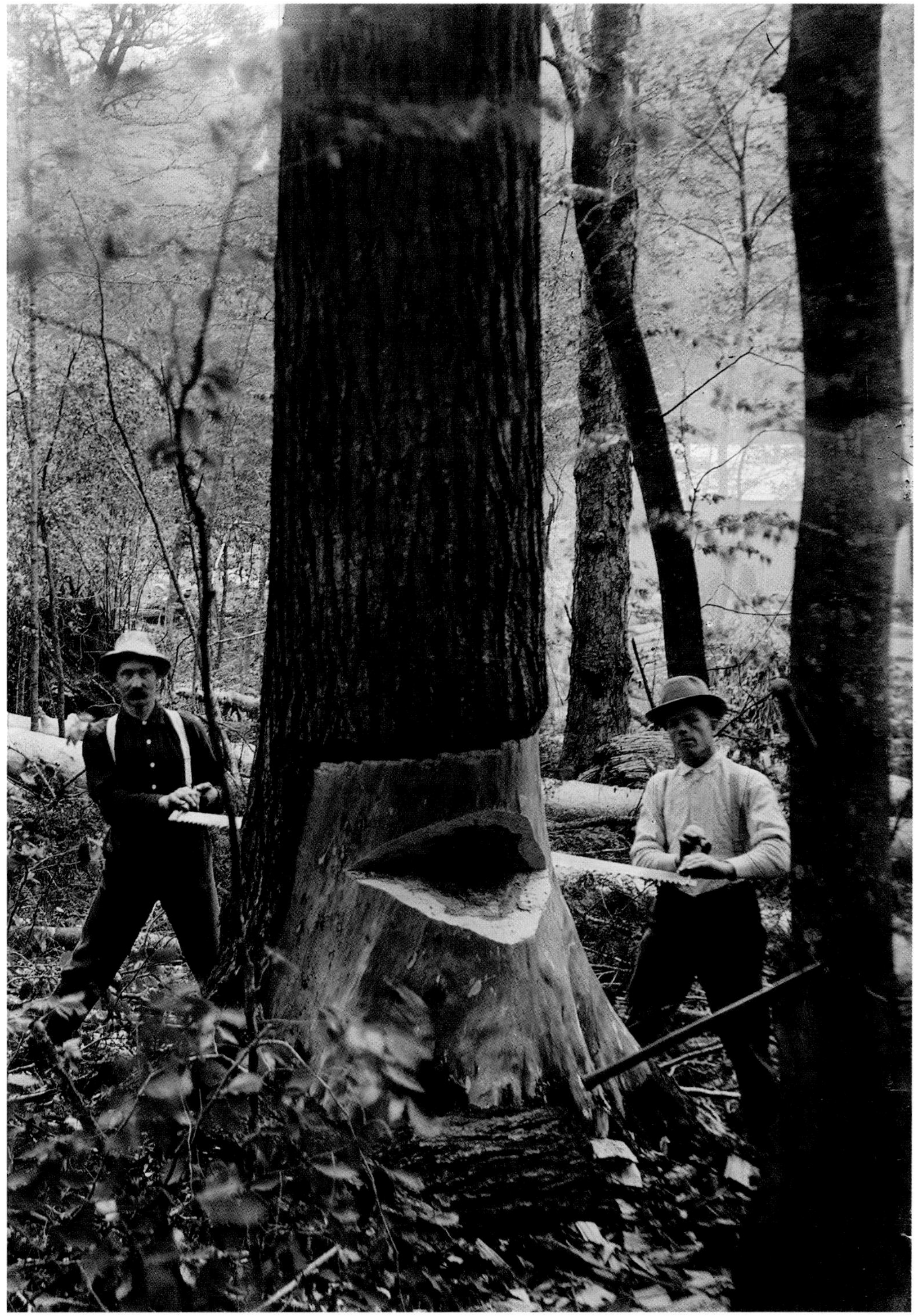

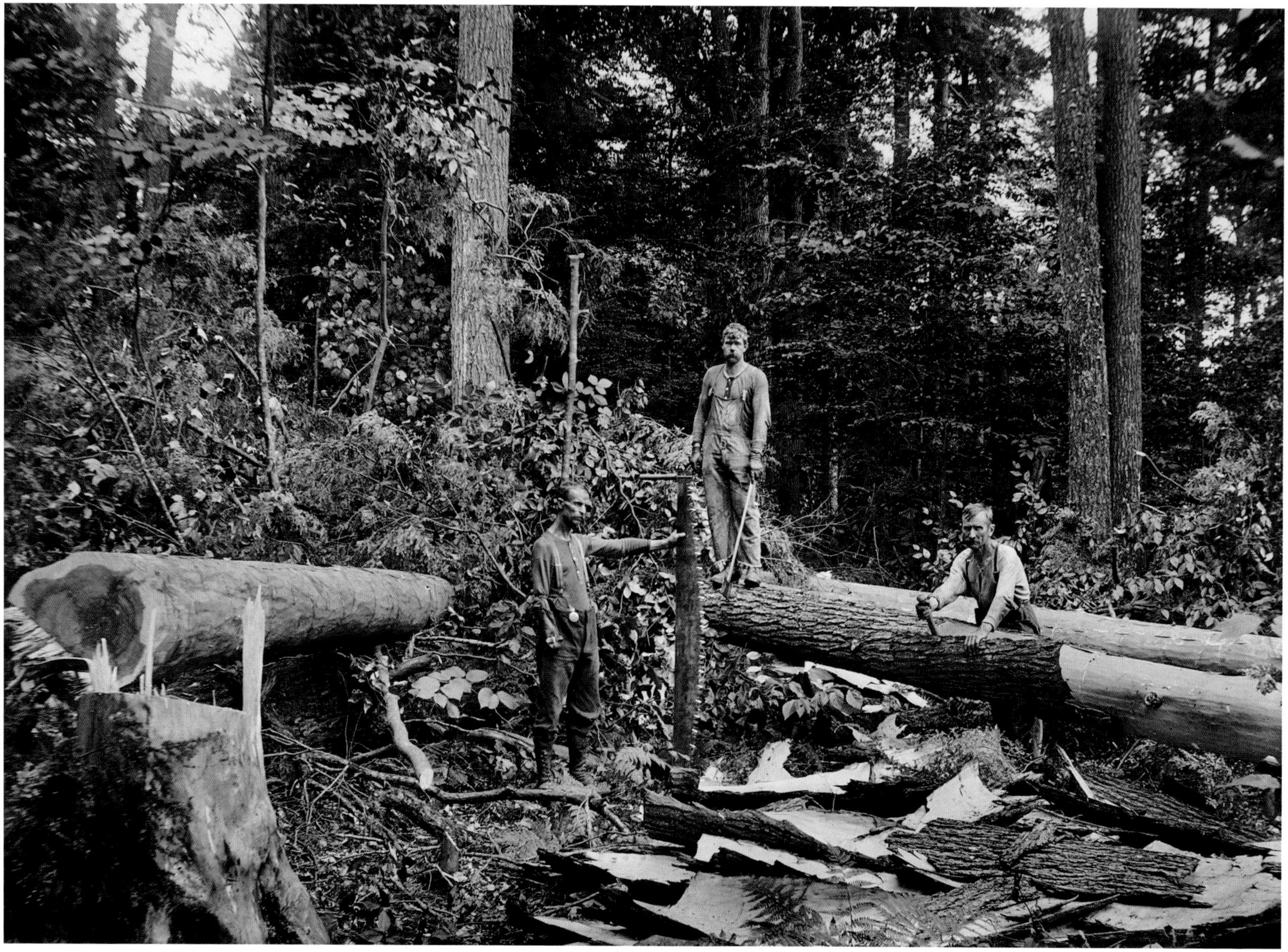

PLATE 26

Sawing hemlock, peeling hemlock bark, Lyman Run, Potter County

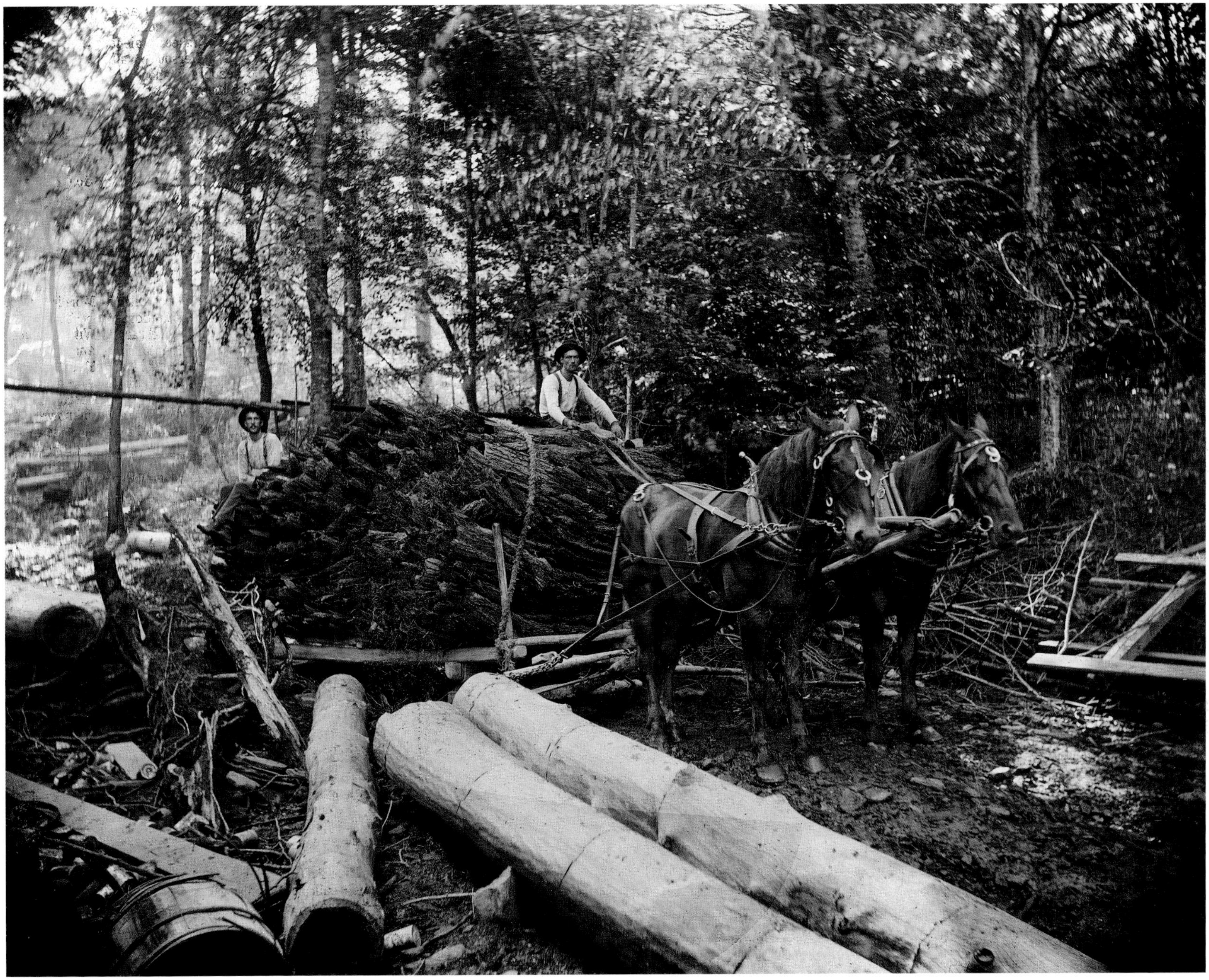

PLATE 27

Bark load, Reed Run, near Mina, Potter County

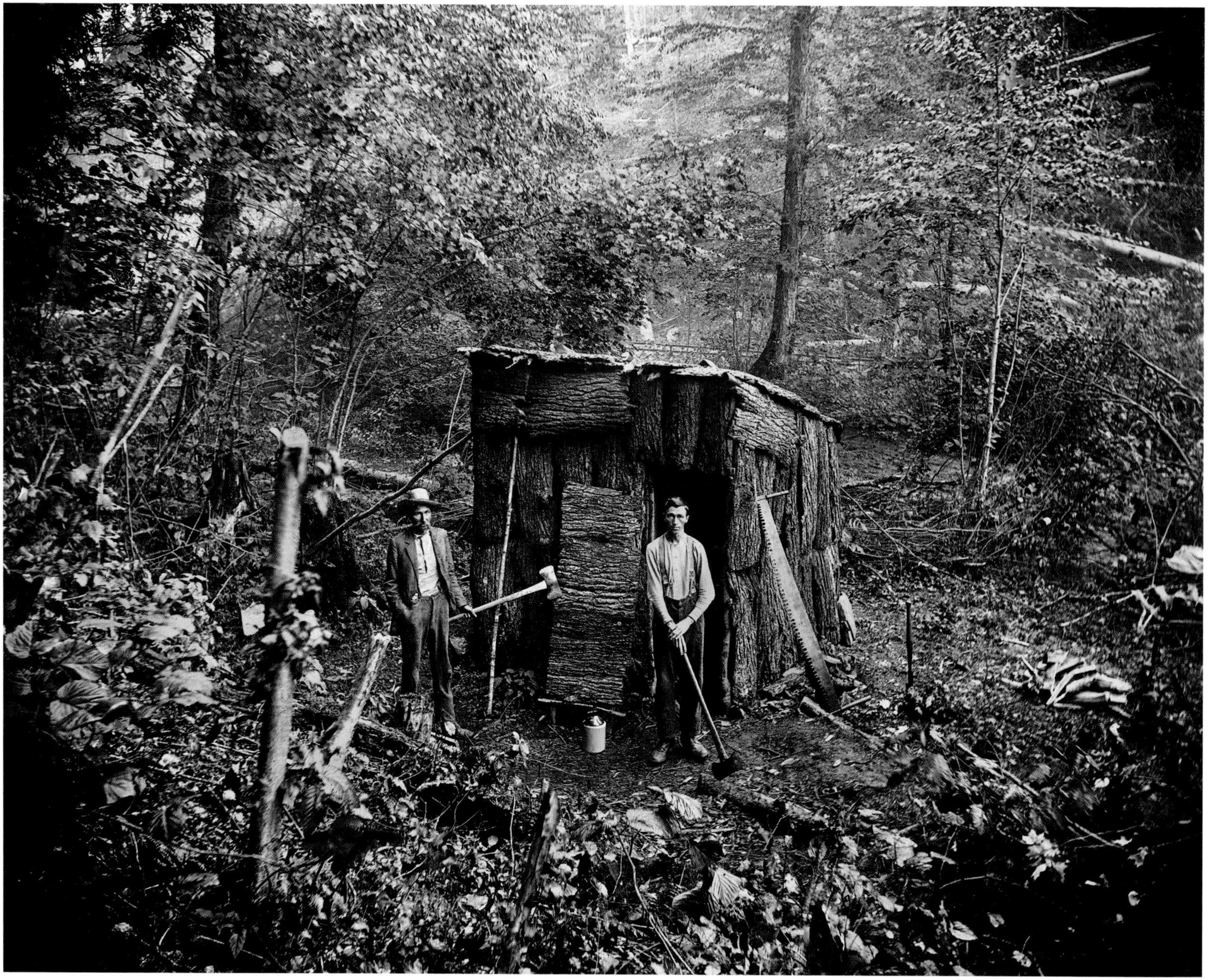

PLATE 28

Wind Fall Run, near Cross Fork, Potter County.

Bark peelers' temporary bark hut

PLATE 29

Buckers sawing hemlock, Nine Mile, Potter County

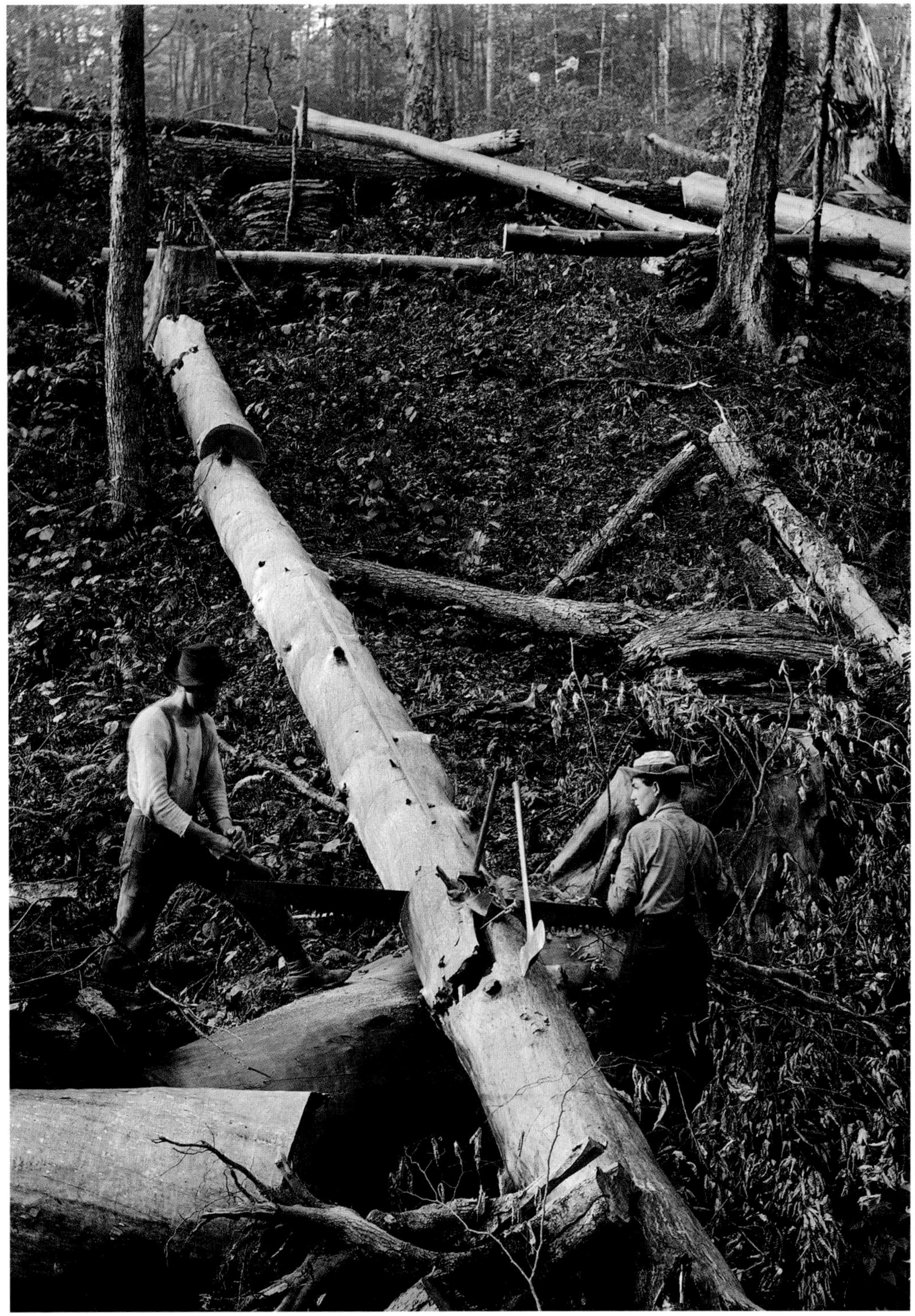

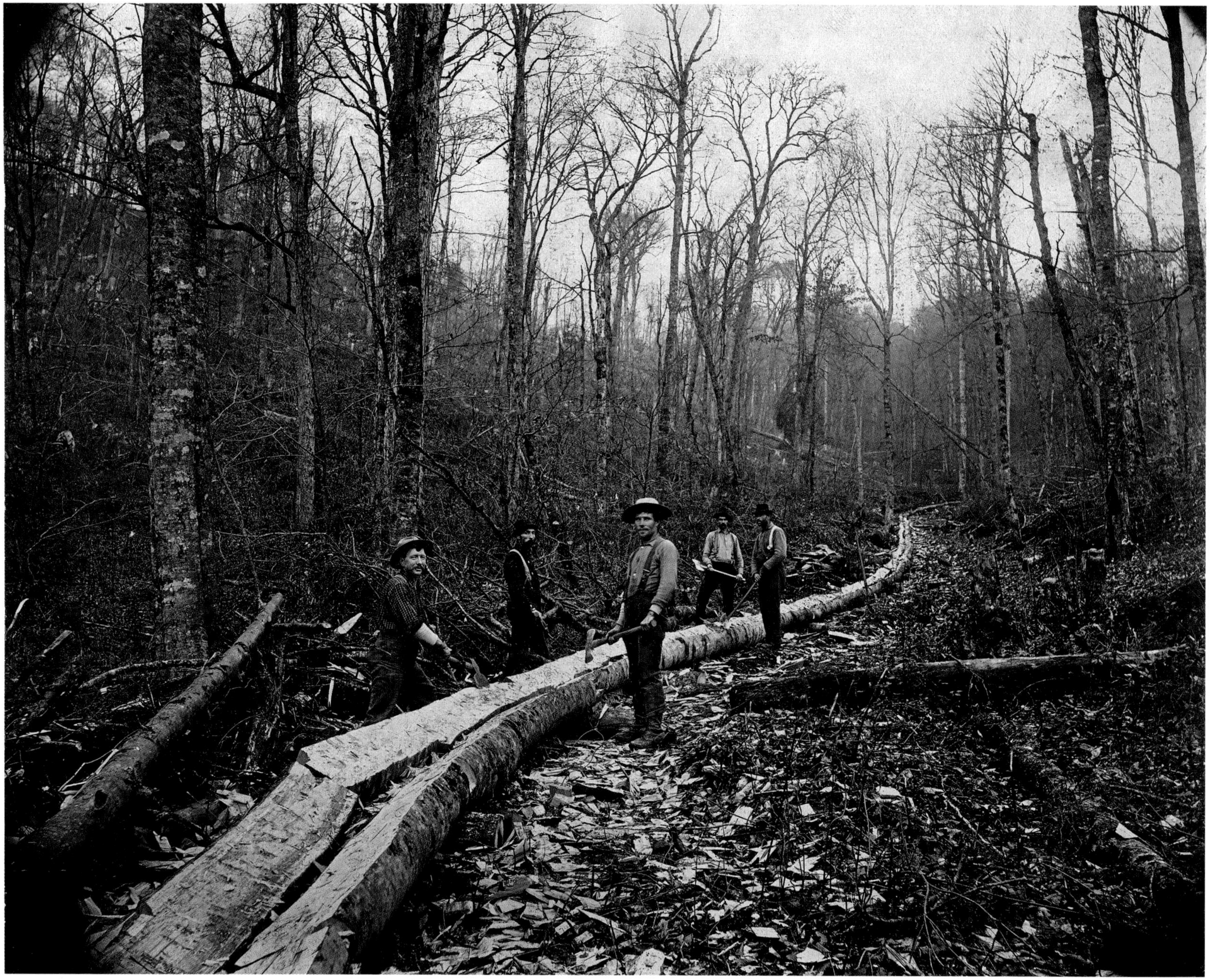

PLATE 30

Building a log slide near Corbett

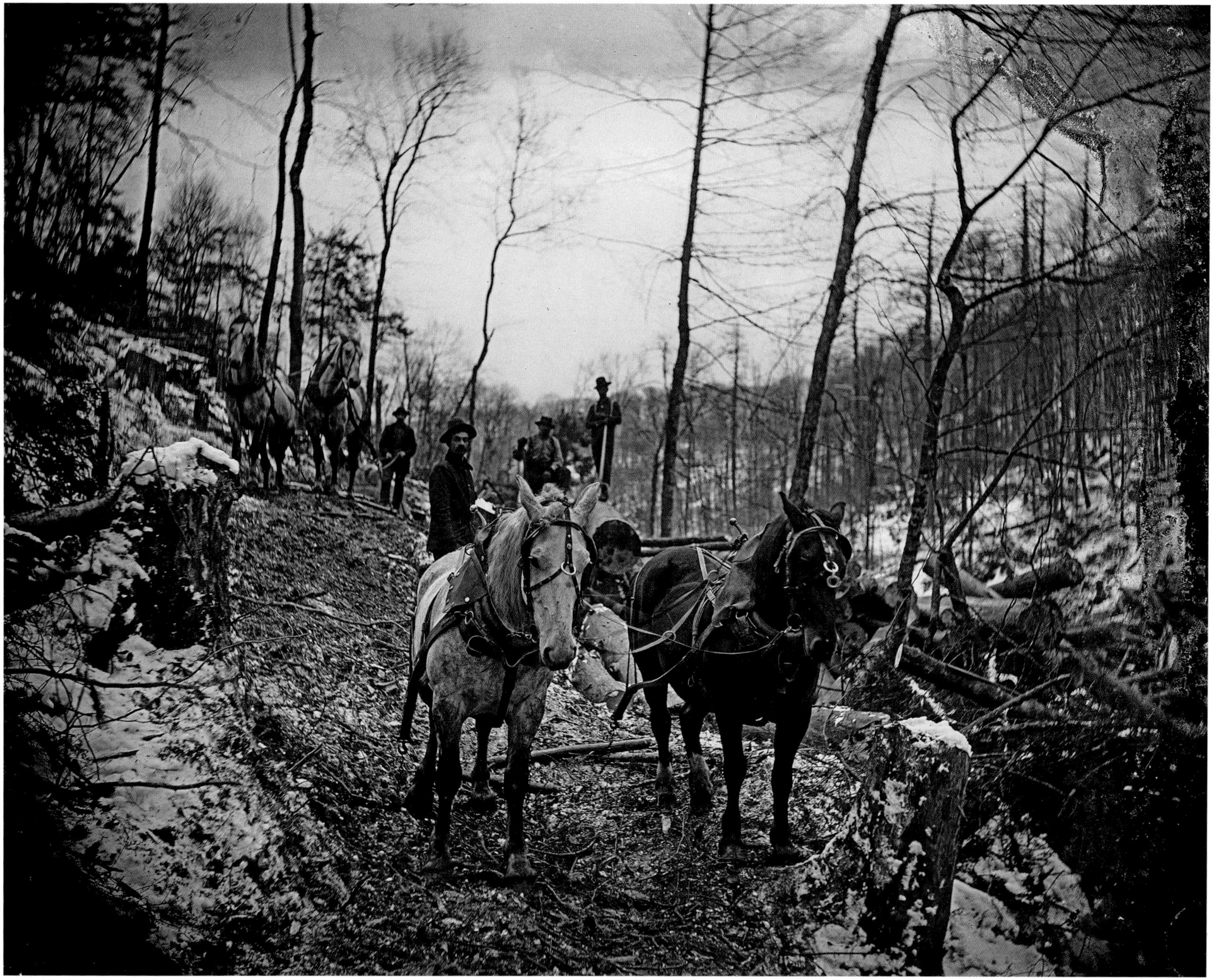

PLATE 31

Horse team at a typical work site

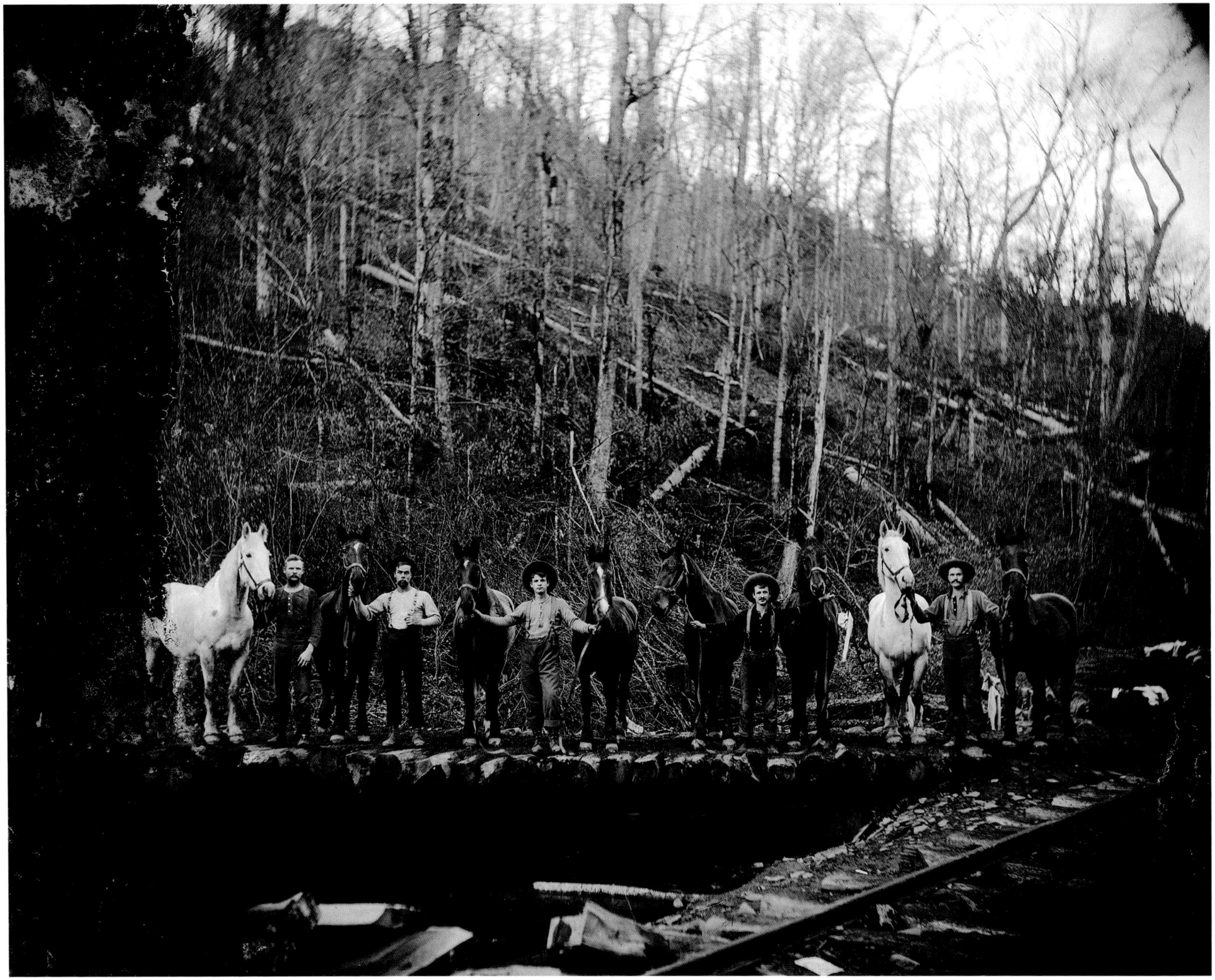

PLATE 32

Five men with eight horses on a log bridge at a railroad spur

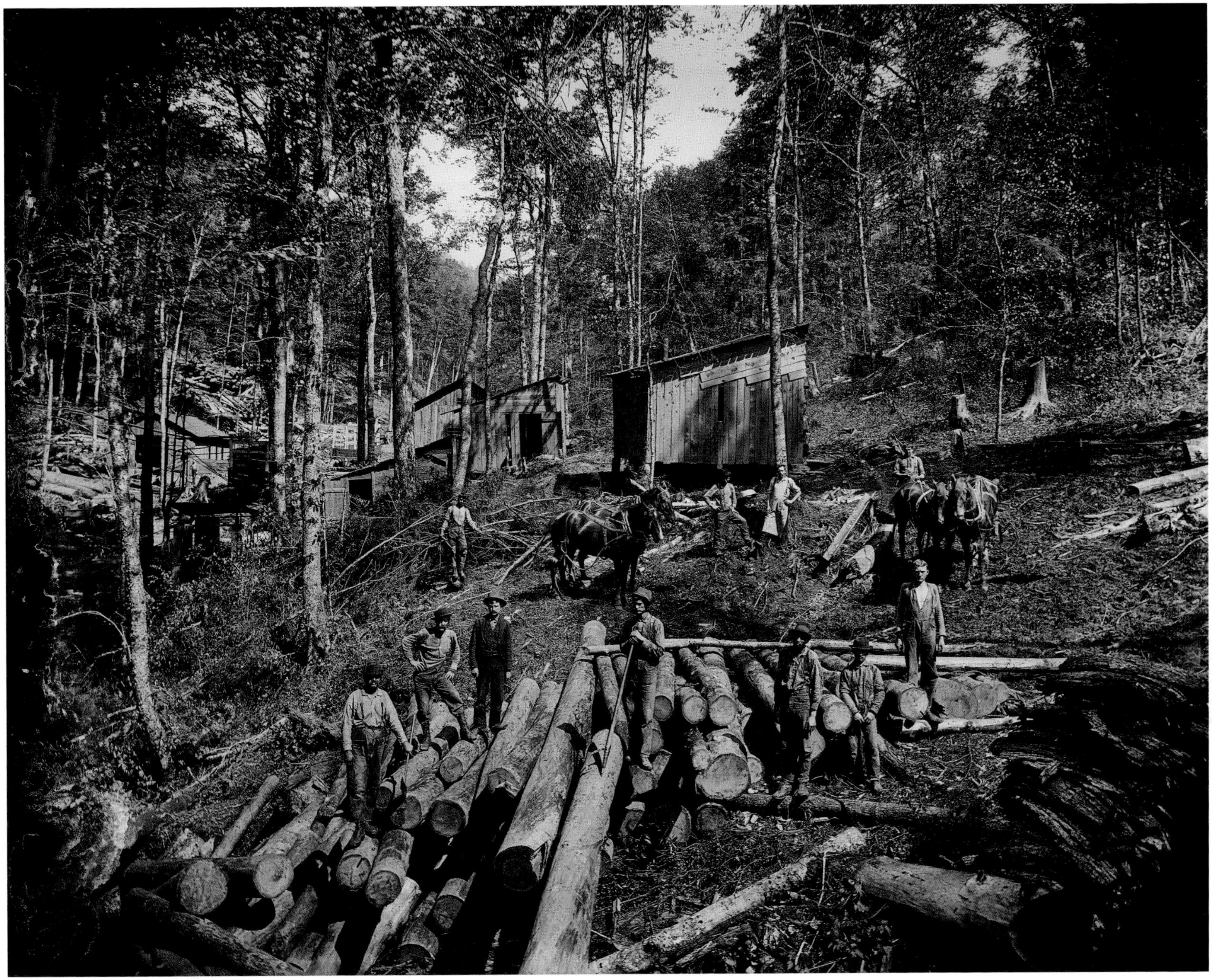

PLATE 33

Logging camp overview

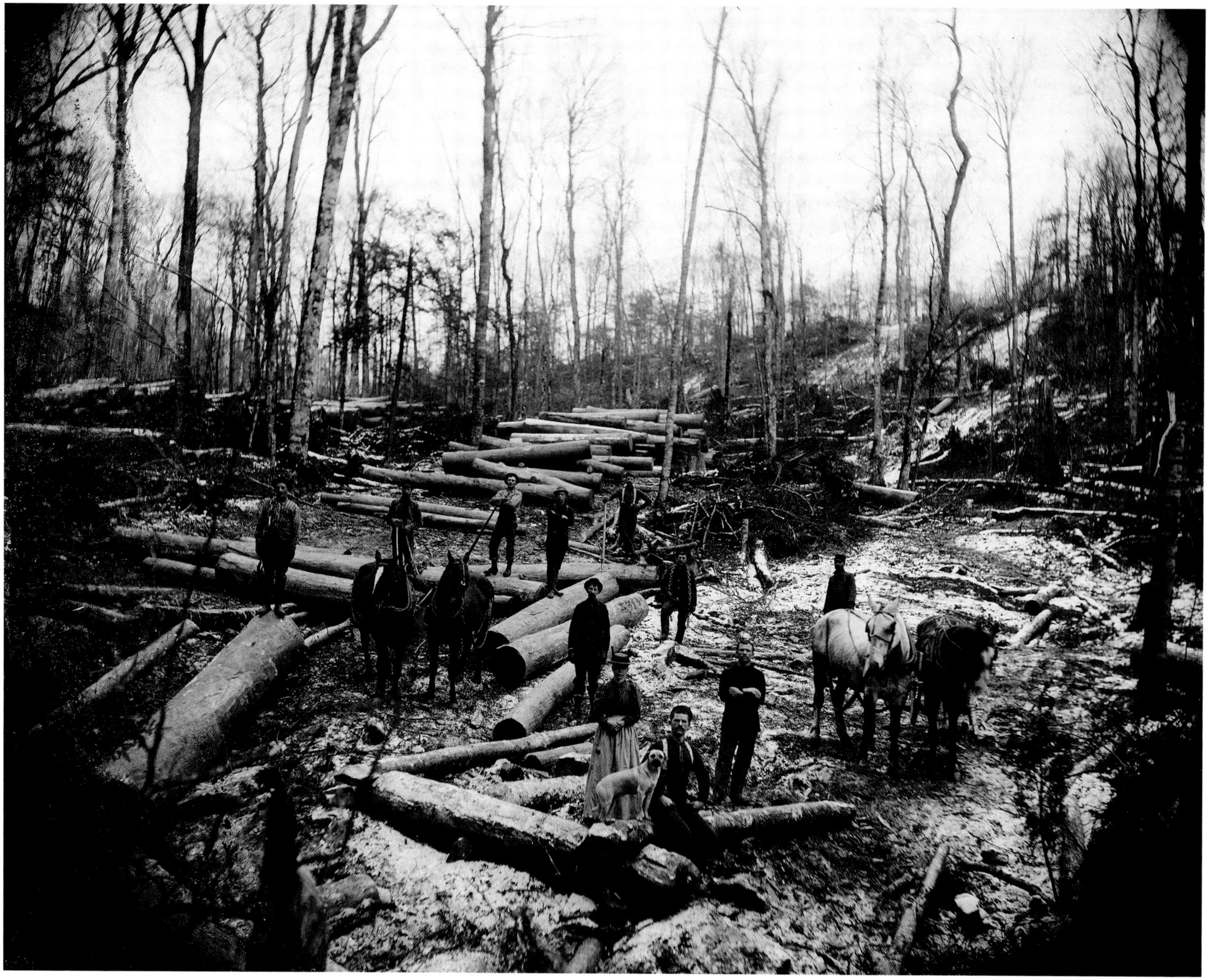

PLATE 34

Moving logs

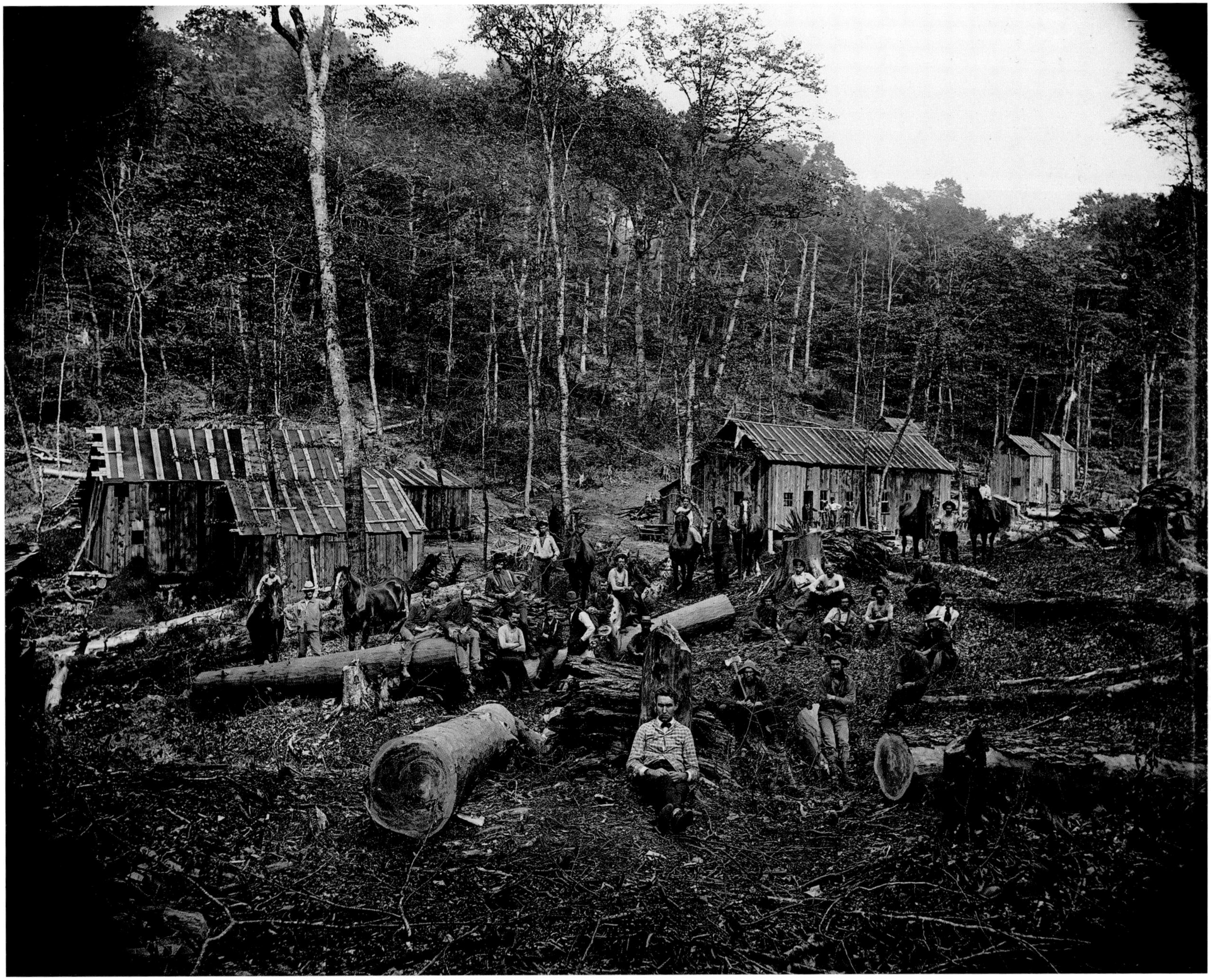

PLATE 35

Logging camp landscape

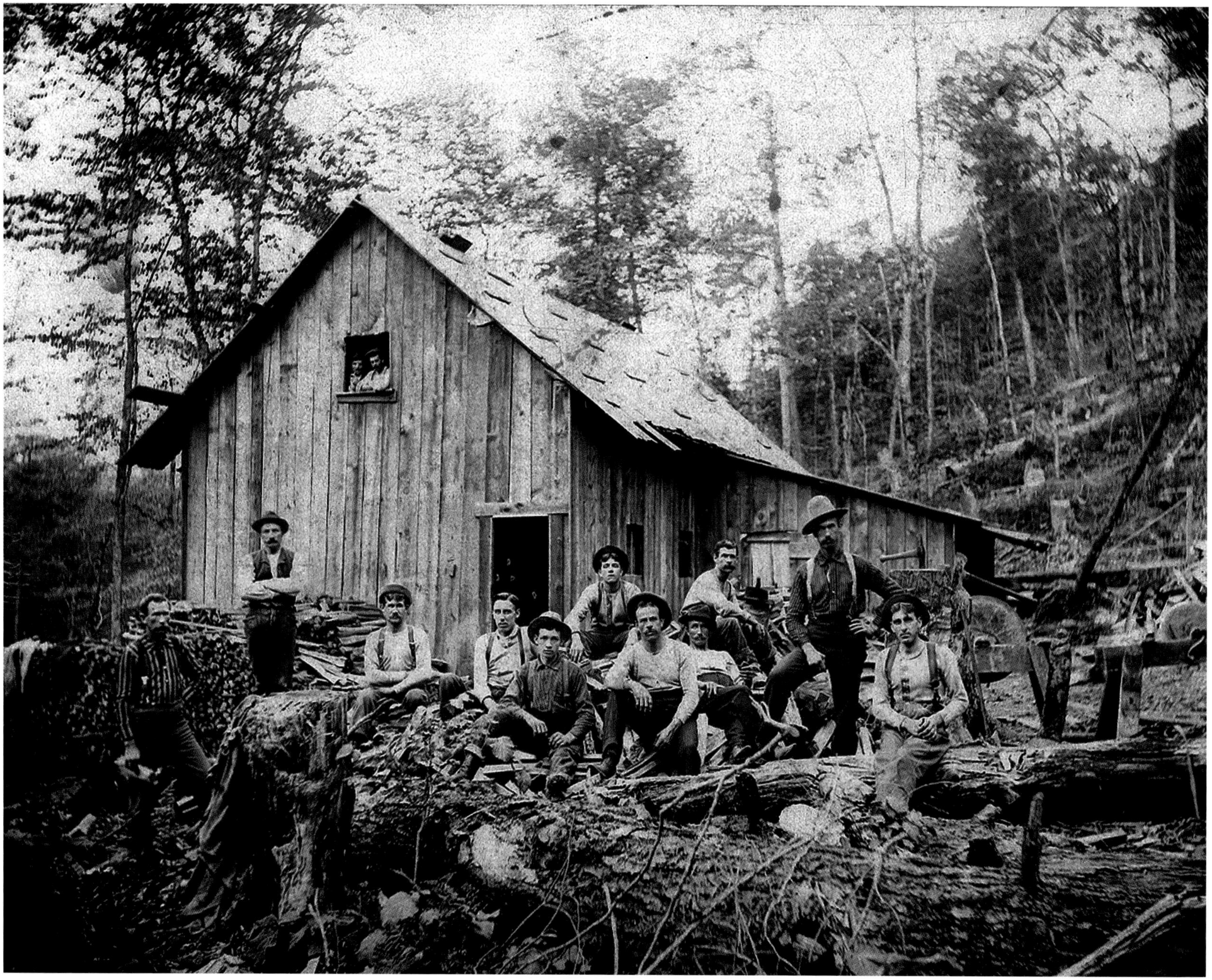

PLATE 36

Have muscles and ax, will travel

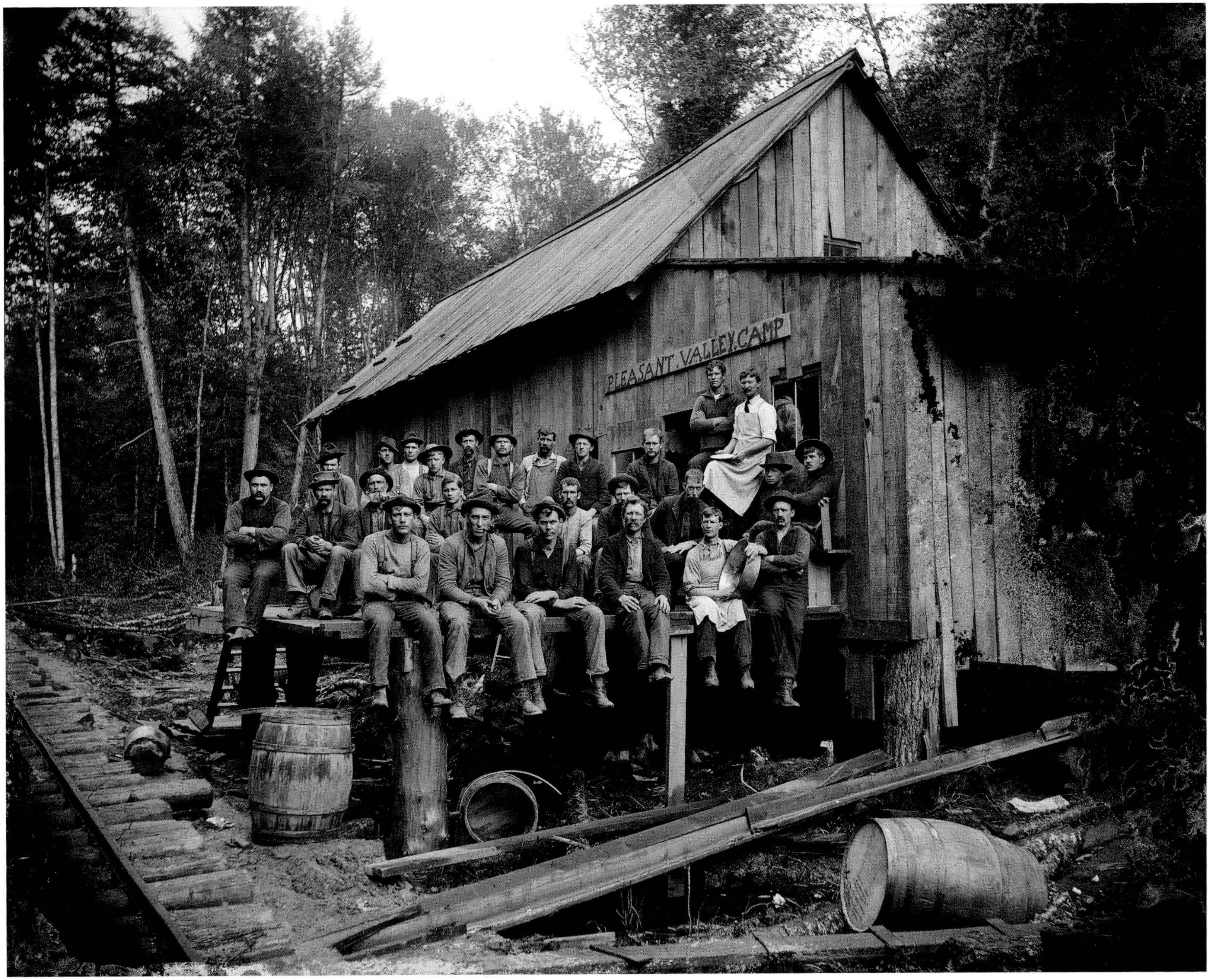

PLATE 37

Small logging camp crew. Cook and cookee in white aprons

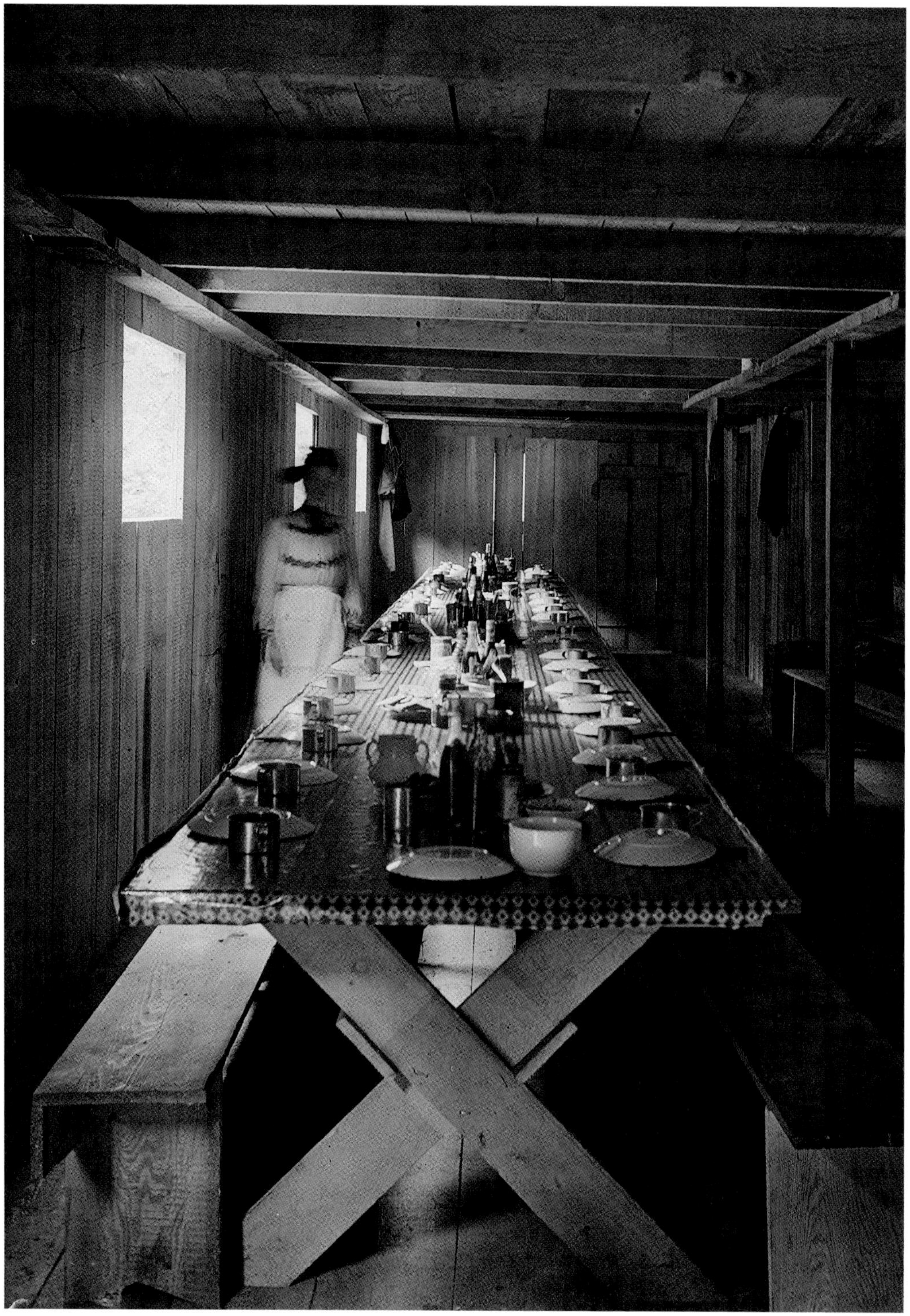

PLATE 38

Dining room interior, tables set.
Hammersley Boarding House,
Hammersley region, Clinton County

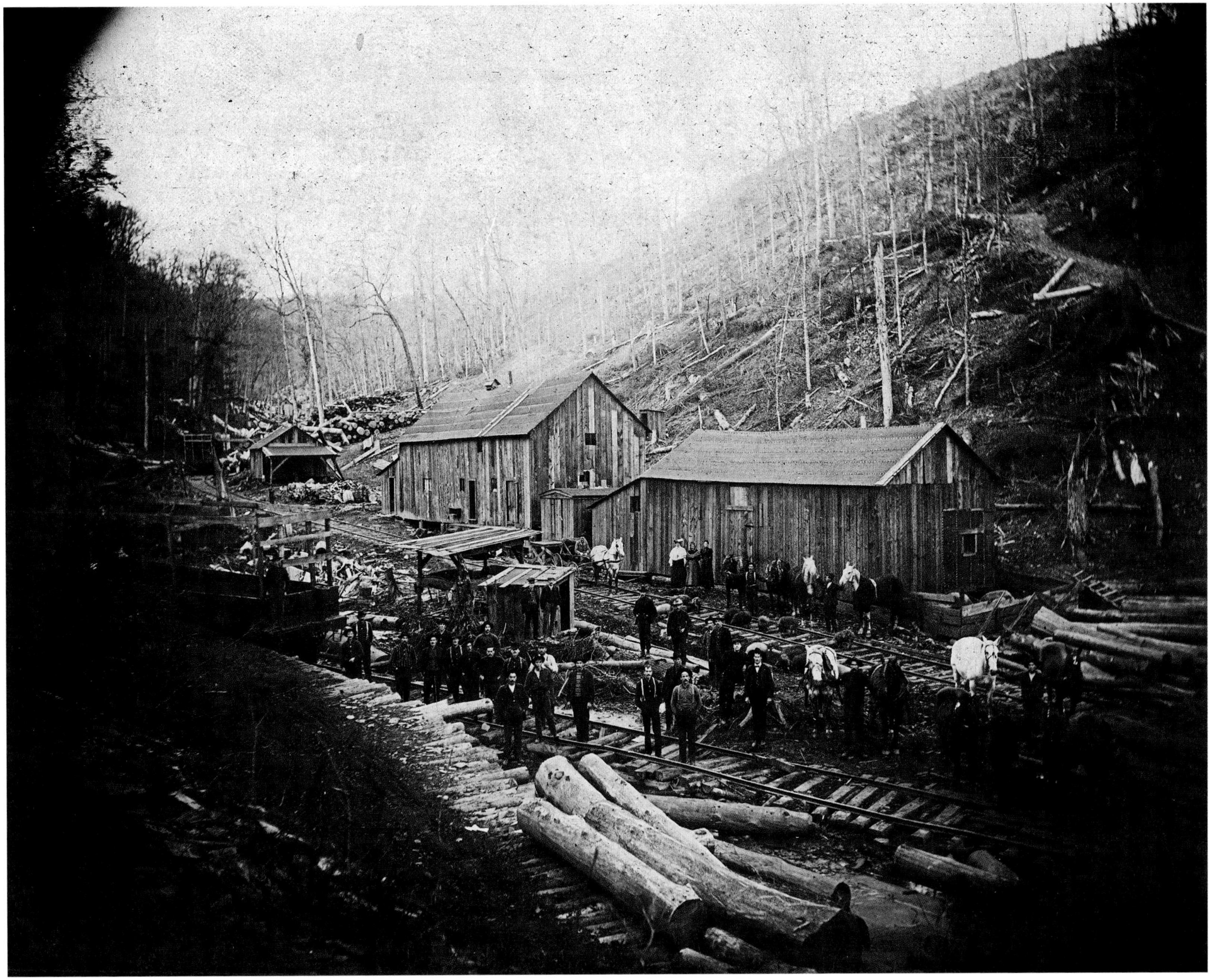

PLATE 39

Jake Swanson's camp

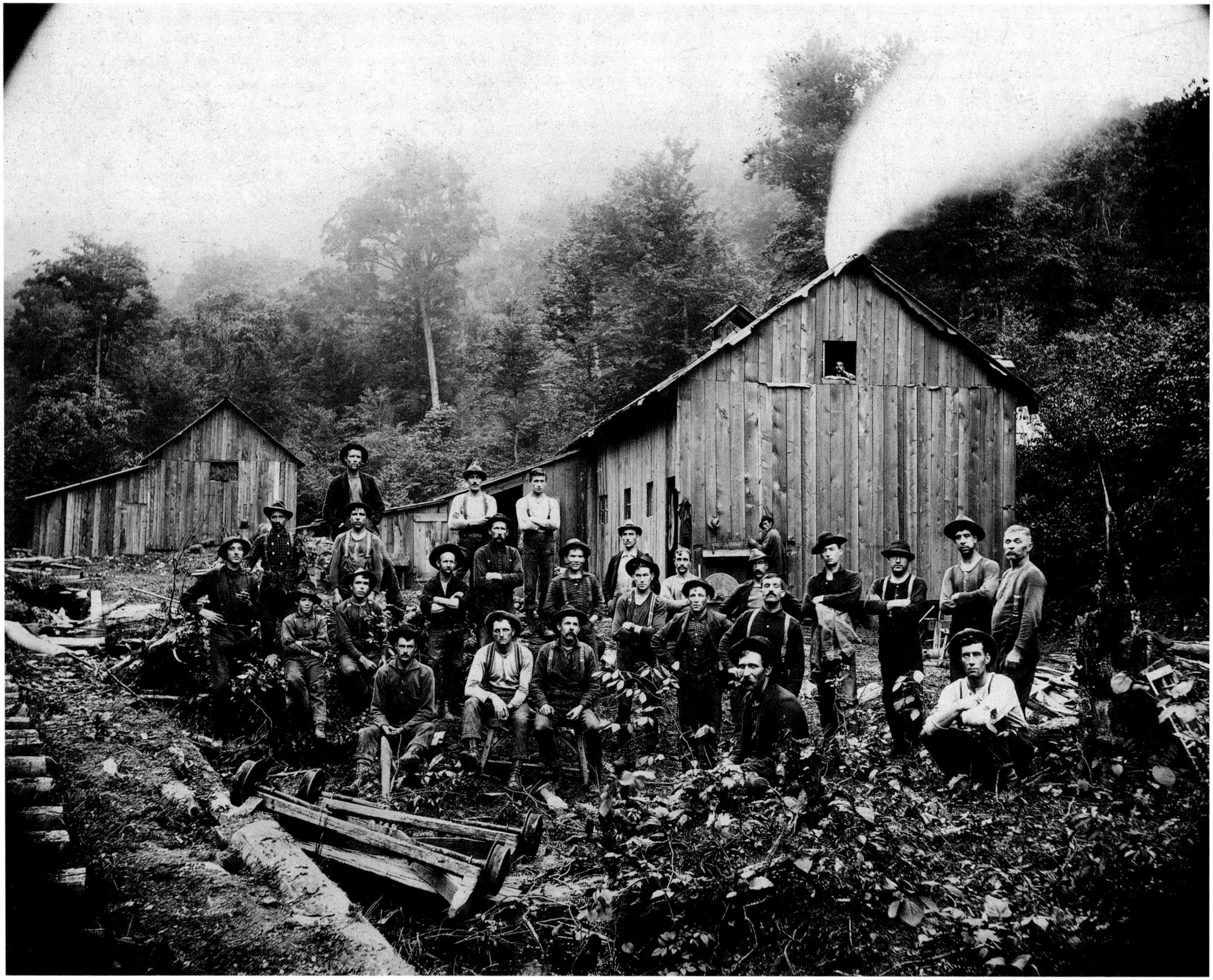

PLATE 40

Logging camp with saw sharpener at his work station (left corner, building)

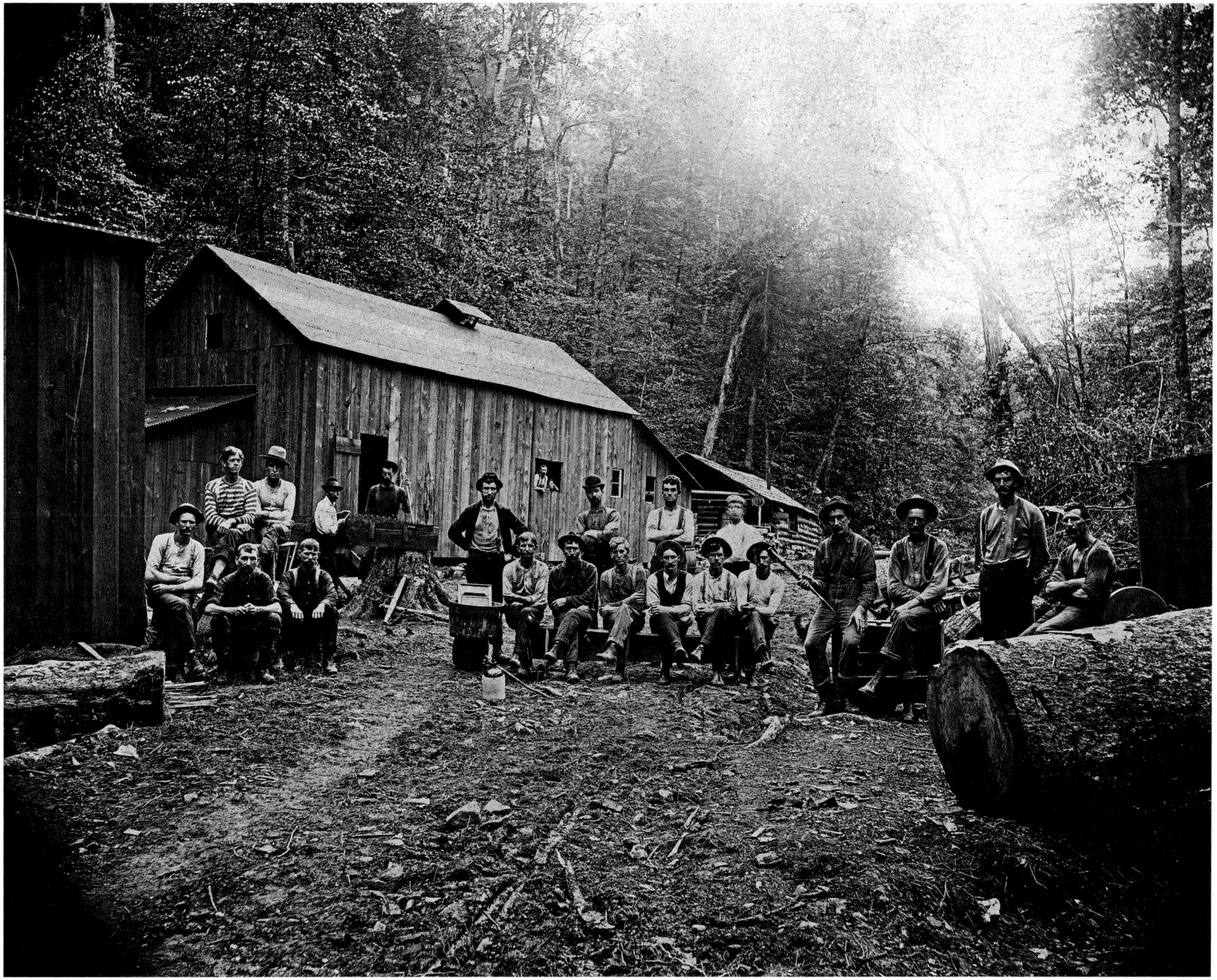

PLATE 41

A saw-filing station nailed to the top of a tree stump, saw sharpener nearby

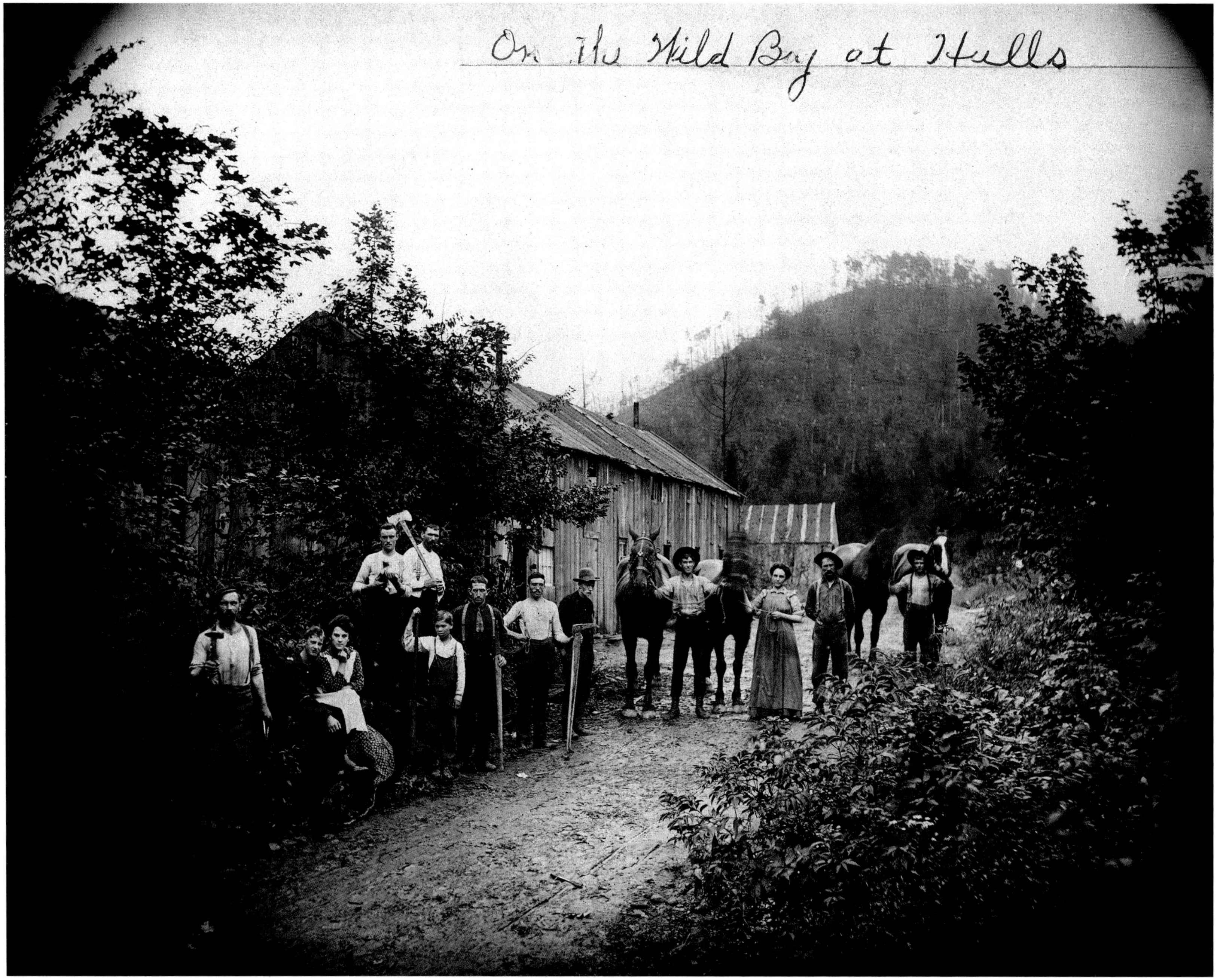

PLATE 42

"On the Wild Boy at Hulls"

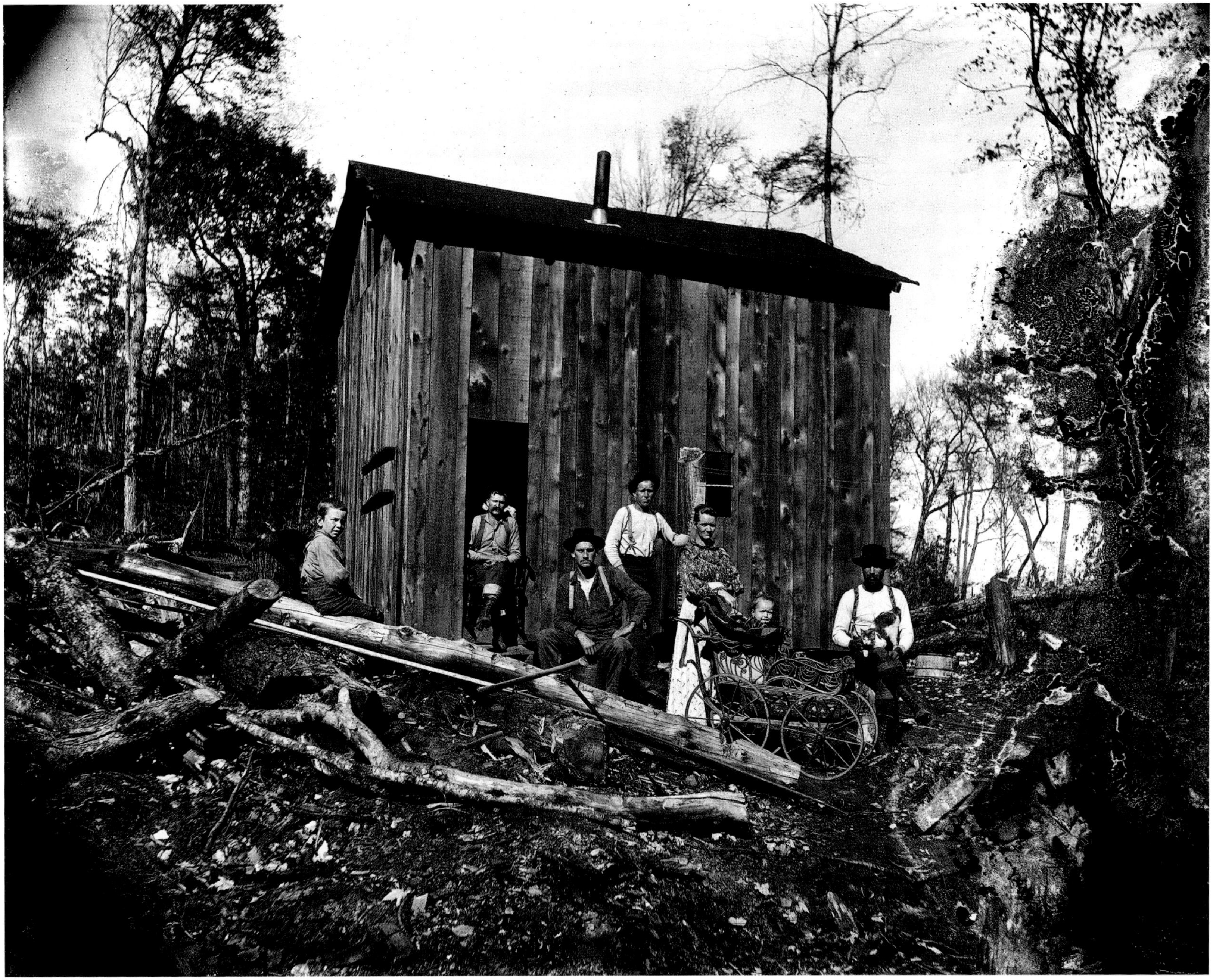

PLATE 43

Family outdoor photo

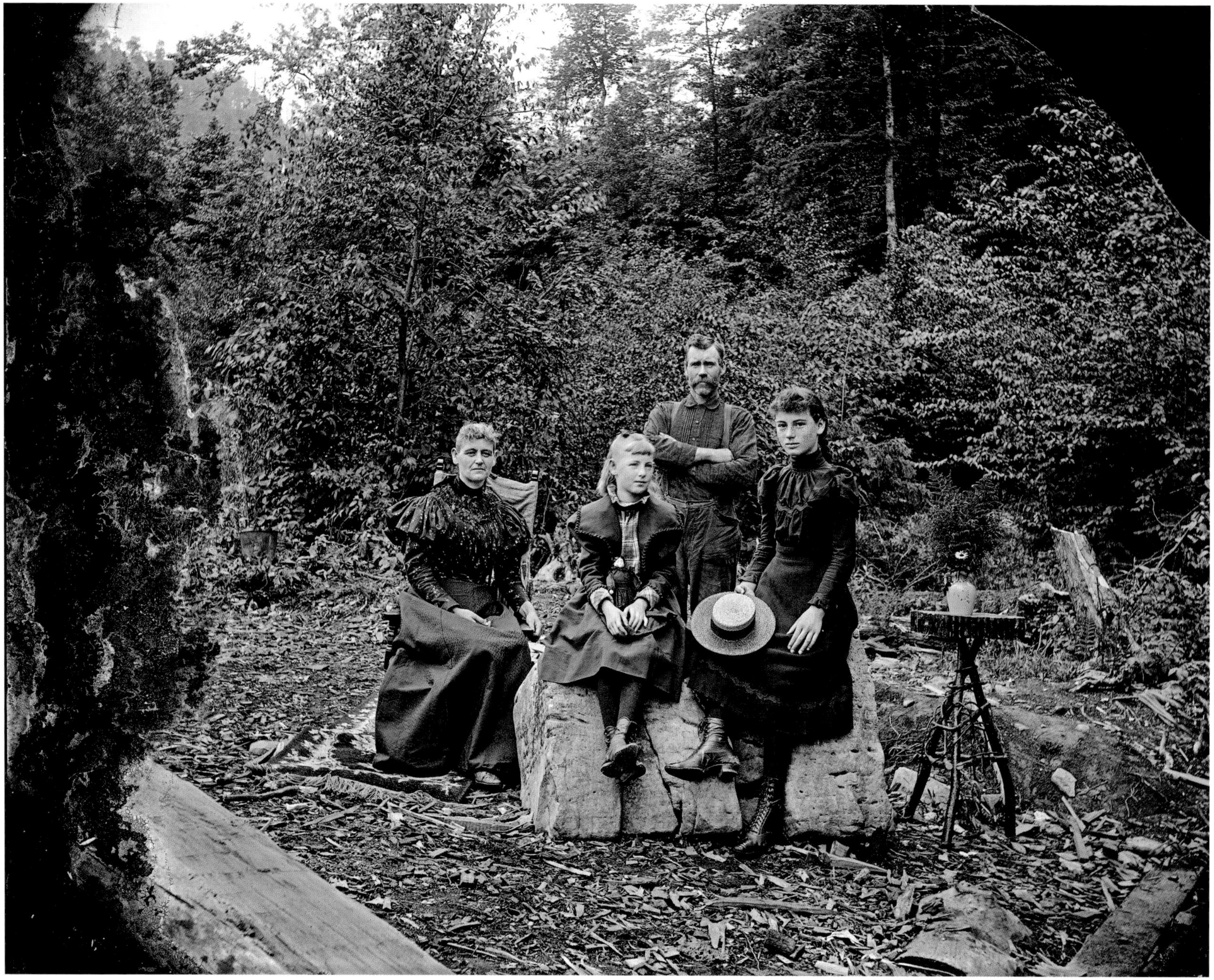

PLATE 44

Jobber Allison's family

PLATE 45

Bill Brewer, "hick preacher."

Hammersley region, Clinton County

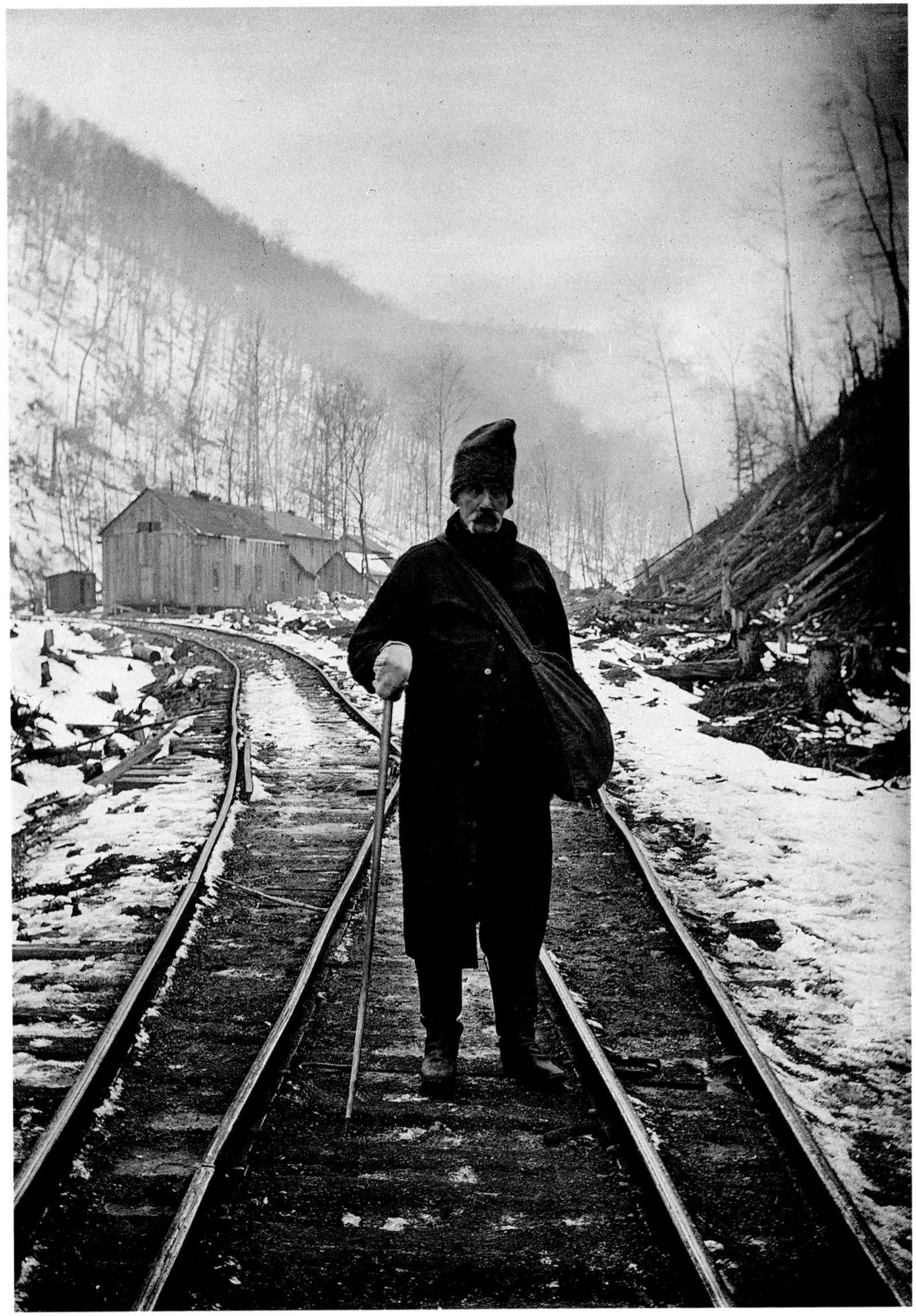

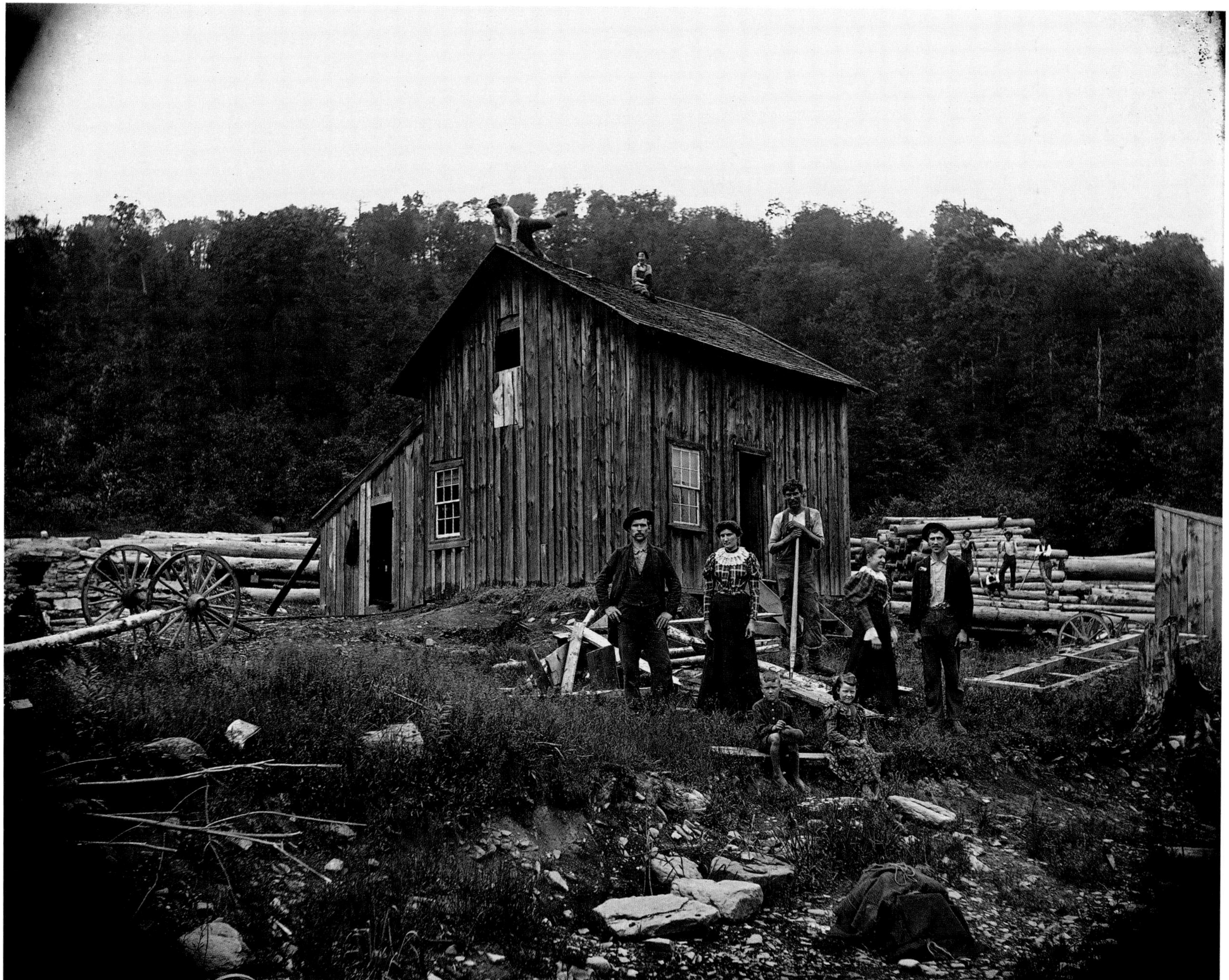

PLATE 46

Horseplay (men on the roof)

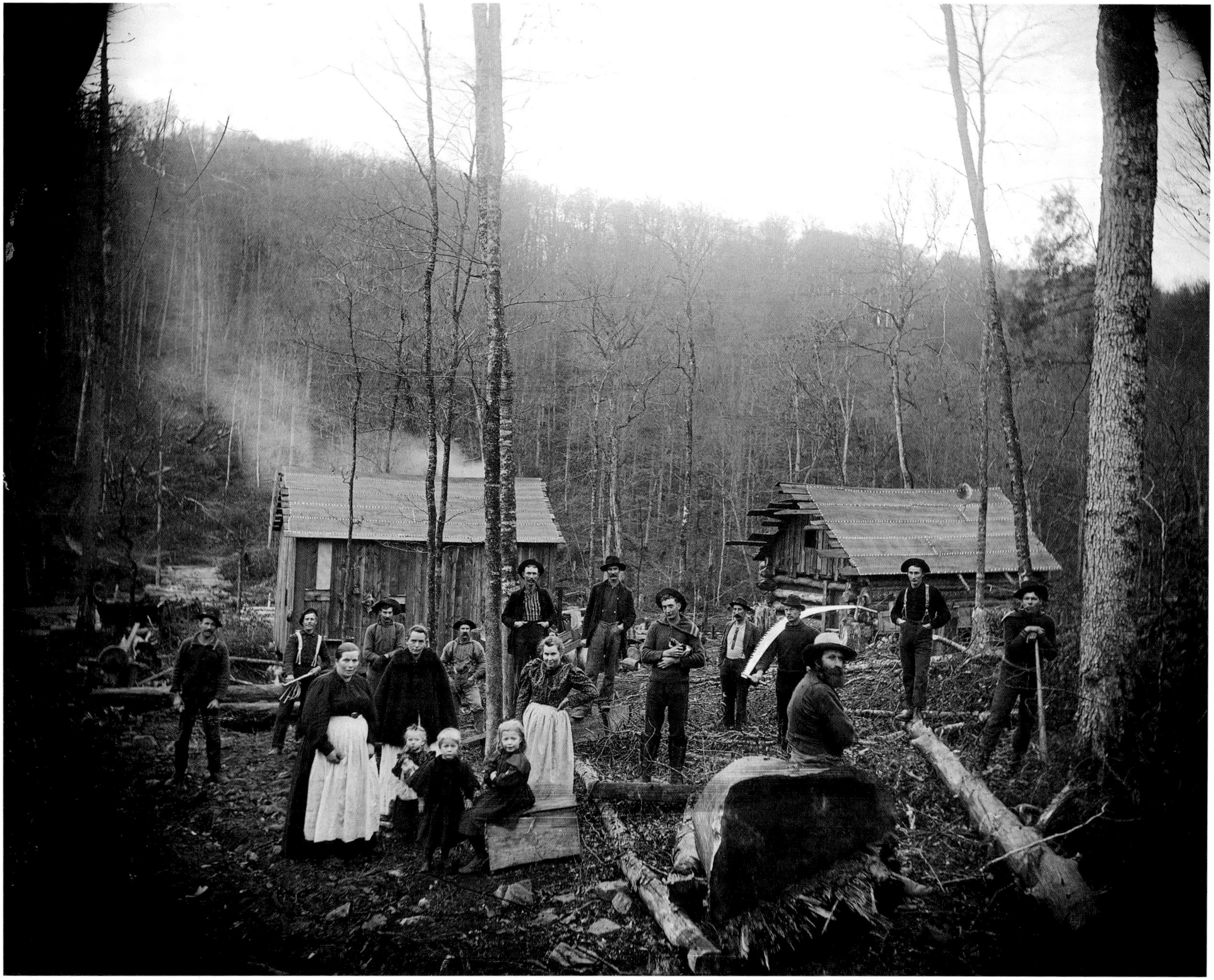

PLATE 47

Logging community, man holding cat

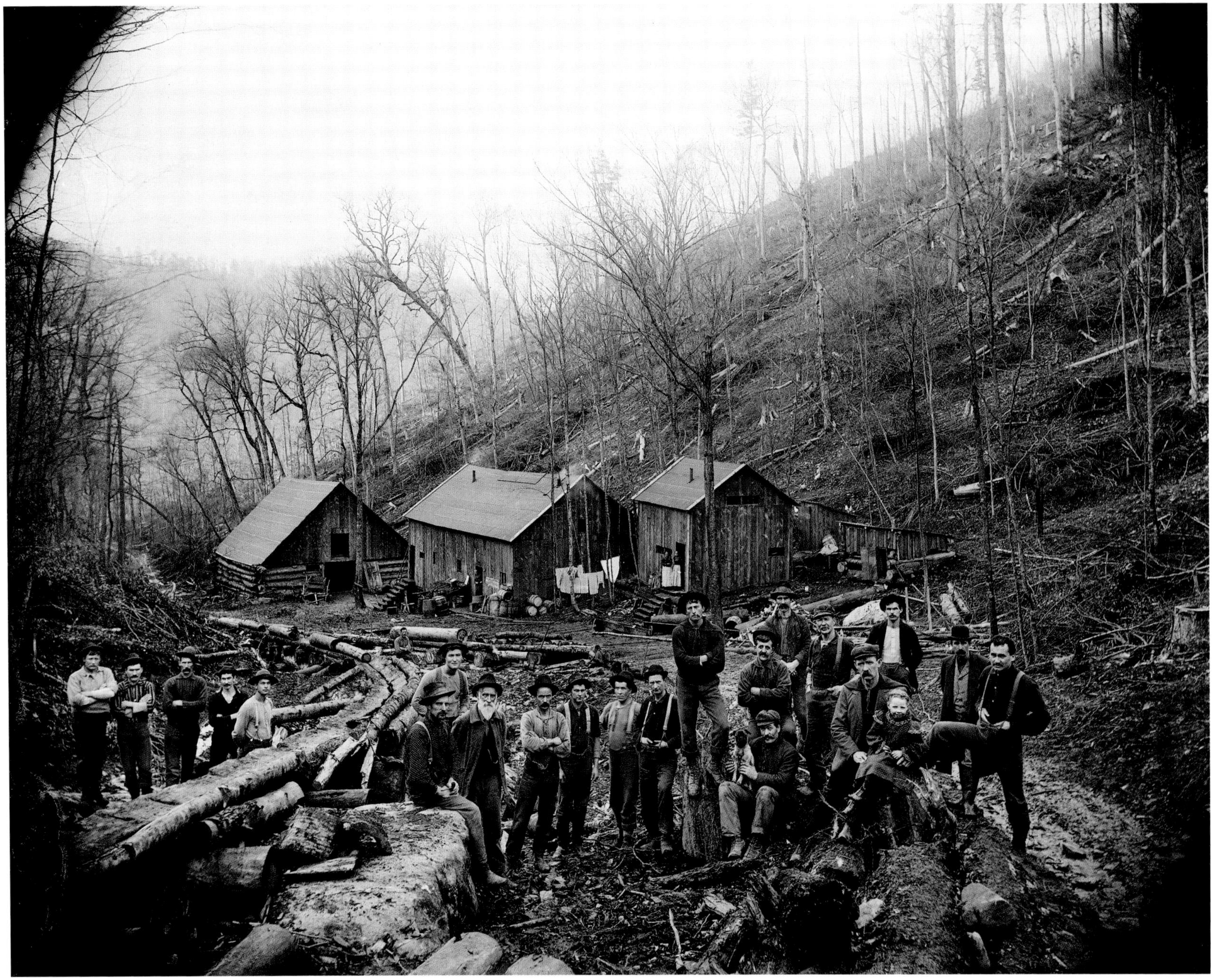

PLATE 48

Gravity slide in logging camp, man holding dog

PLATE 49

Boiling up. Cooperative laundry, Lyman Run lumber camp

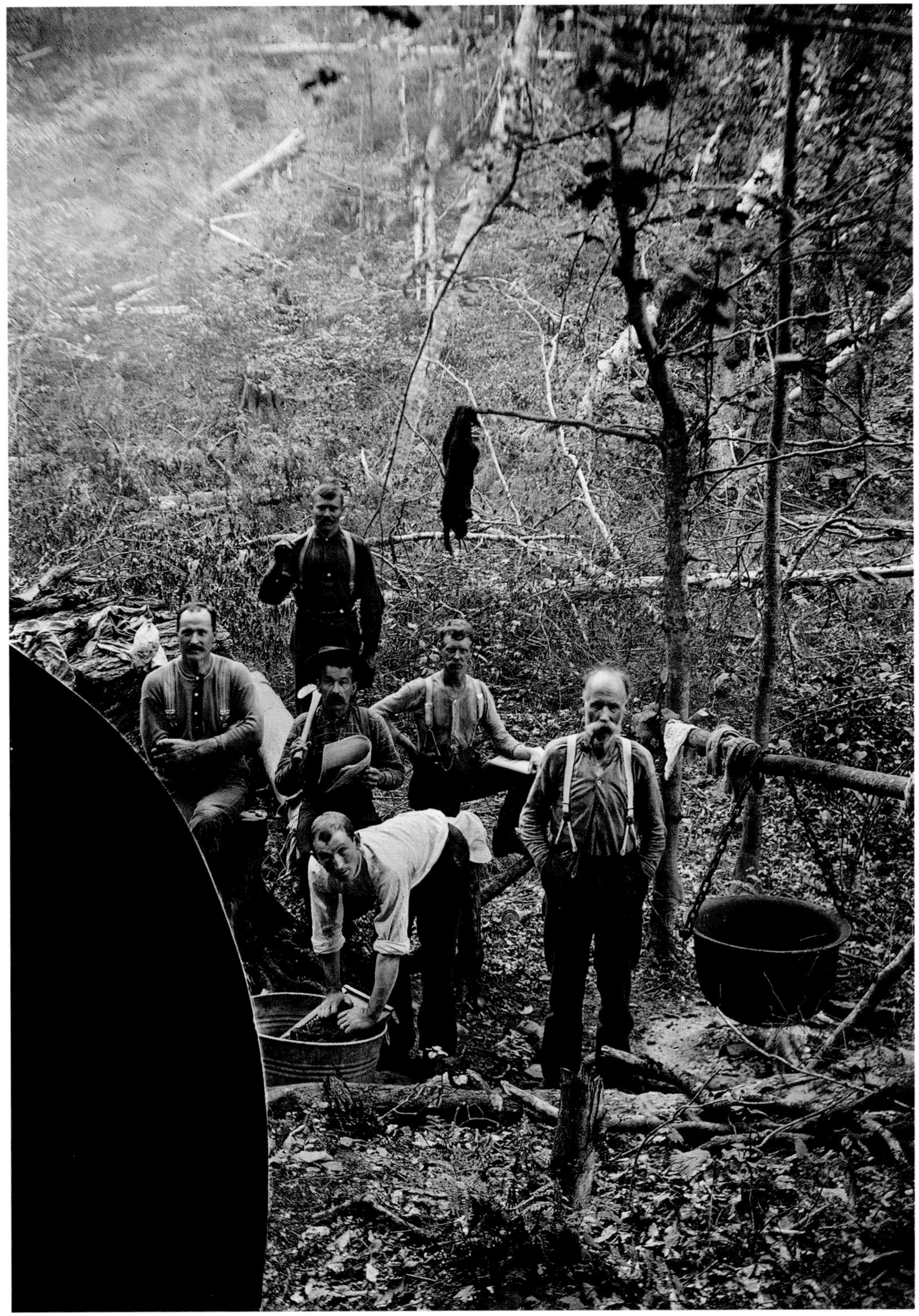

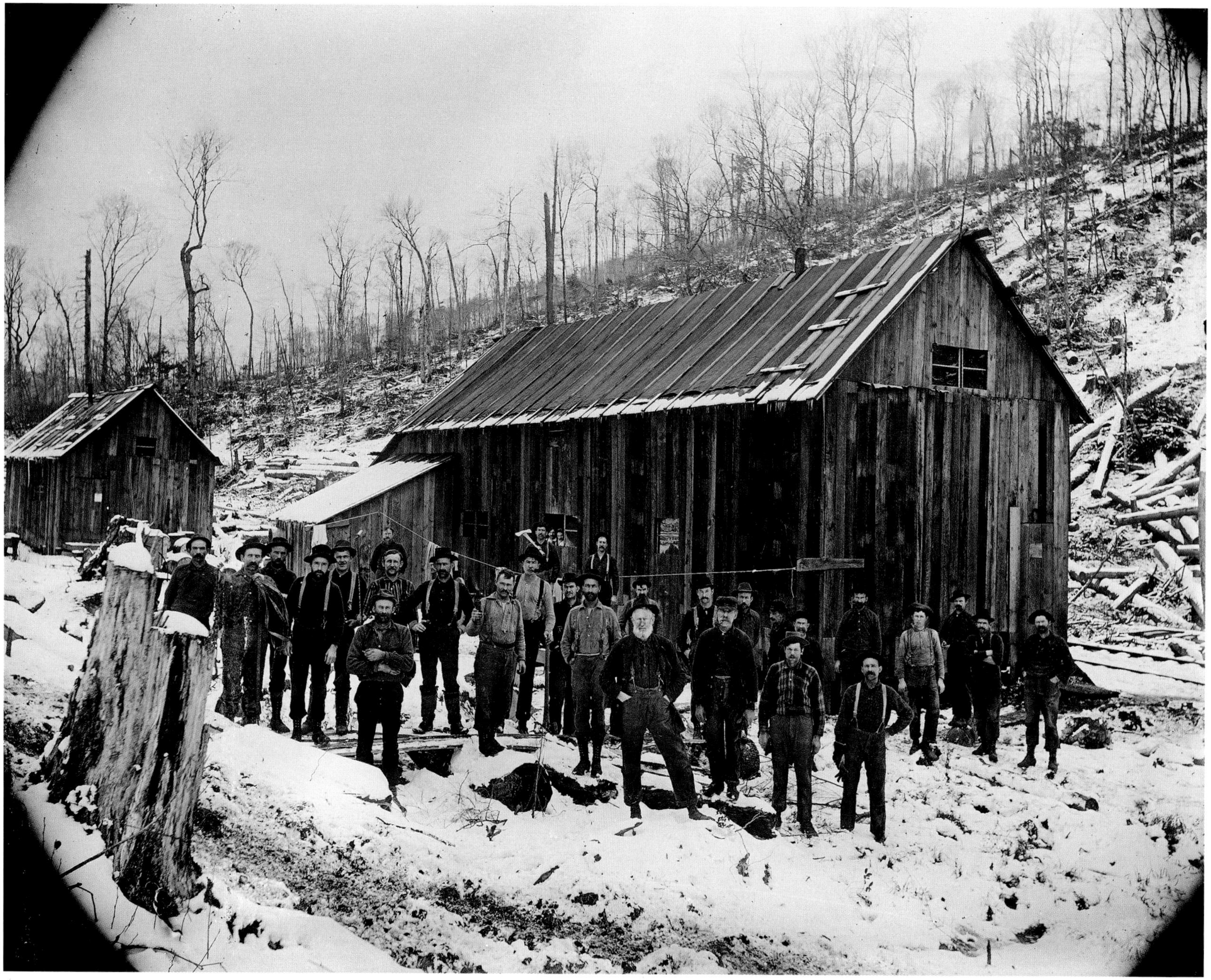

PLATE 50

Twenty-nine men at winter camp

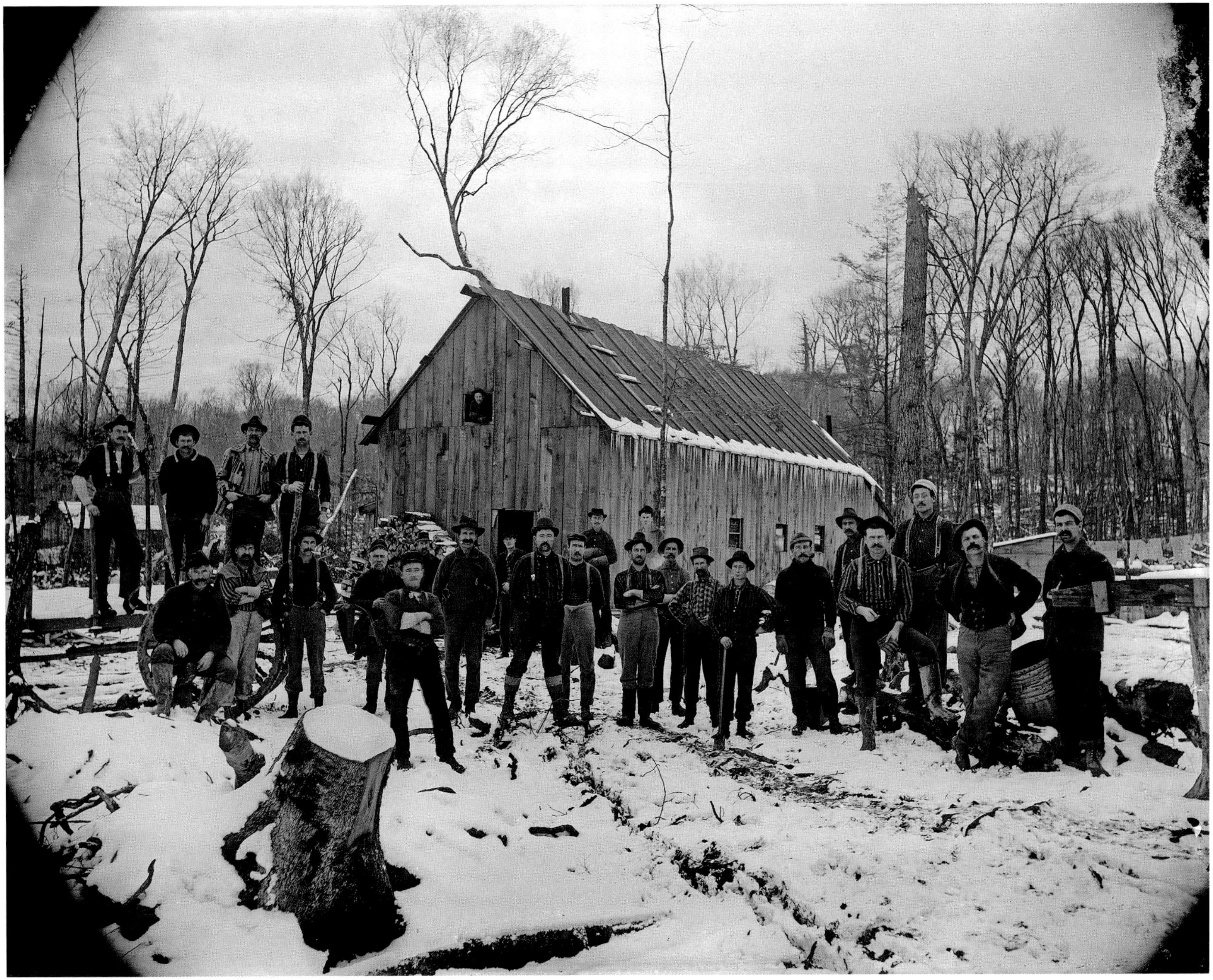

PLATE 51

Men outside winter camp

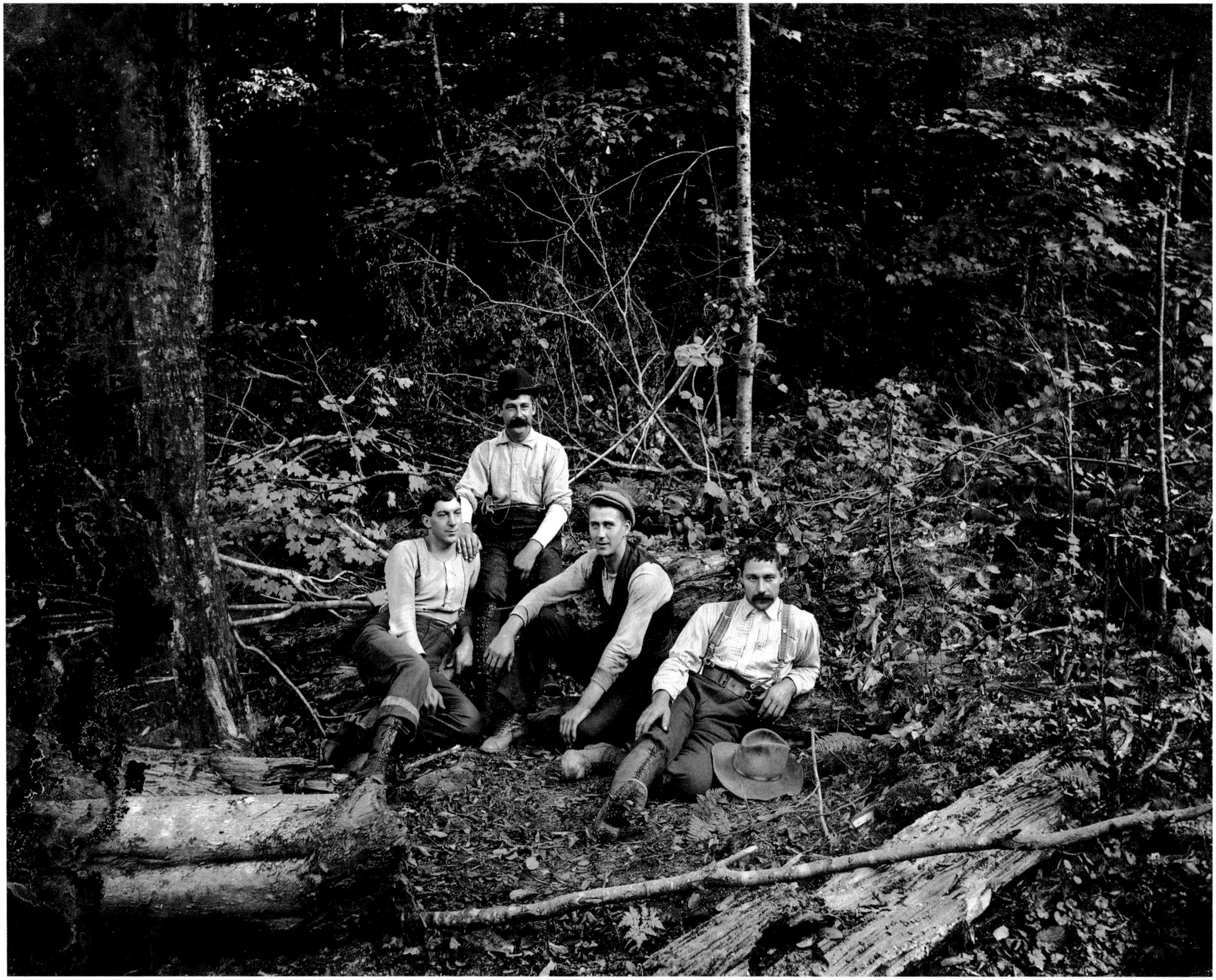

PLATE 52

Celebration ahead. Four men all dressed up

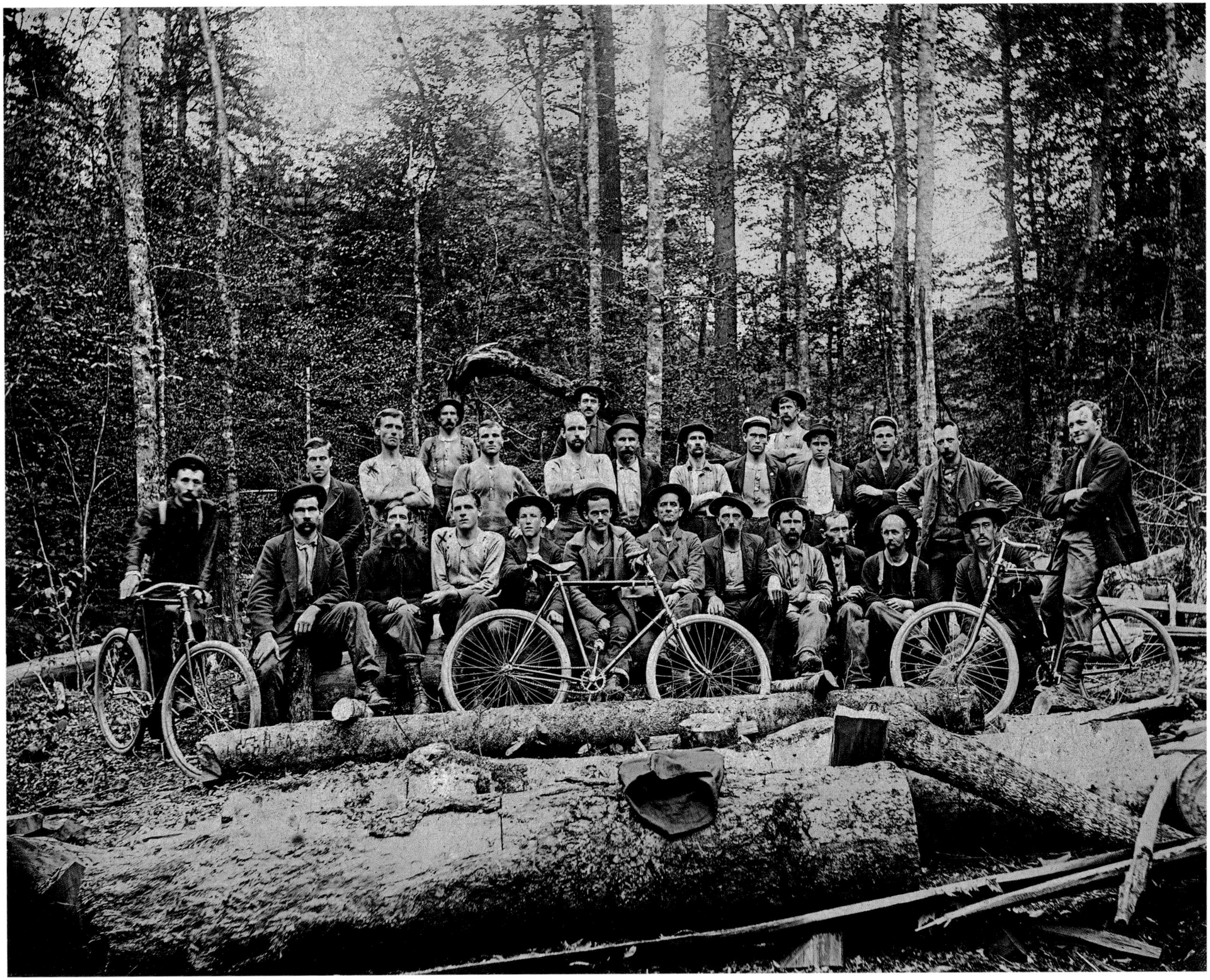

PLATE 53

Men and three bicycles

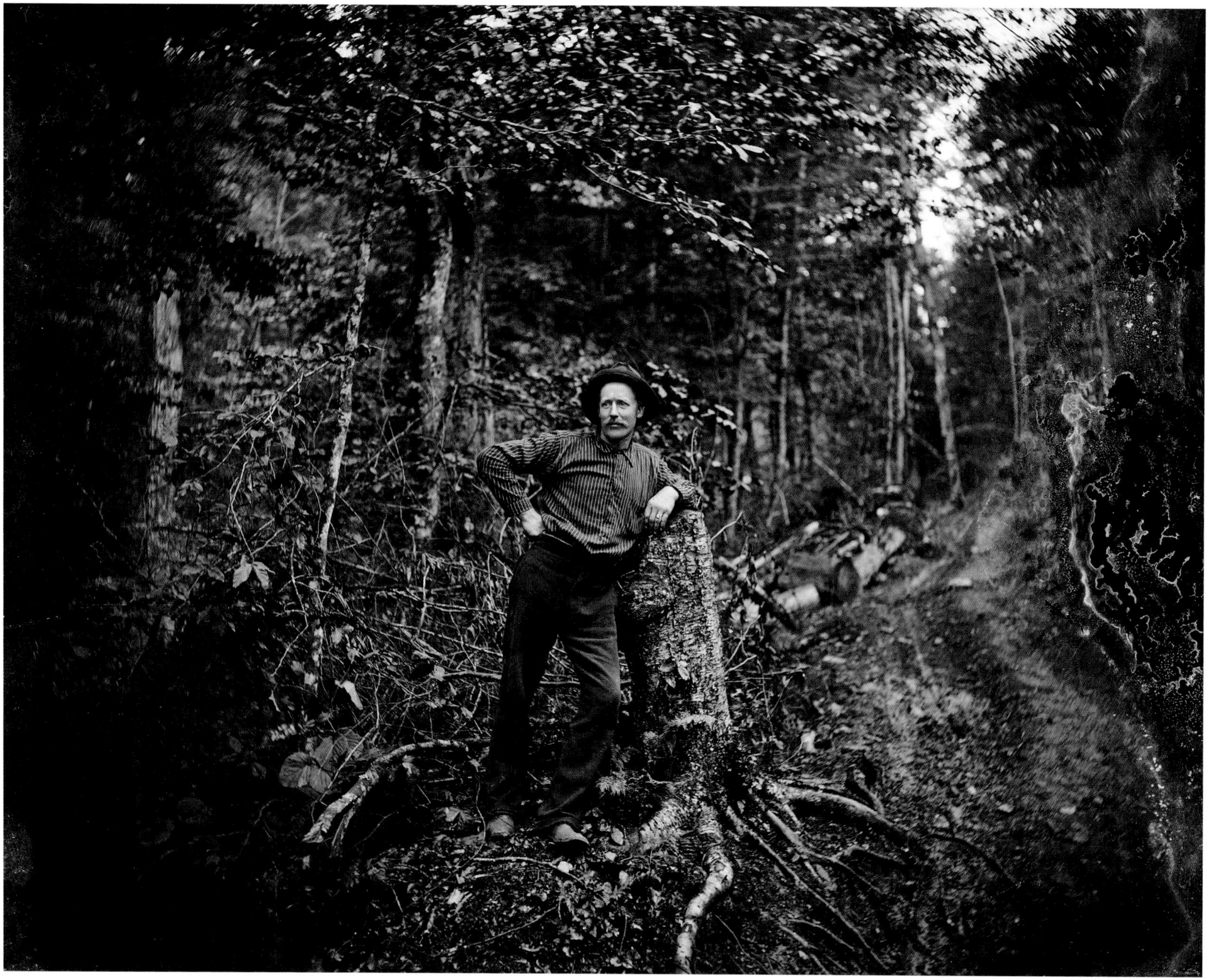

PLATE 54

Man leaning on stump near rough road

PLATE 55

Possible “lobby hog” (detail of plate 50)

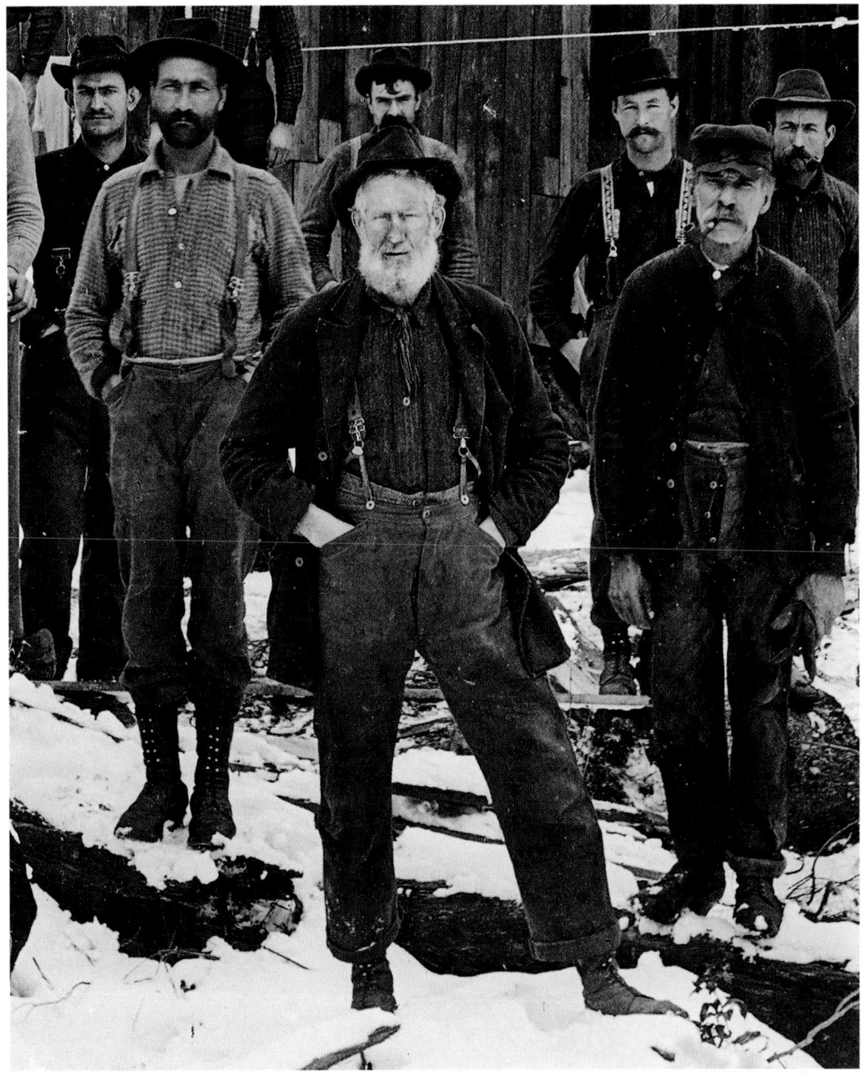

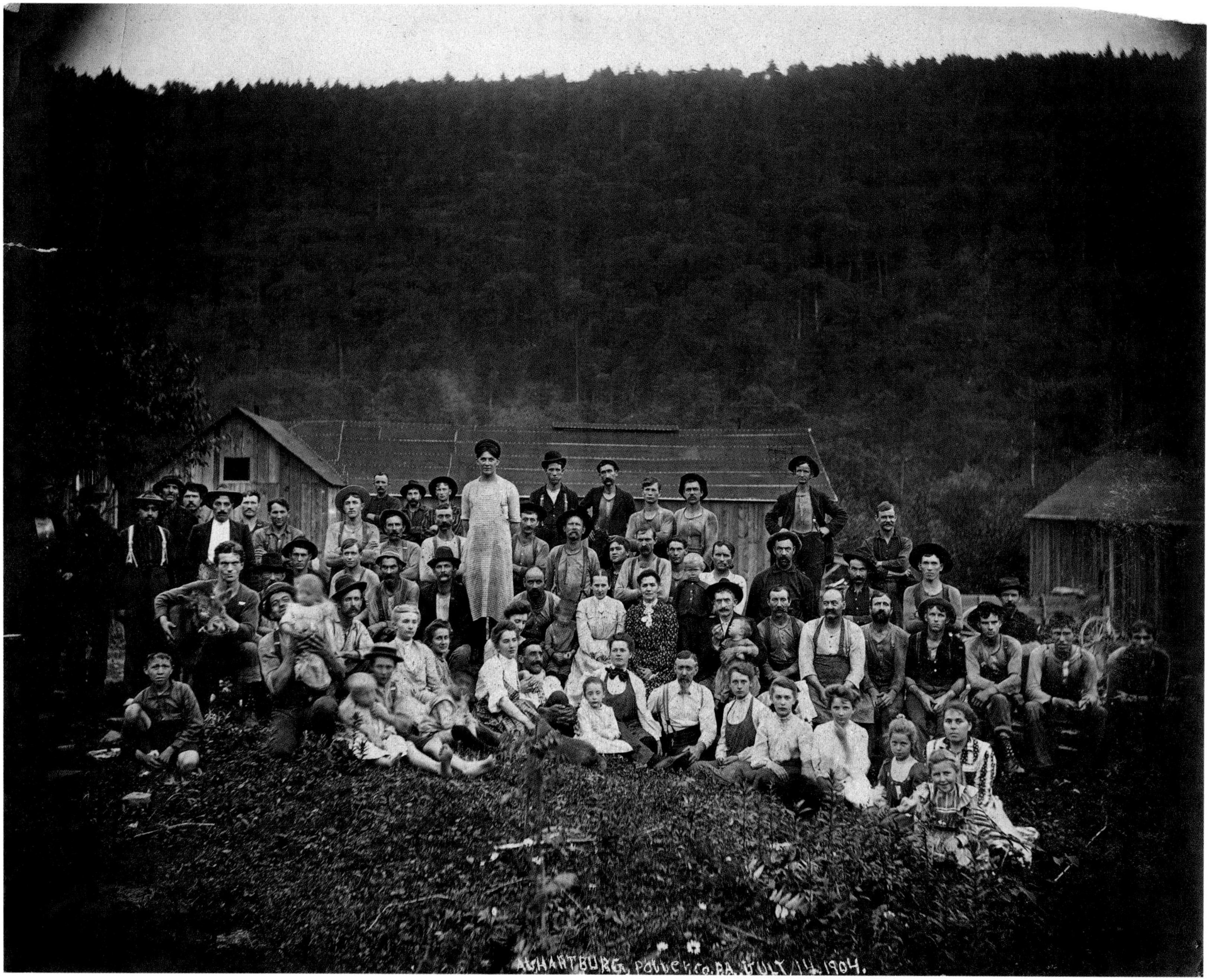

PLATE 56

Hartsburg, Potter County, July 4, 1904

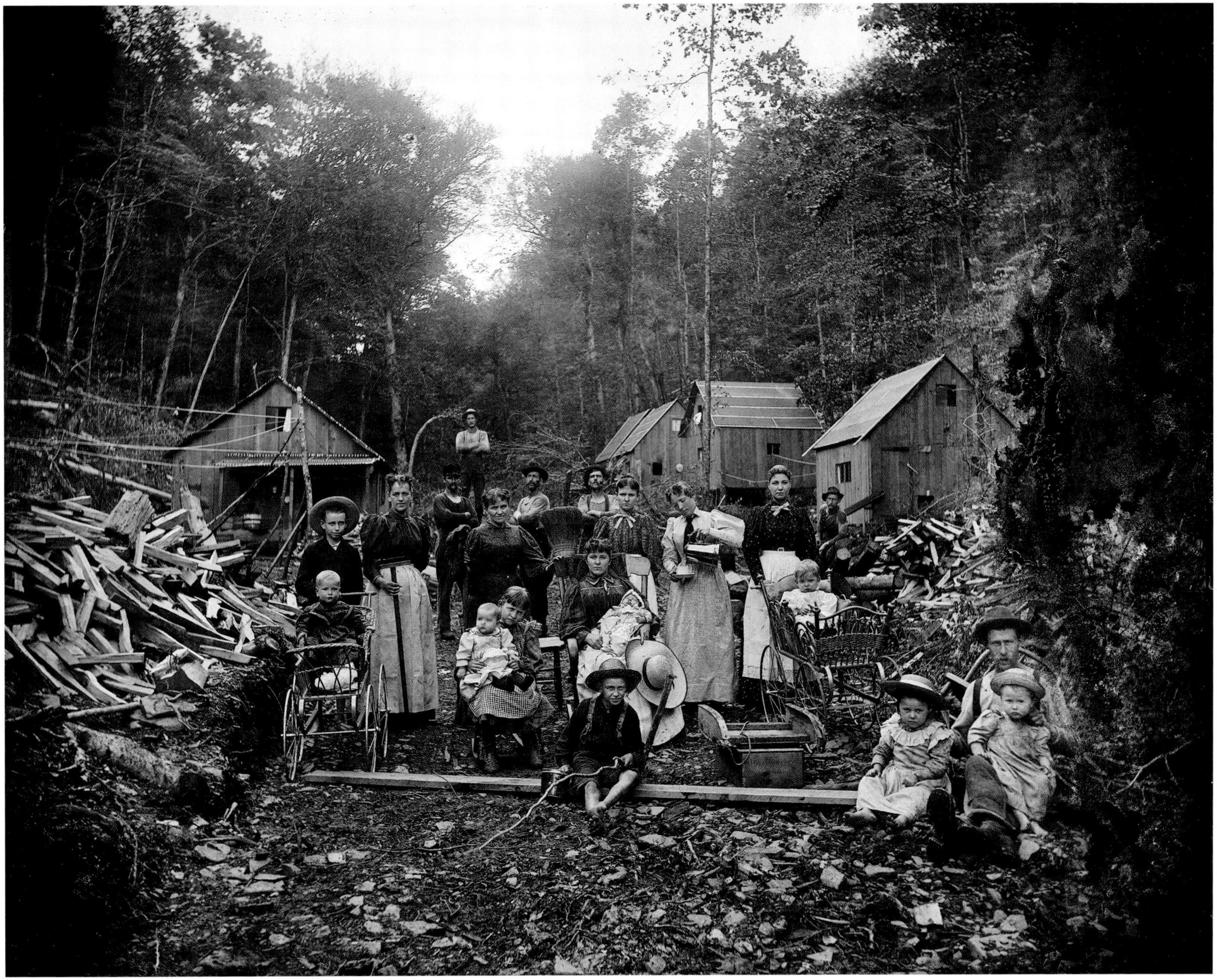

PLATE 57

A fledgling community with aspirations

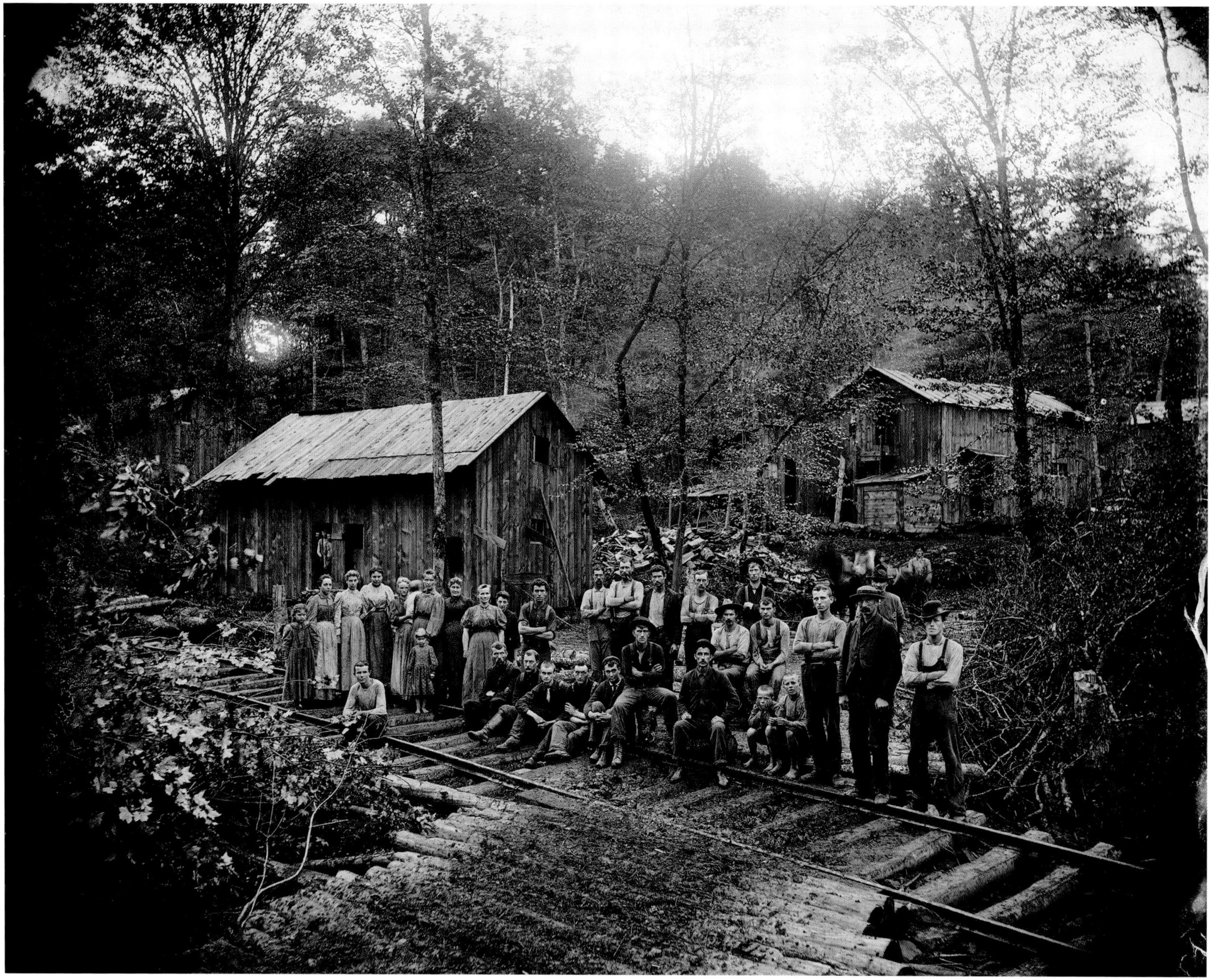

PLATE 58

Communities formed where transportation existed

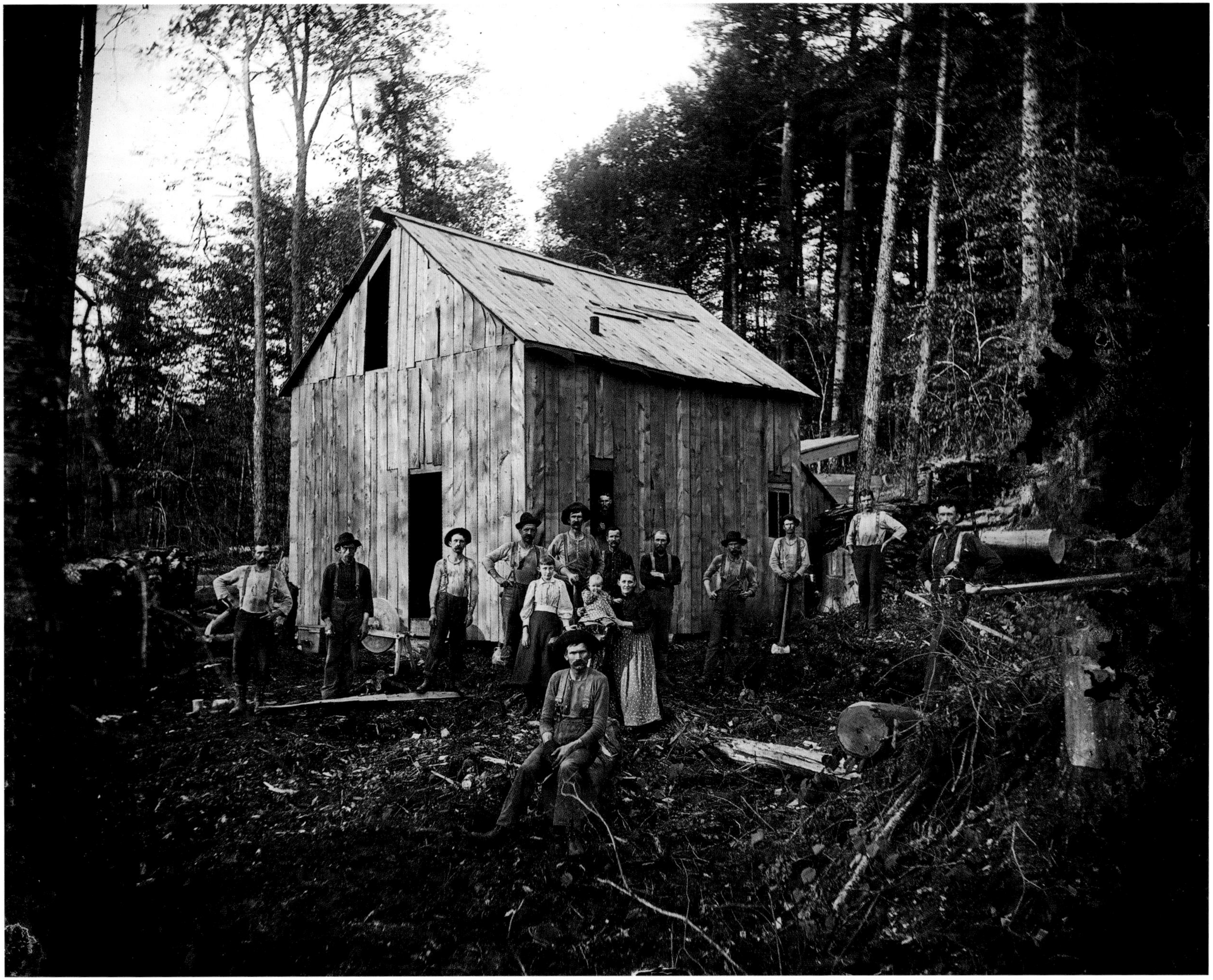

PLATE 59

Generations and extended family roots

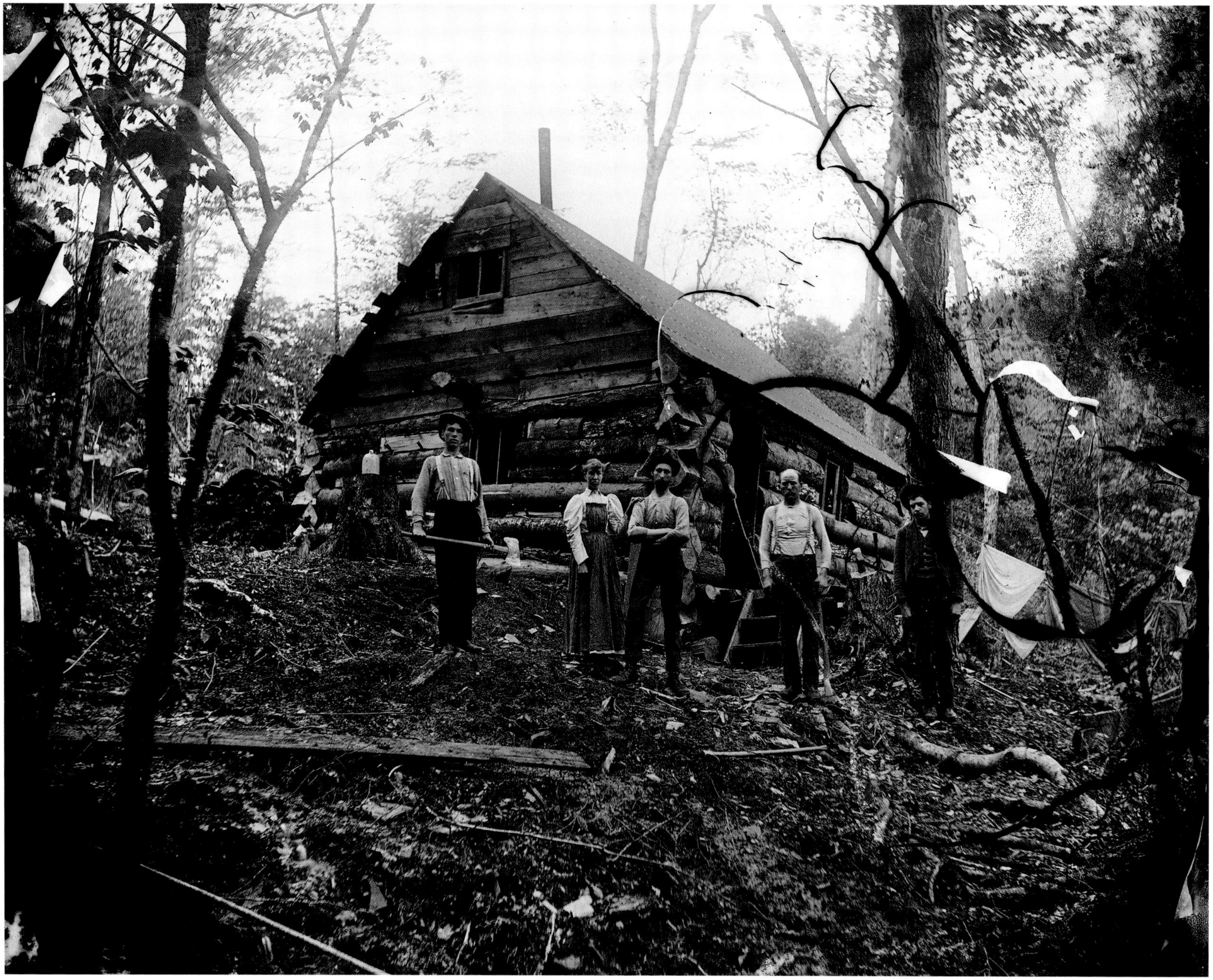

PLATE 60

The body language of commitment

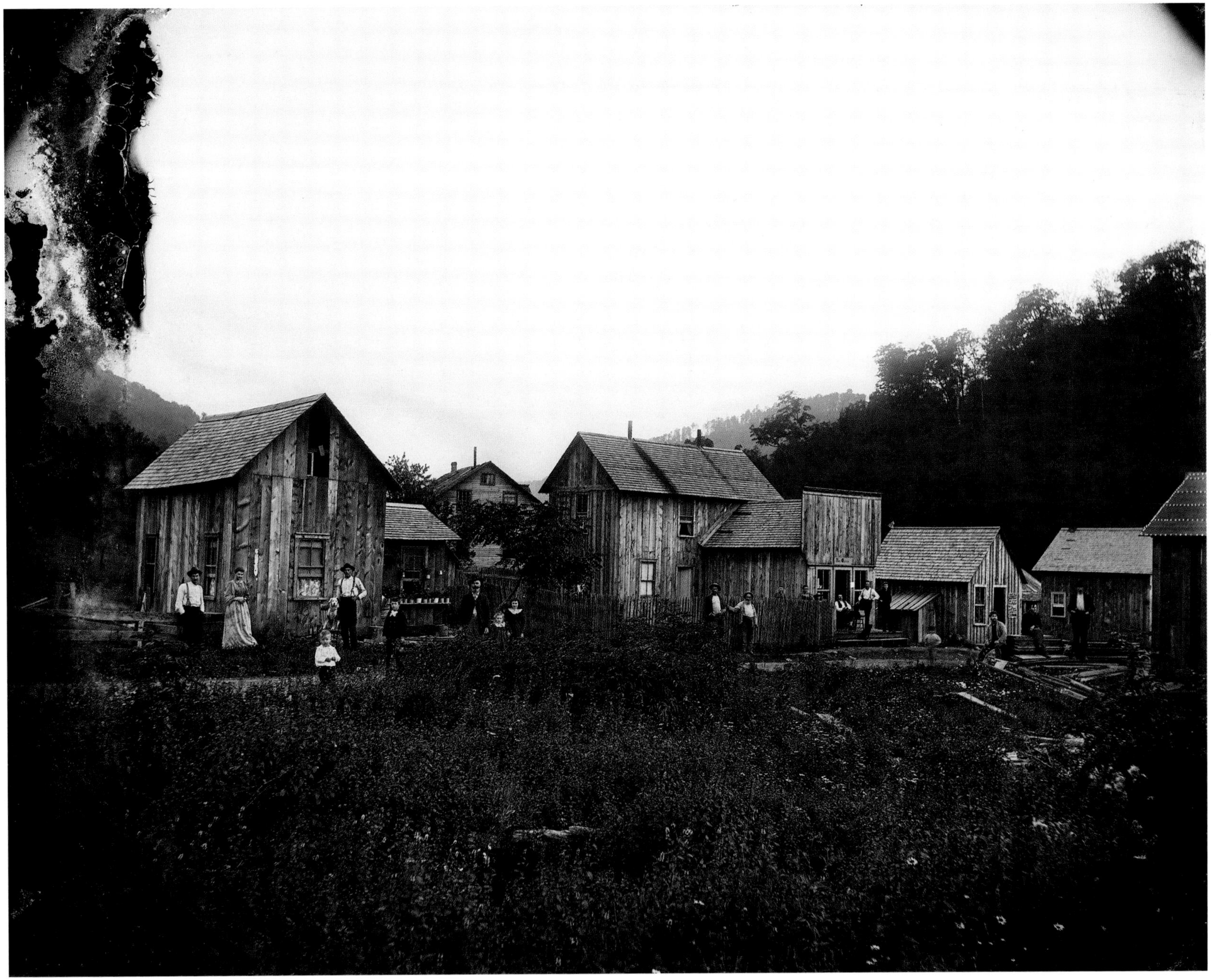

PLATE 61

Baileyville, 1897. At the peak of logging operations

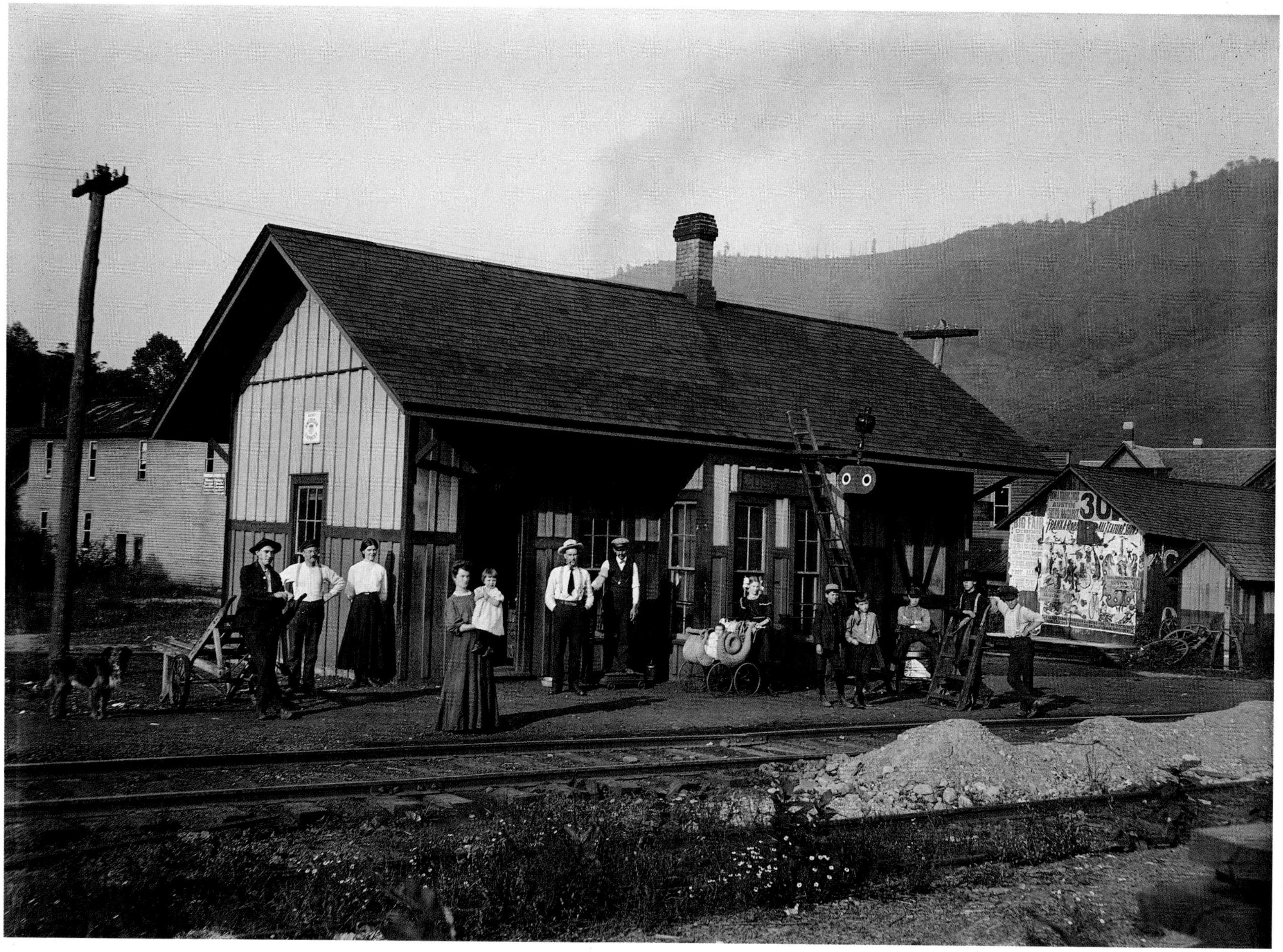

PLATE 62

Costello Railroad Station, Potter County

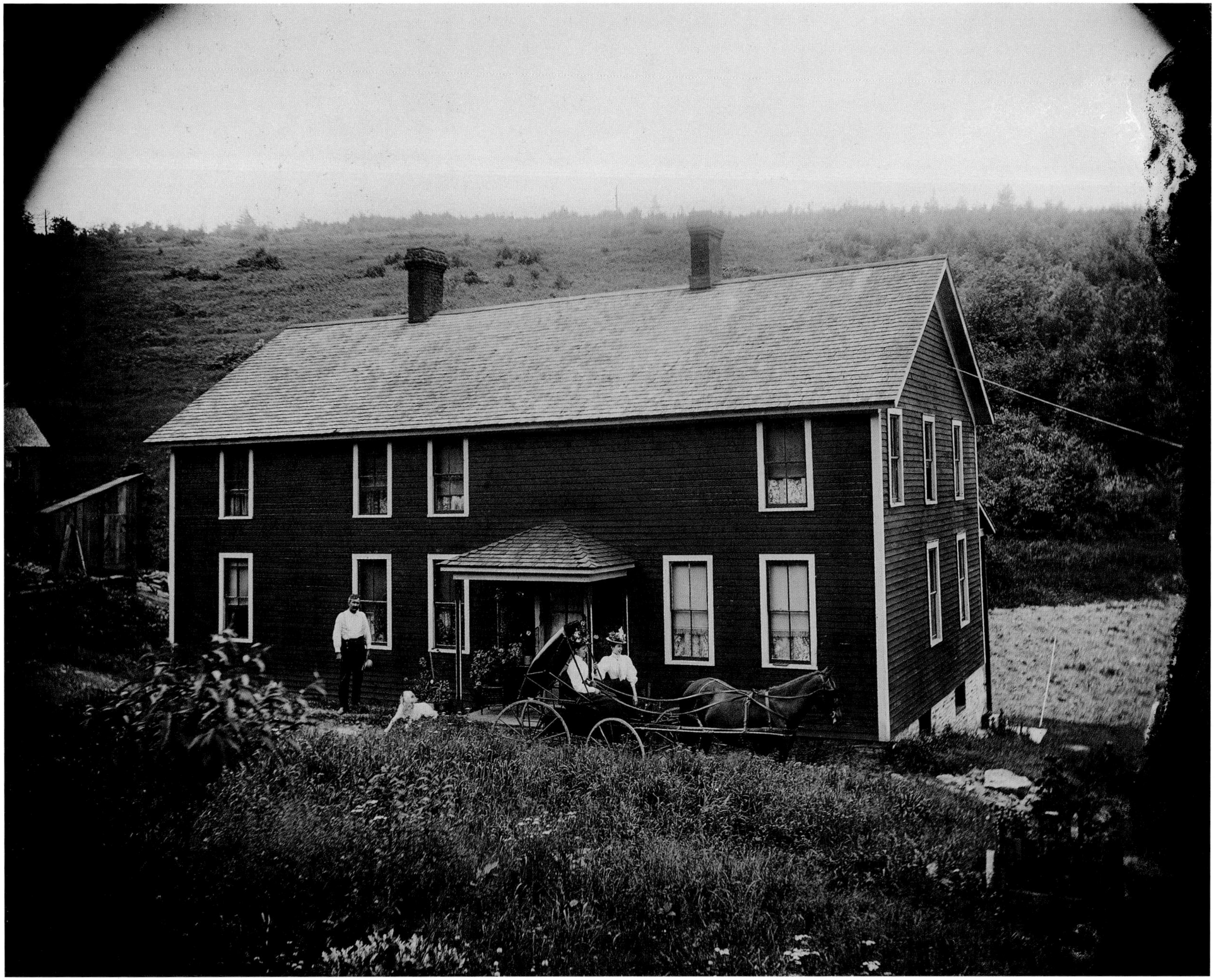

PLATE 63

Genteel living

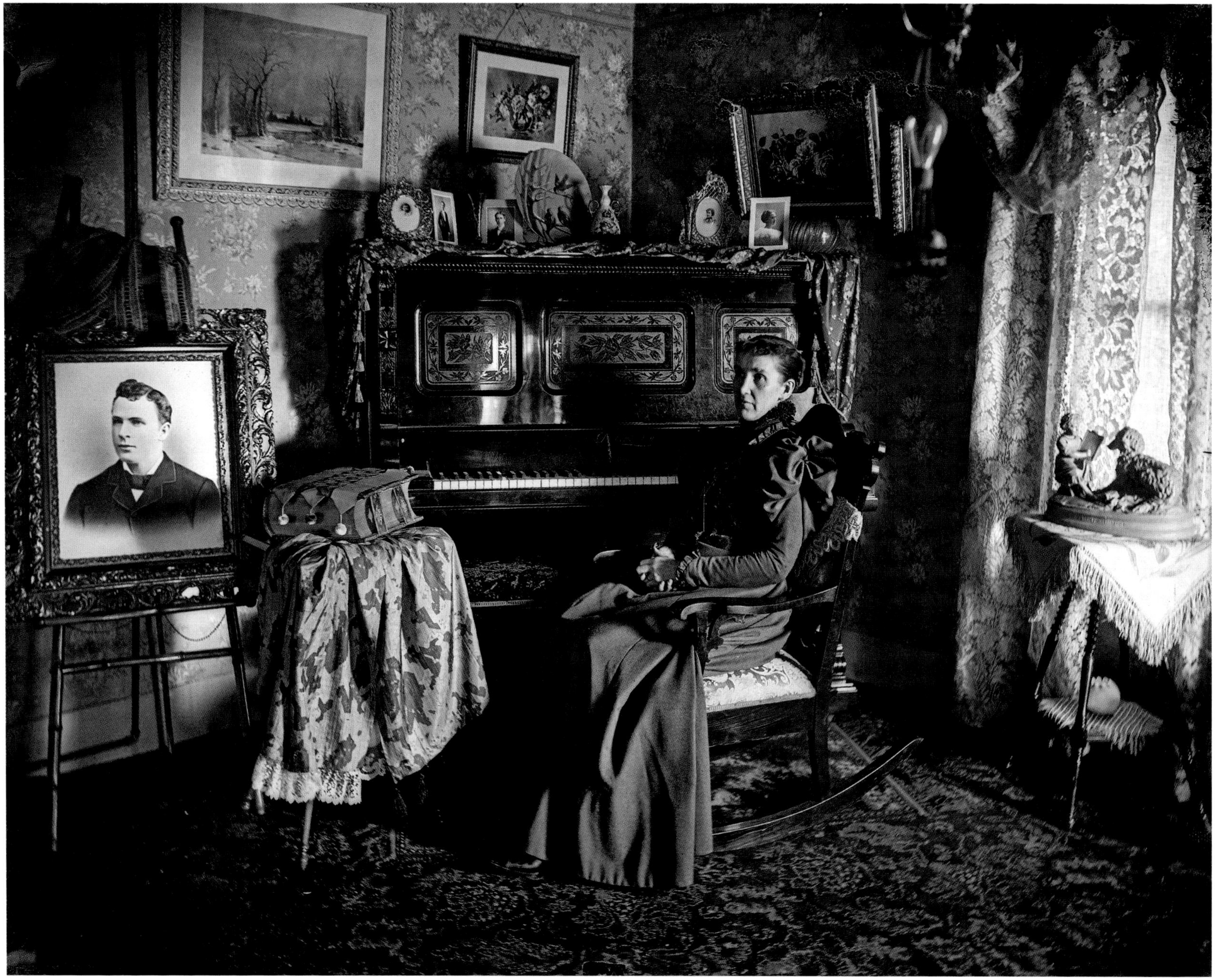

PLATE 64

Victorian interior, well-to-do woman, north-central Pennsylvania

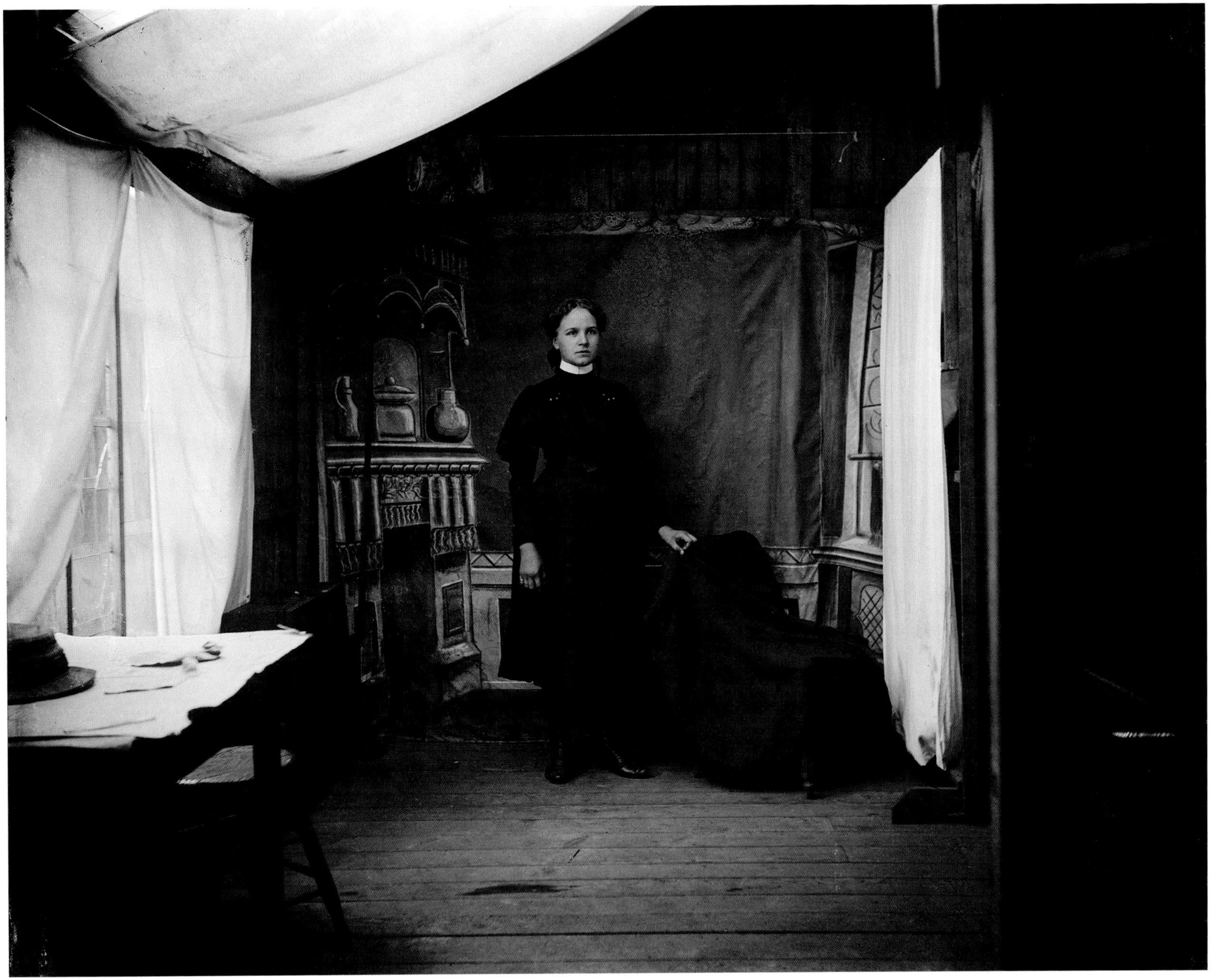

PLATE 65

Young woman's portrait, Clarke's studio

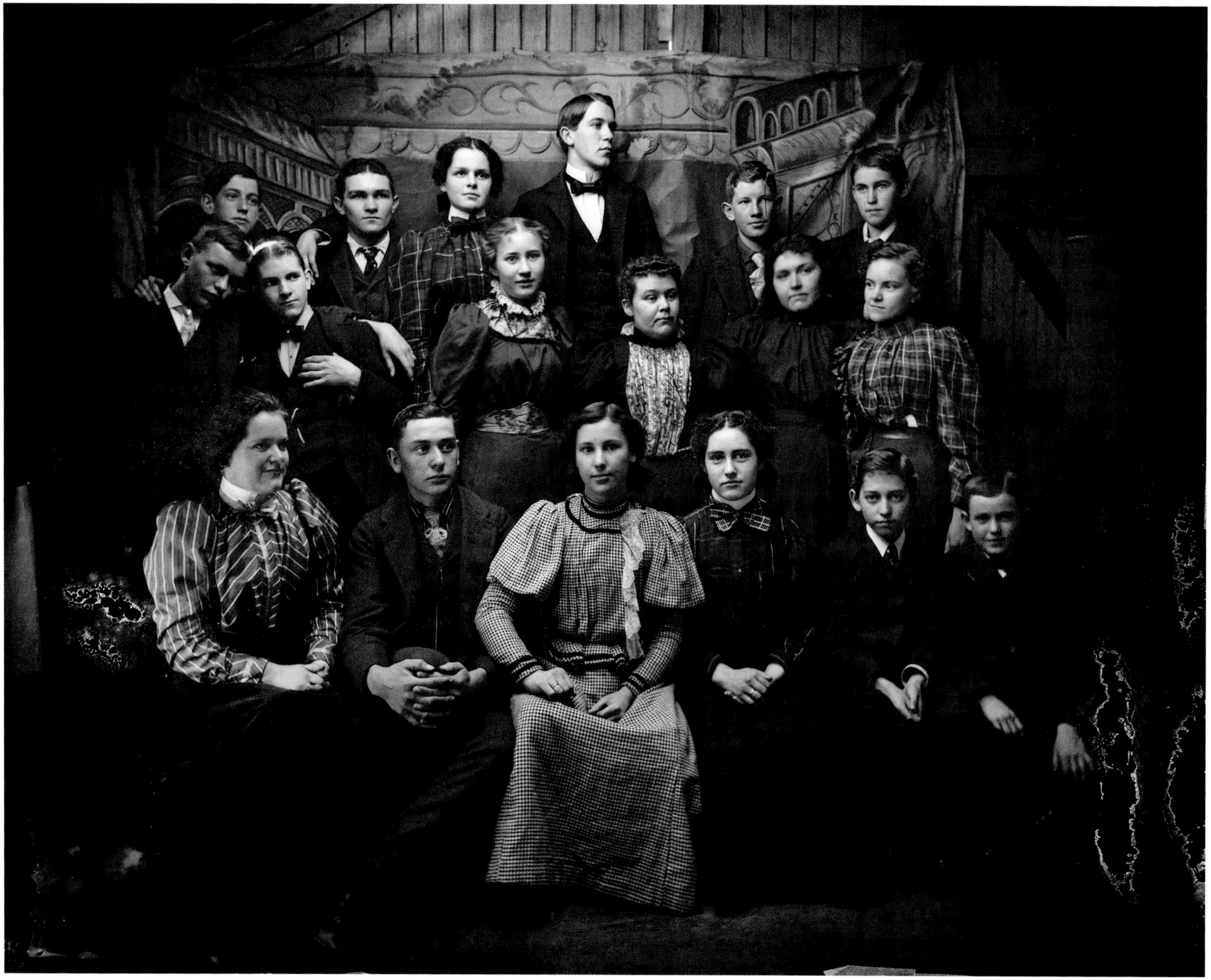

PLATE 66

Youth group packed inside Clarke's studio

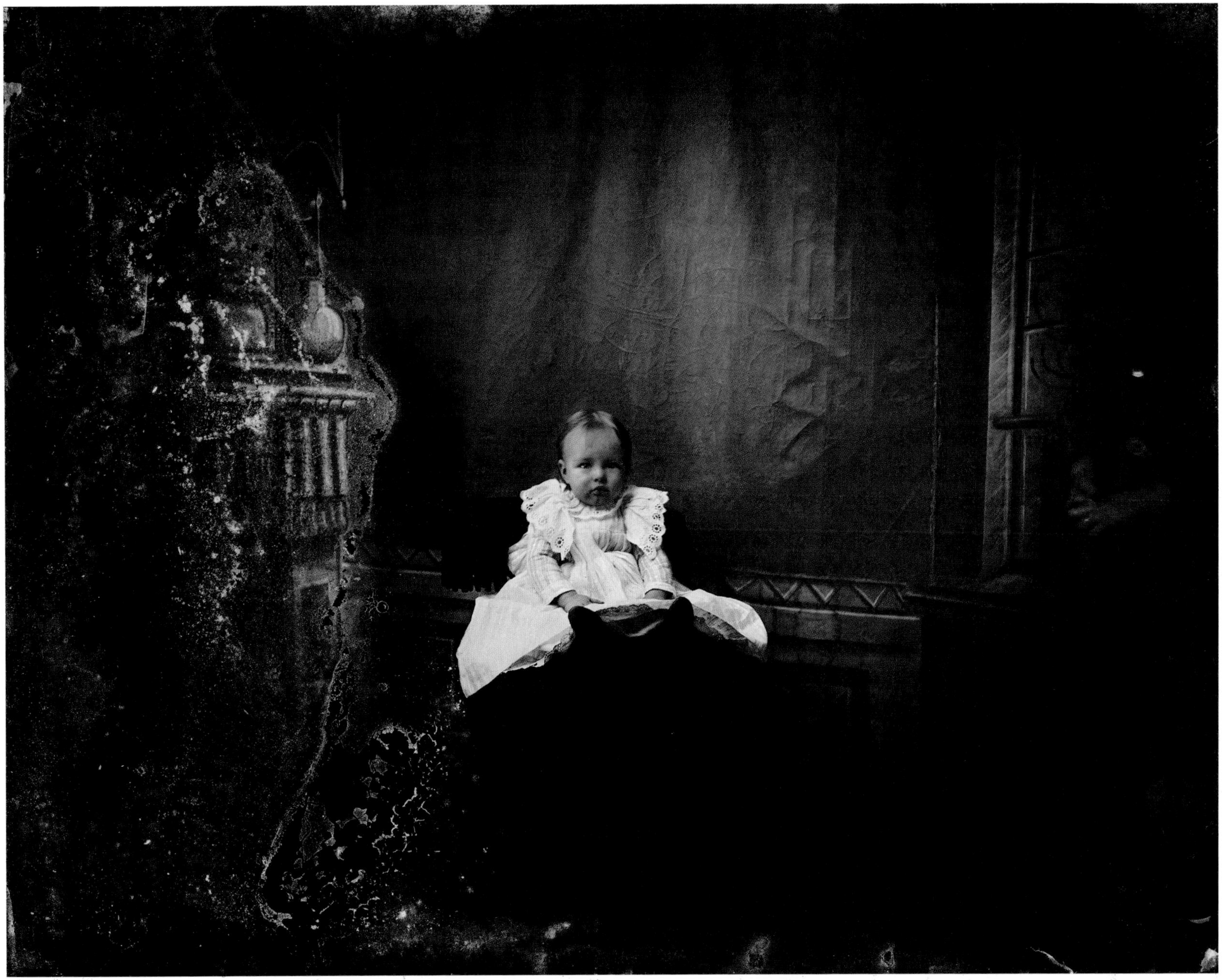

PLATE 67

Baby, studio portrait

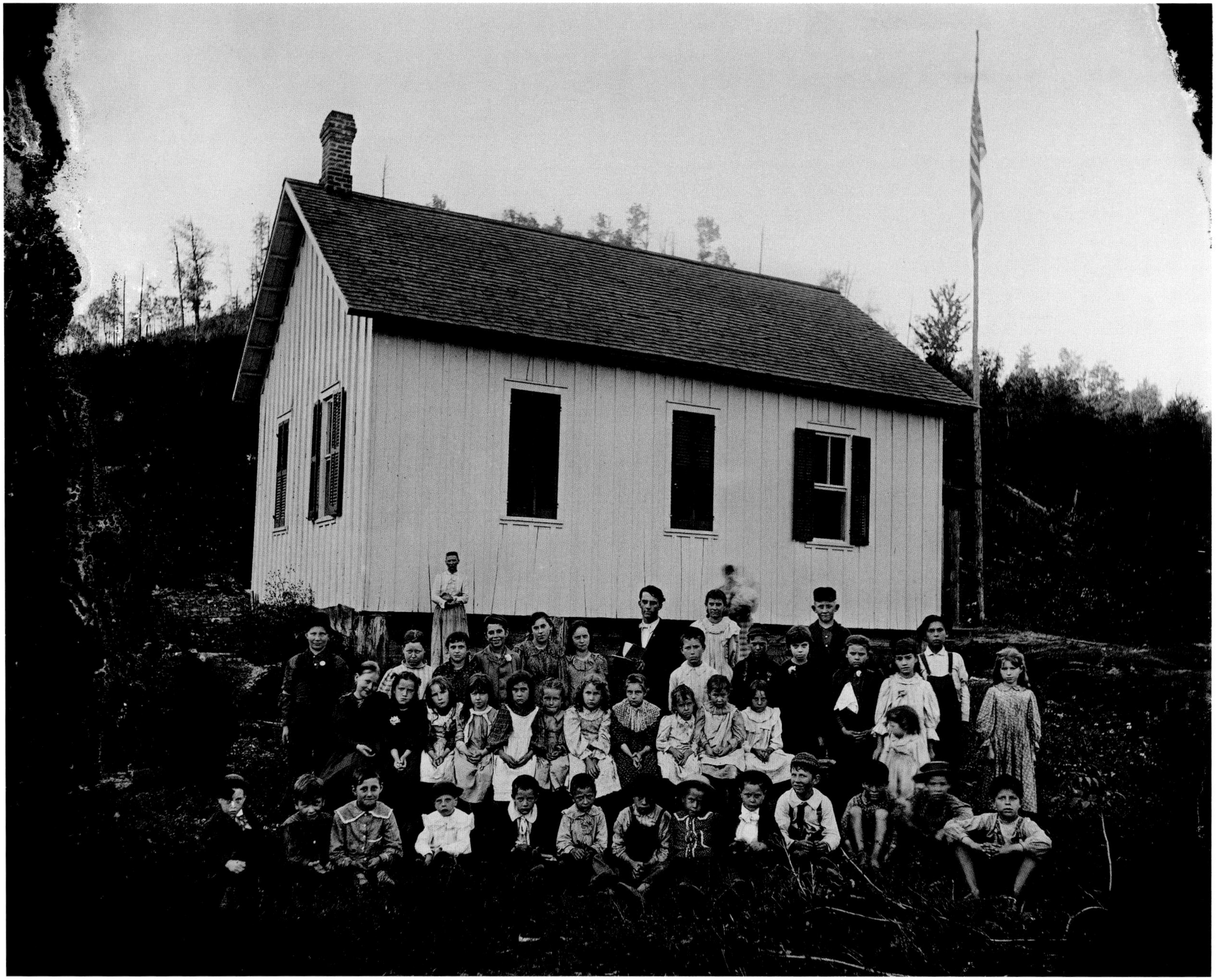

PLATE 68

Typical north-central Pennsylvania school

PLATE 69

Church, Keating Summit,
Potter County

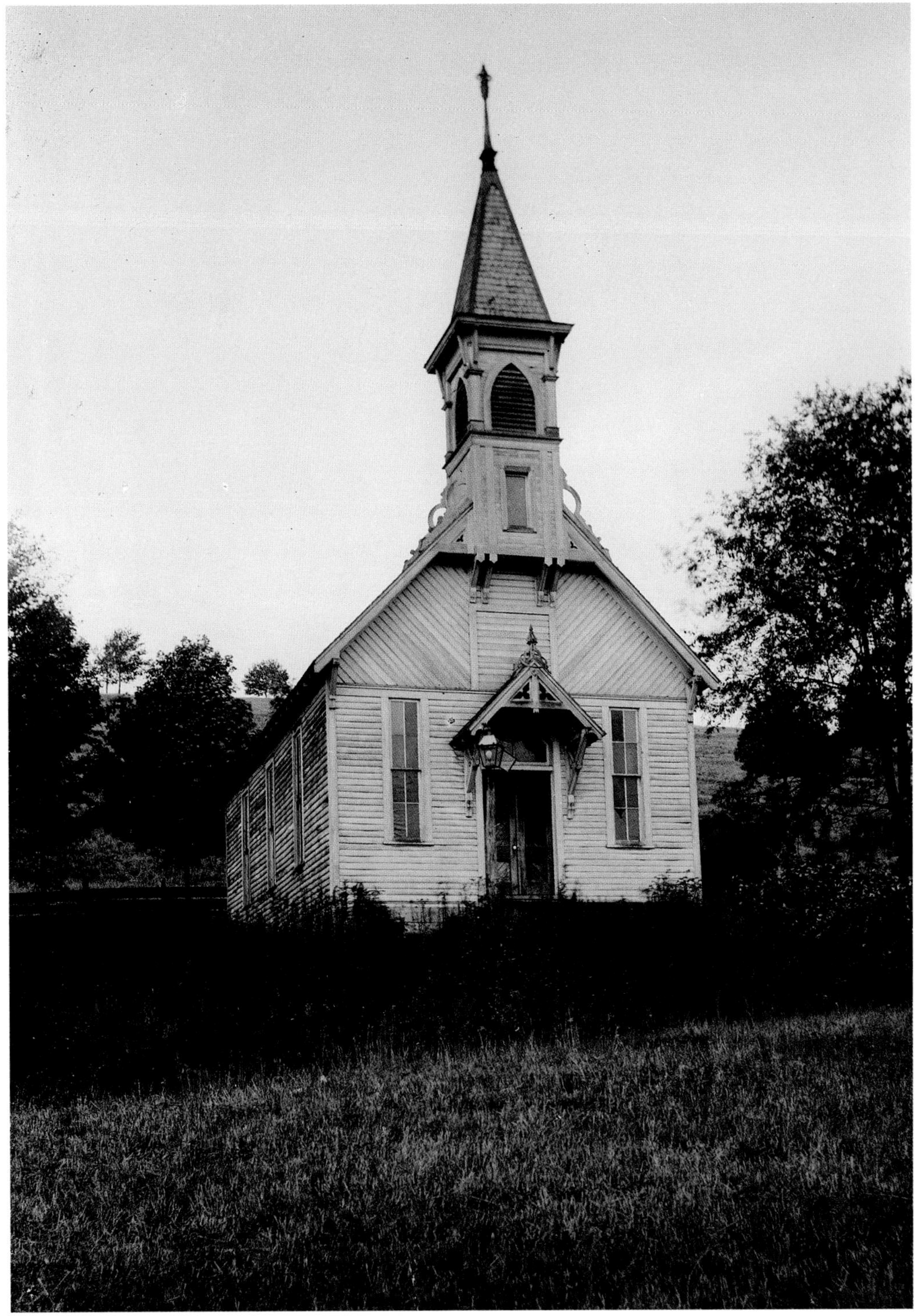

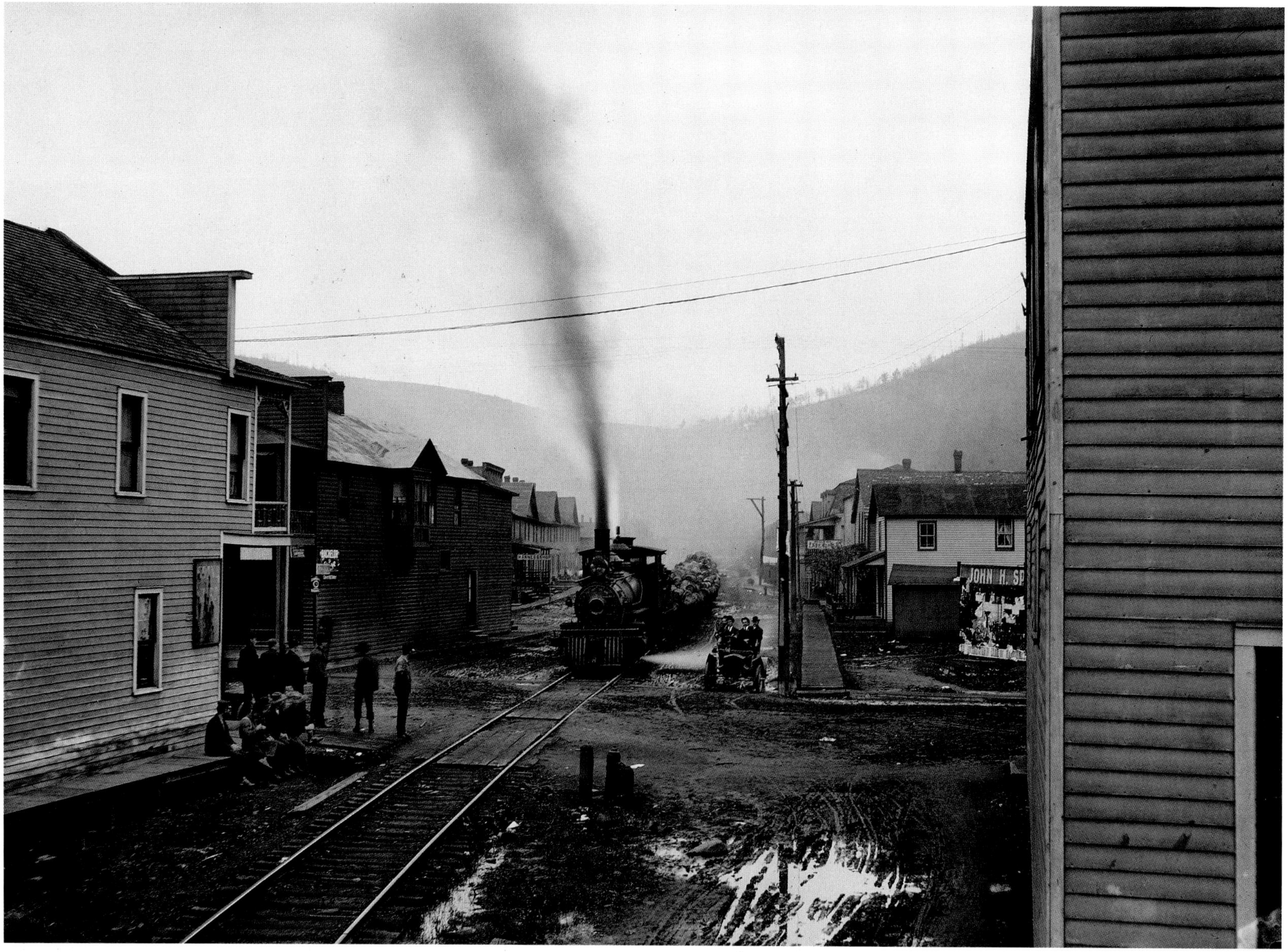

PLATE 70

Locomotive, automobile. Lackawanna Avenue, Cross Fork, Potter County

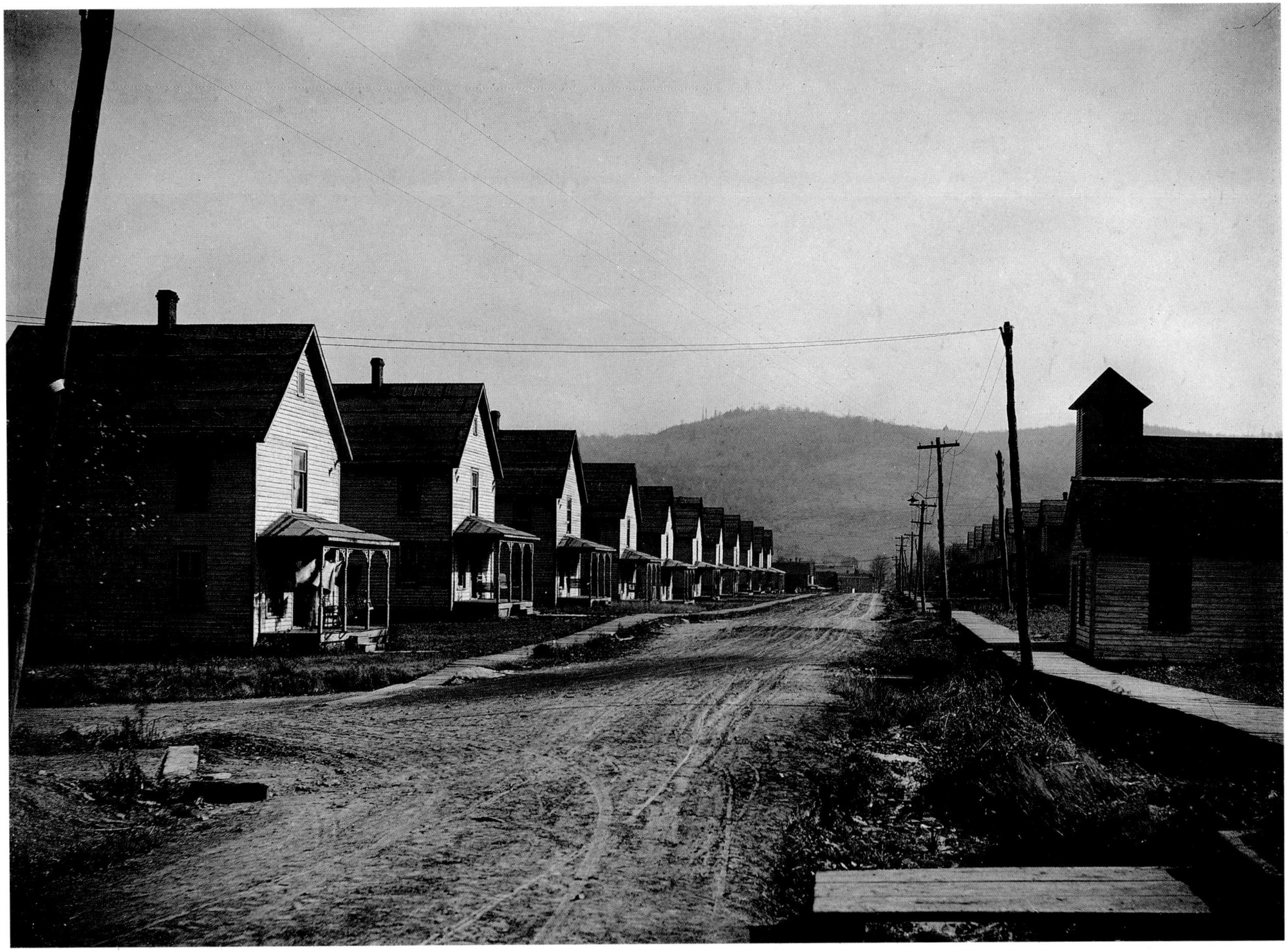

PLATE 71

Residential section, Betula, McKean County

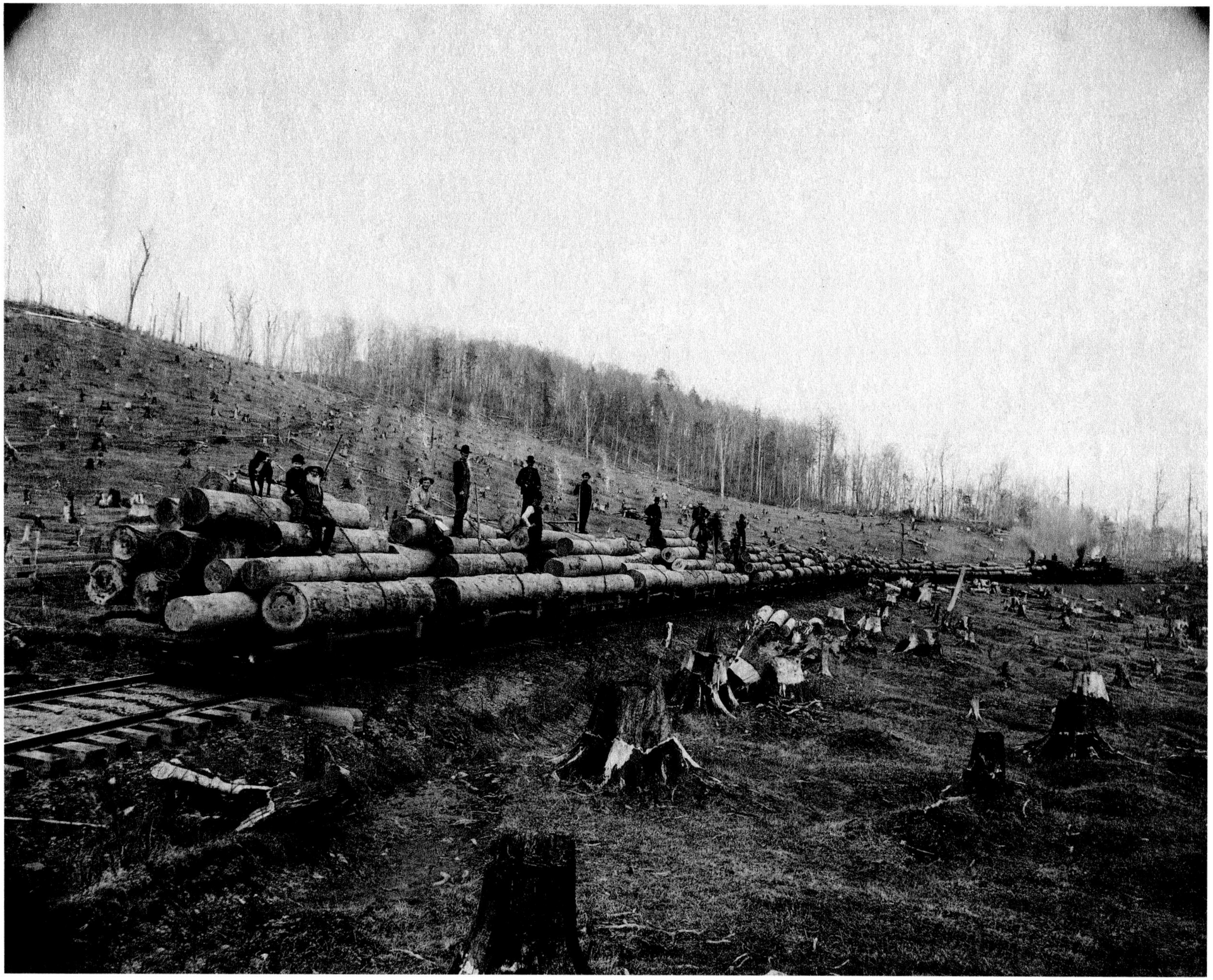

PLATE 72

A Goodyear lumber train of some twenty cars pauses on the Dennis Hall Farm, two miles southeast of Keating Summit

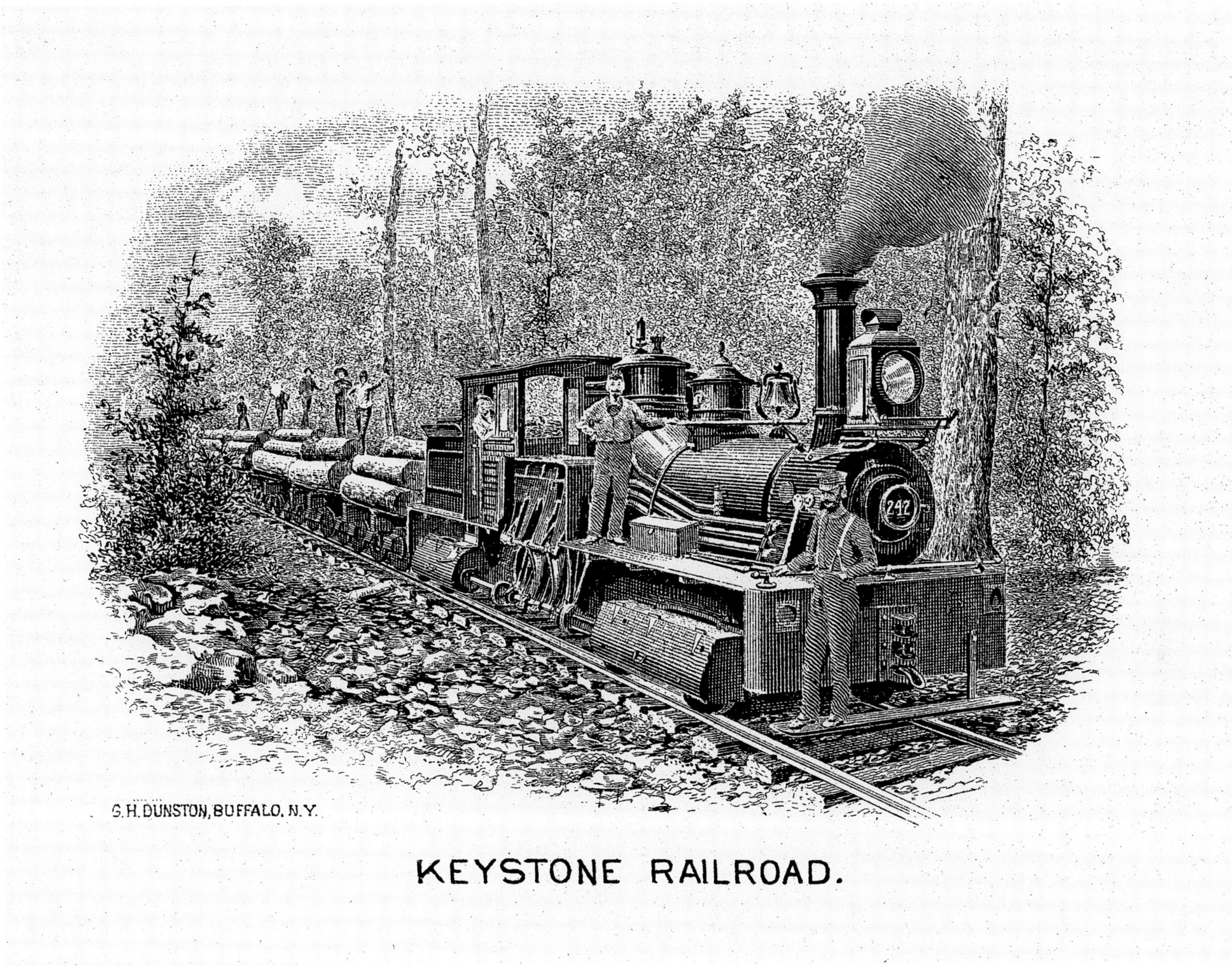

PLATE 73

Rich Lumber detail, Keystone RR stationery

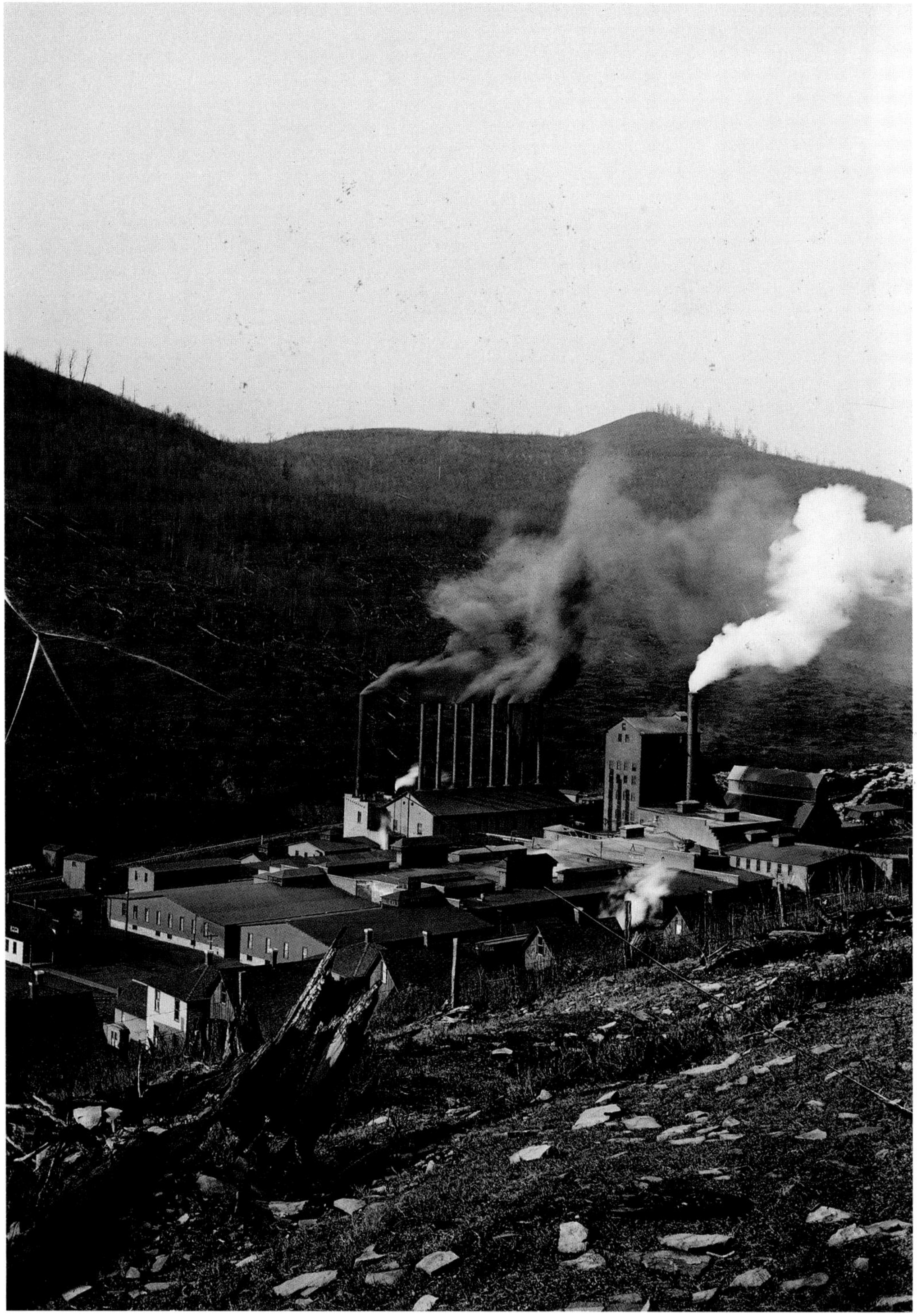

PLATE 74

Austin's Bayless Pulp and Paper Company, Potter County

PLATE 75

One of the last original areas
of forest remnants in north-central
Pennsylvania, 1910

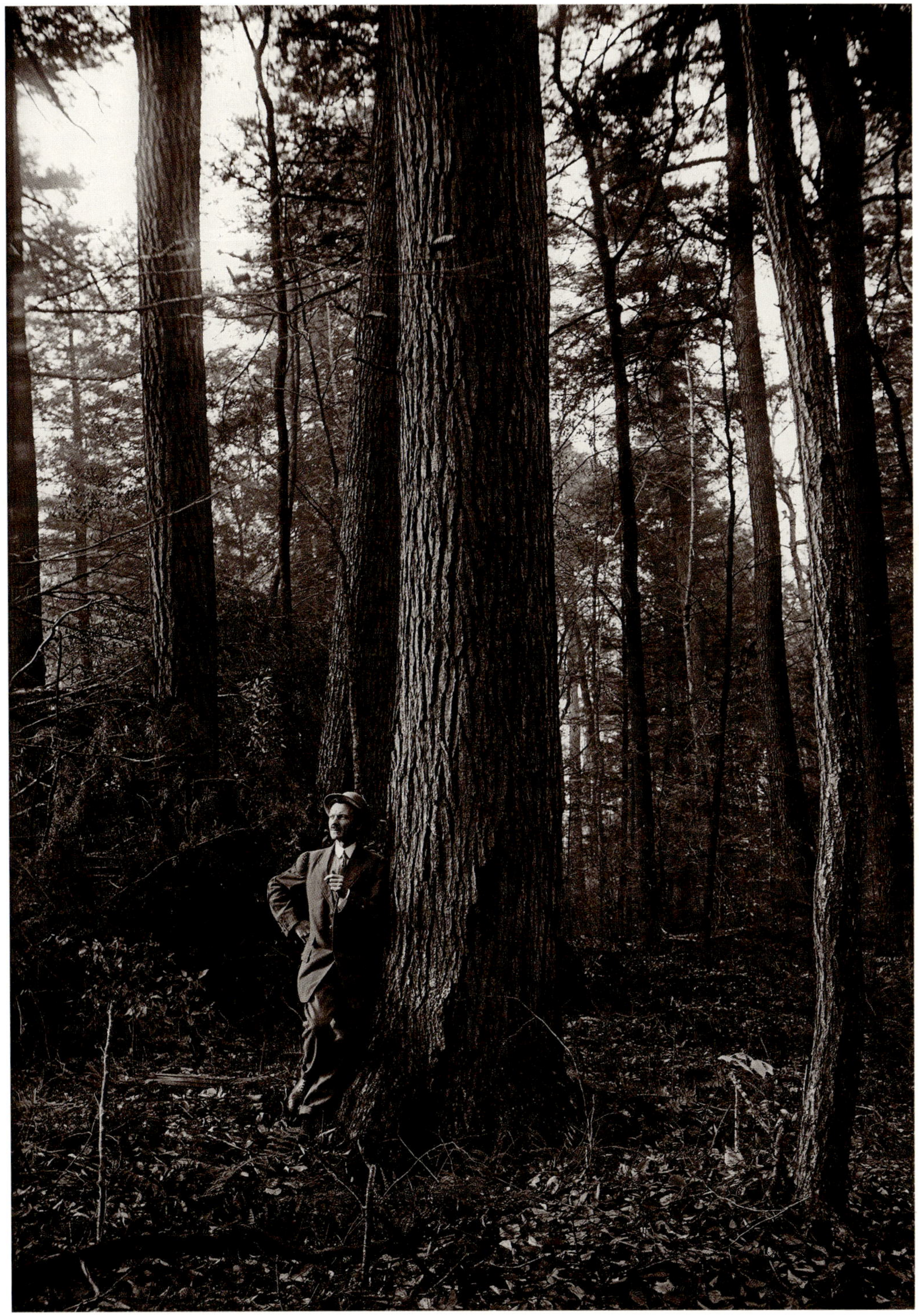

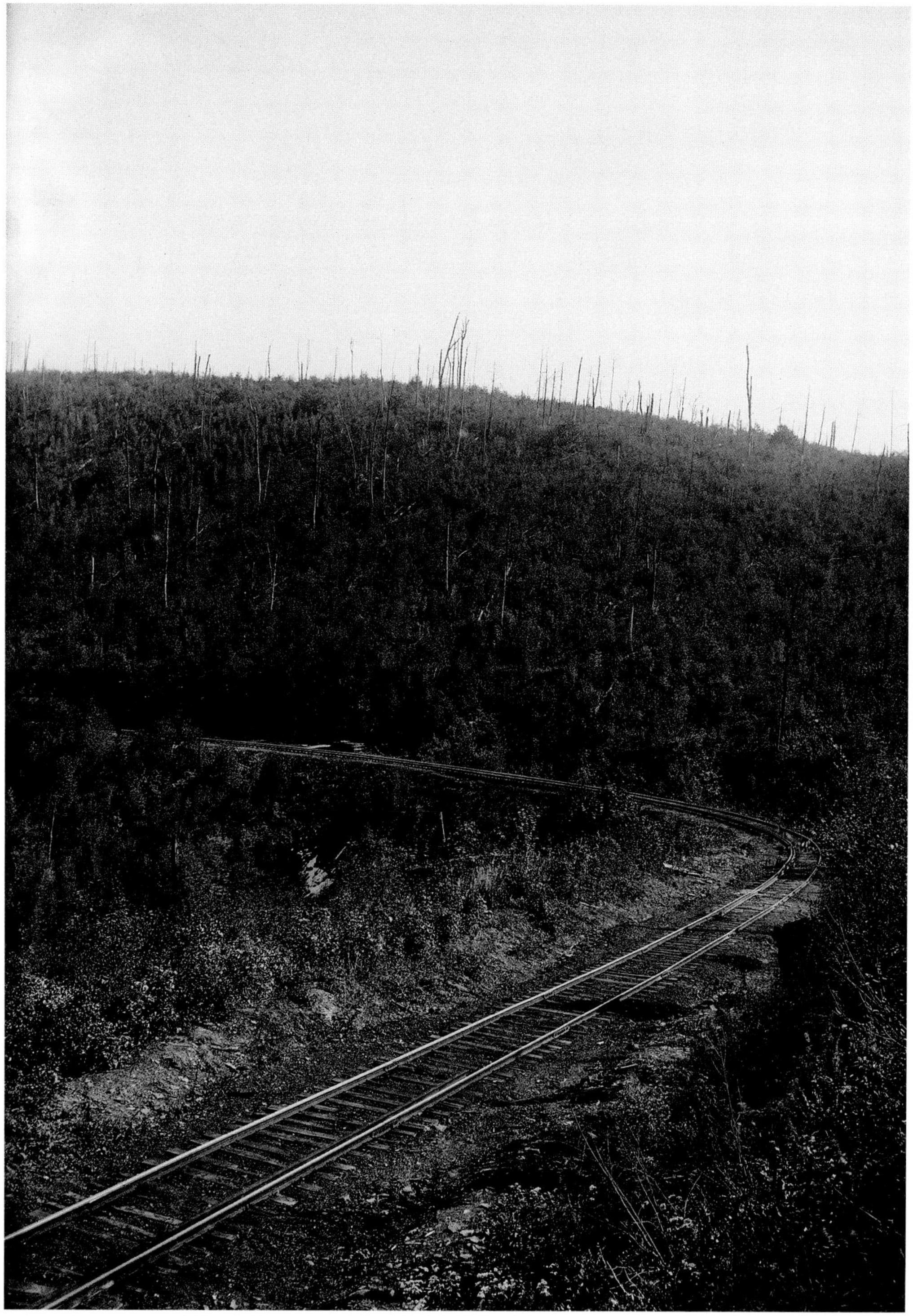

PLATE 76

Desolate skyline

PLATE 77

Railroad near Cross Fork Junction, Potter County

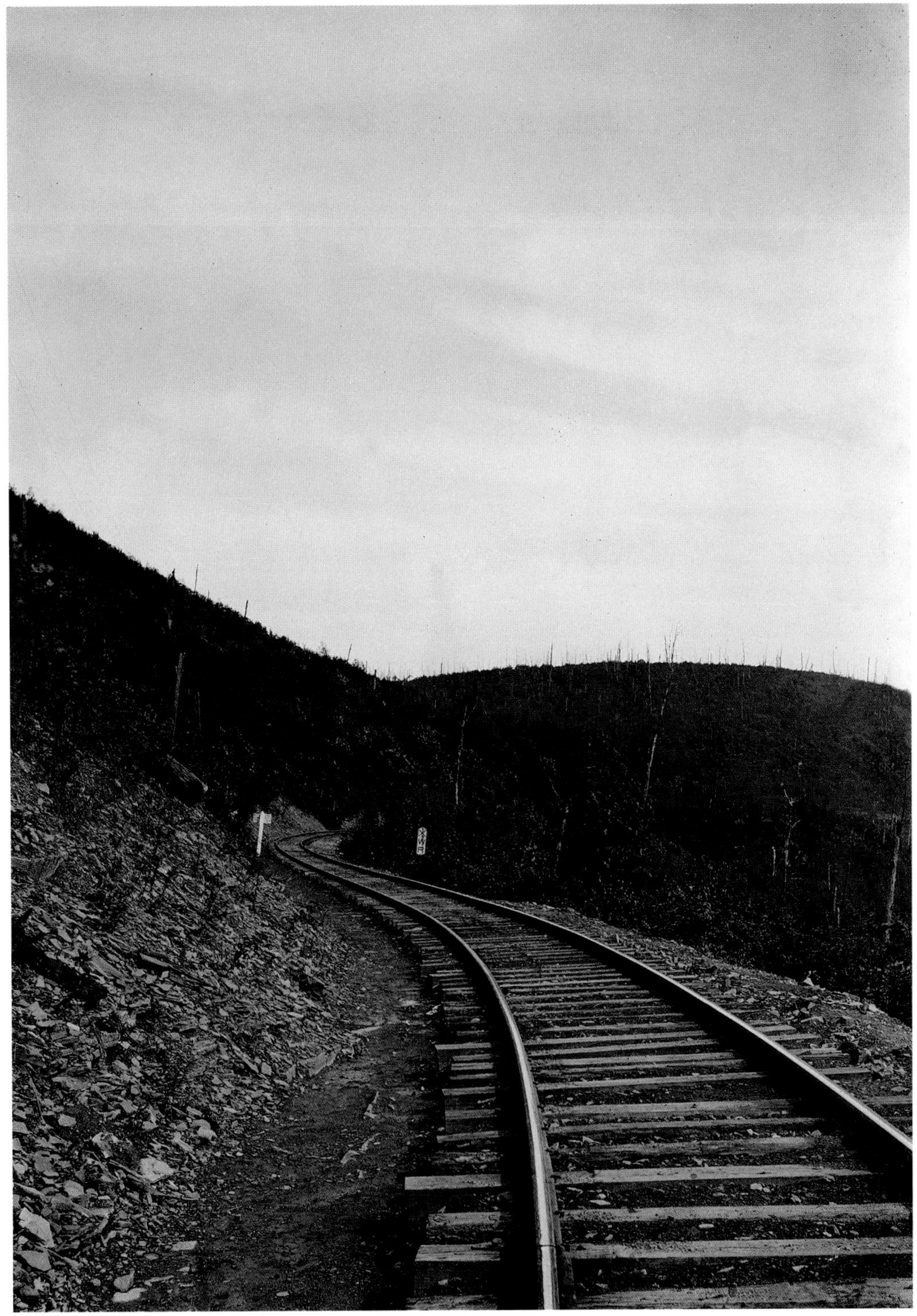

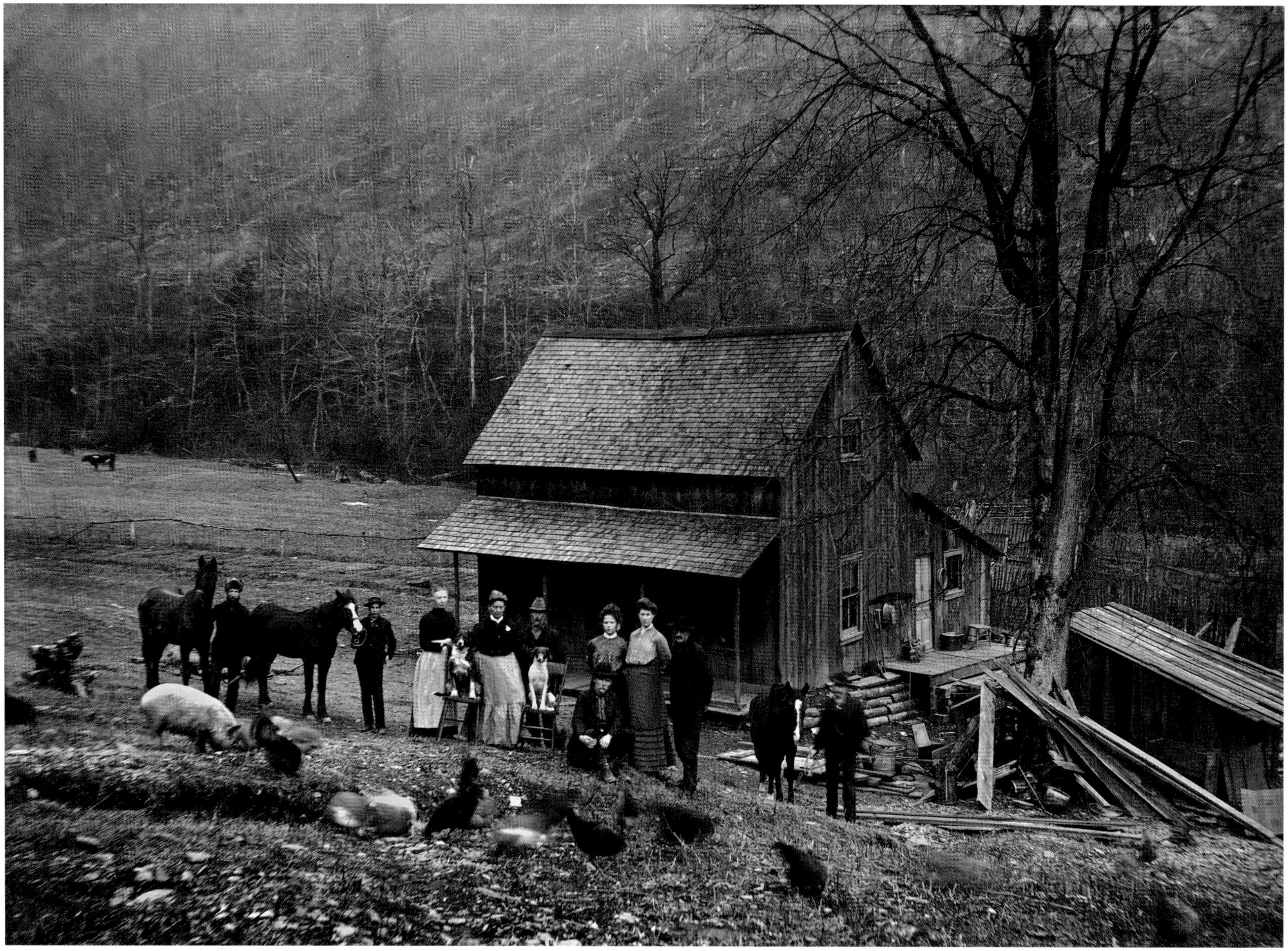

PLATE 78

Family farm

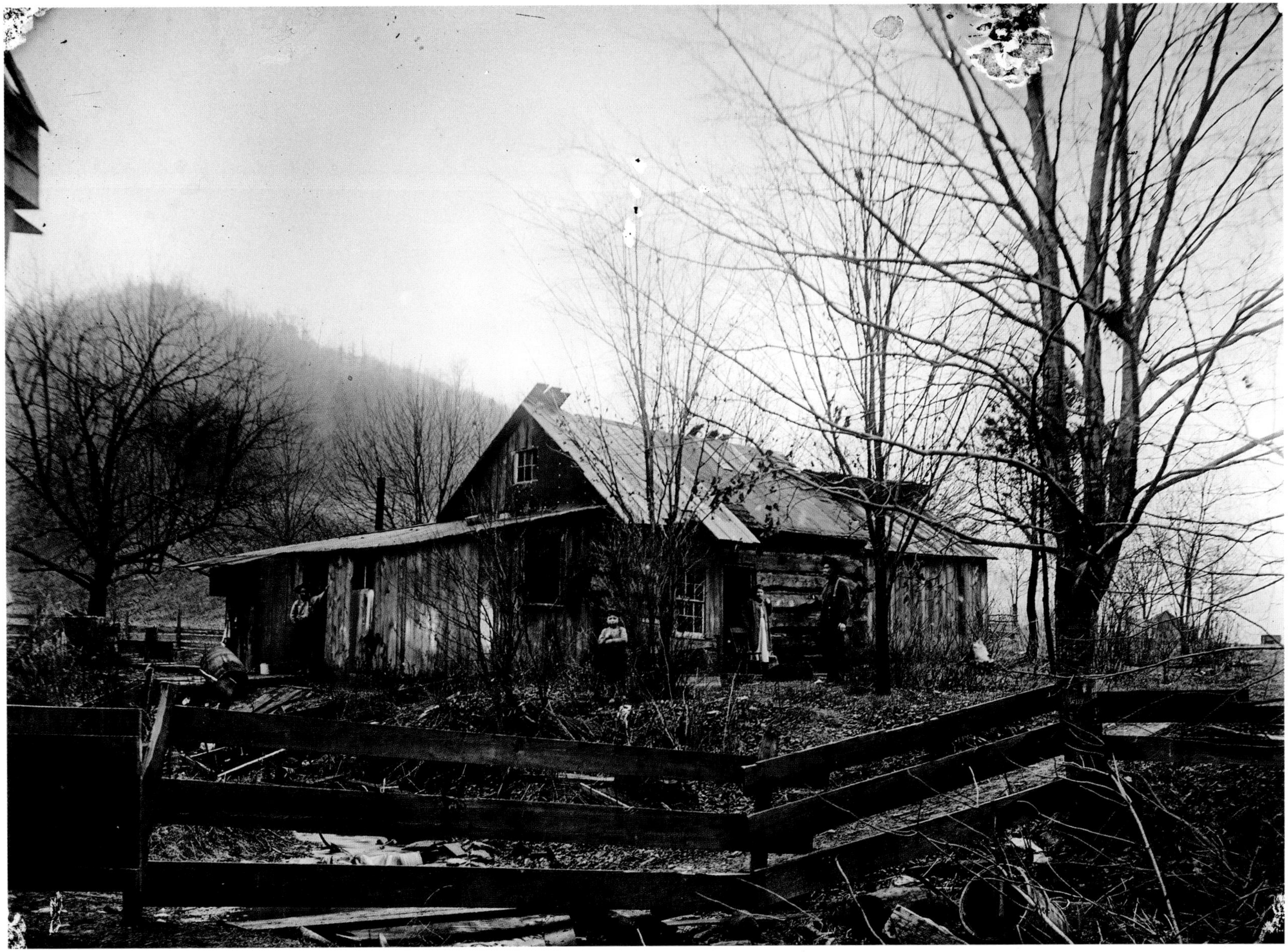

PLATE 79

Home, be it ever so humble, near Hull, Potter County

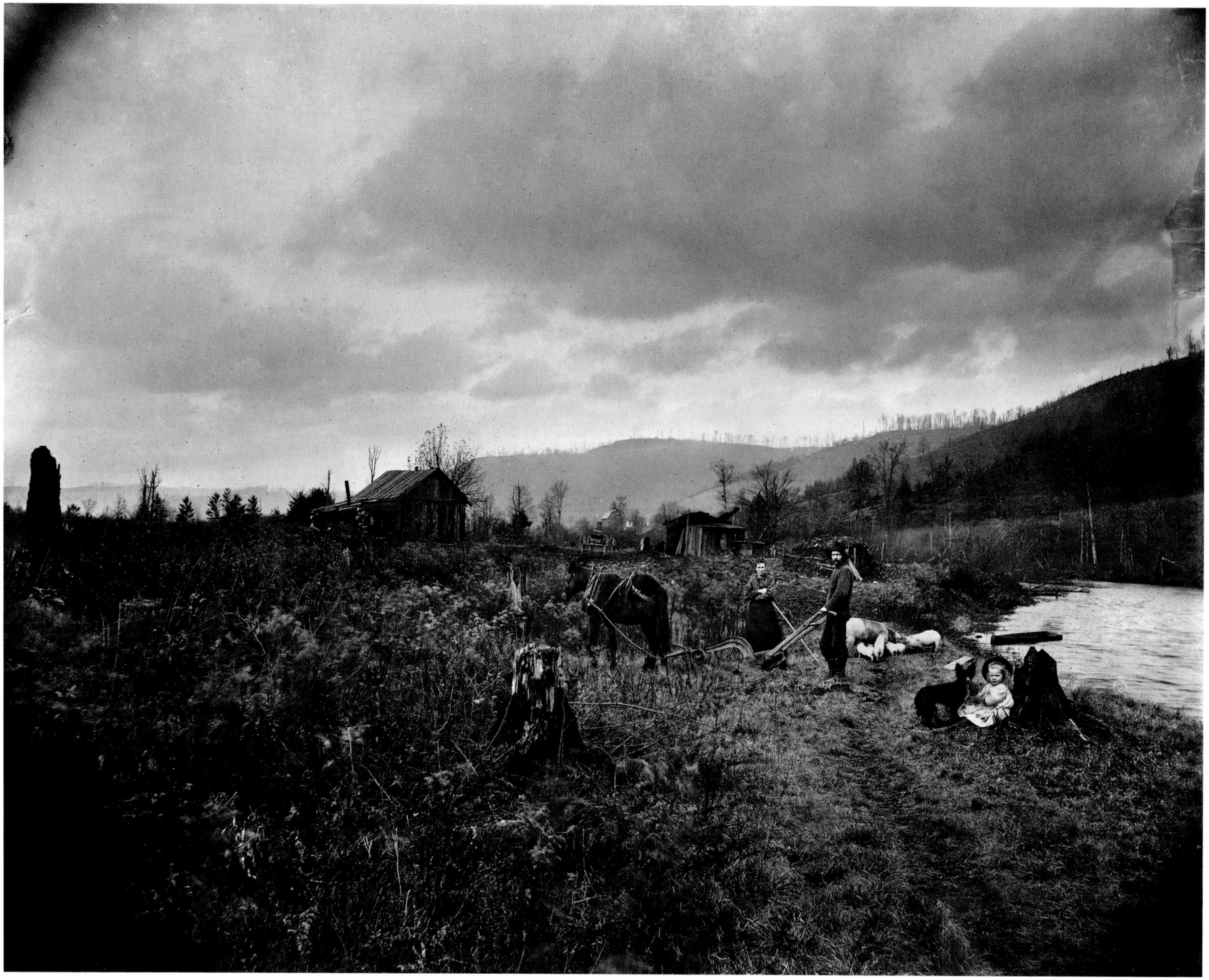

PLATE 80

Desolate Potter County farm near Mina

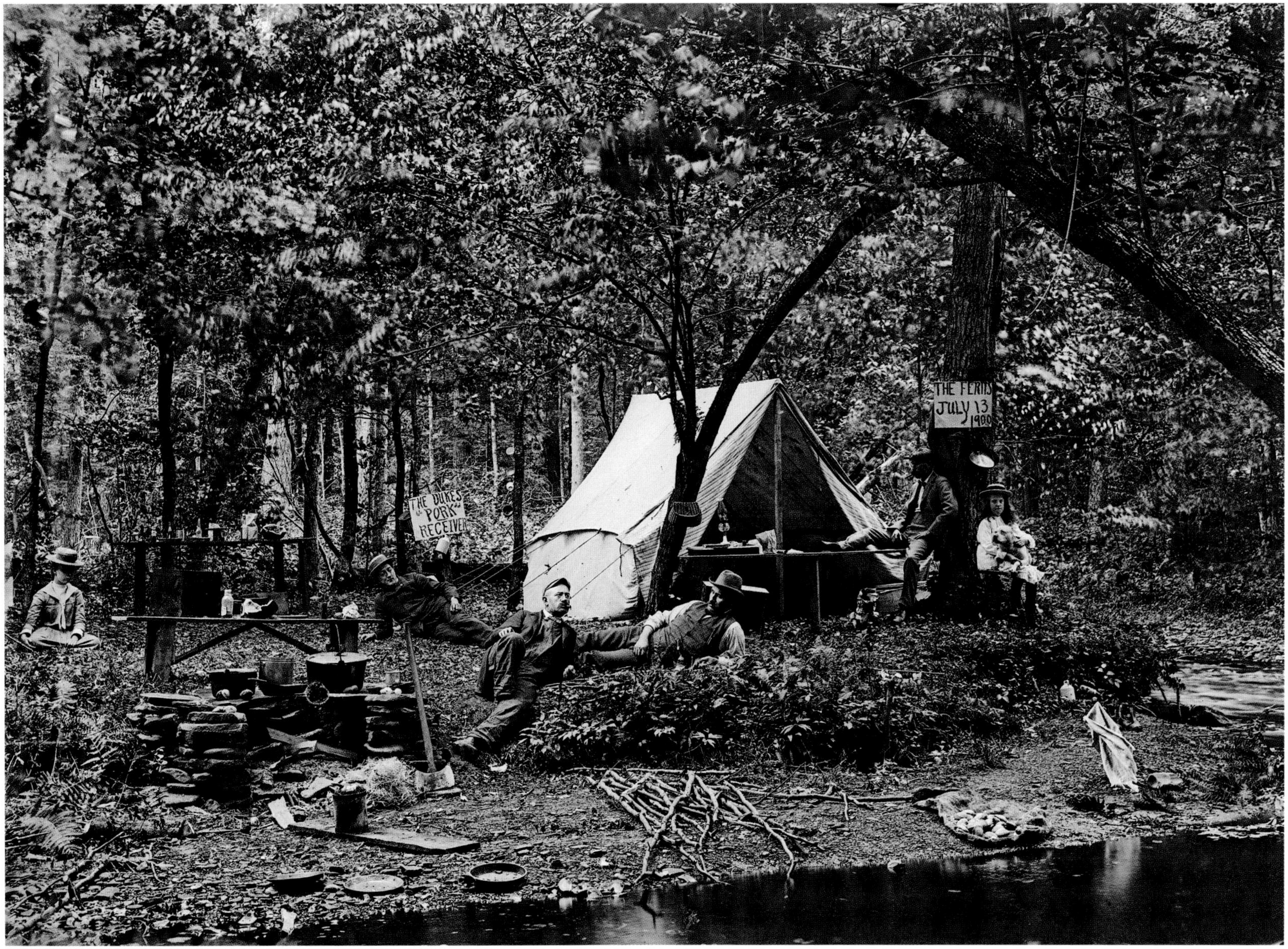

PLATE 81

Camping near Hull, Potter County

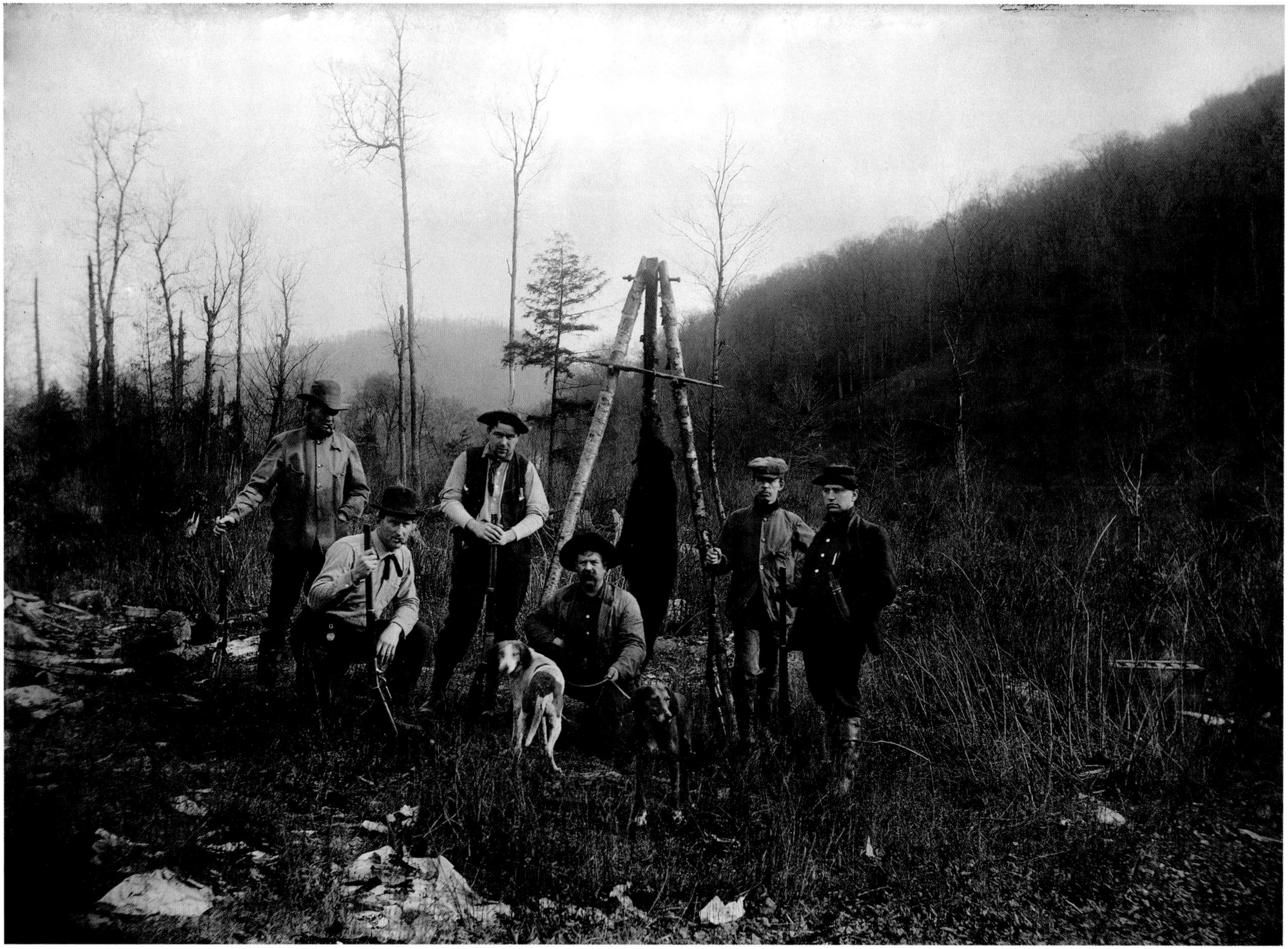

PLATE 82

Hunting scene, near Hull, Potter County

PLATE 83

Oil derrick, near Hull, Potter County

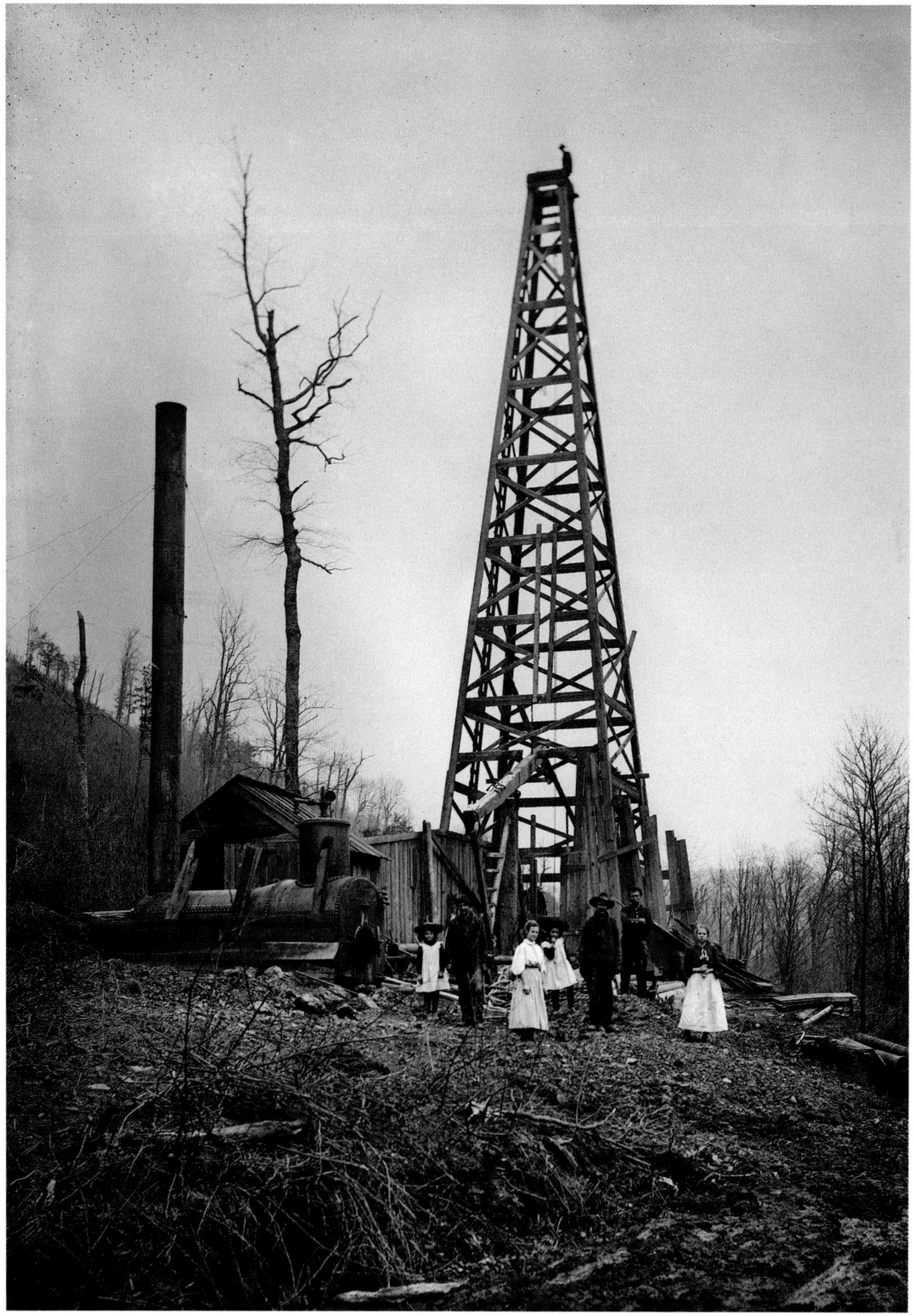

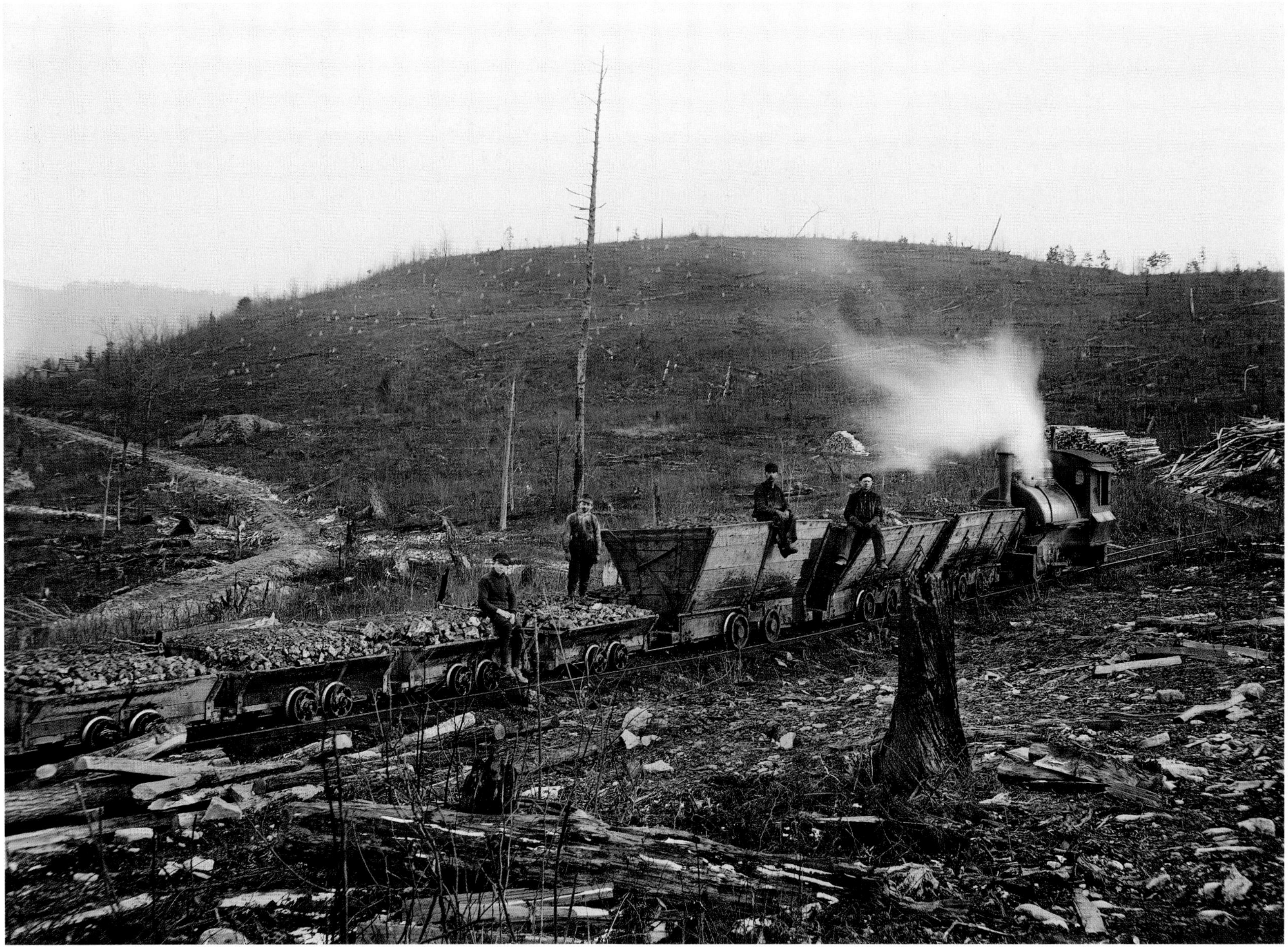

PLATE 84

Railroad hauling coal, Sinnemahoning

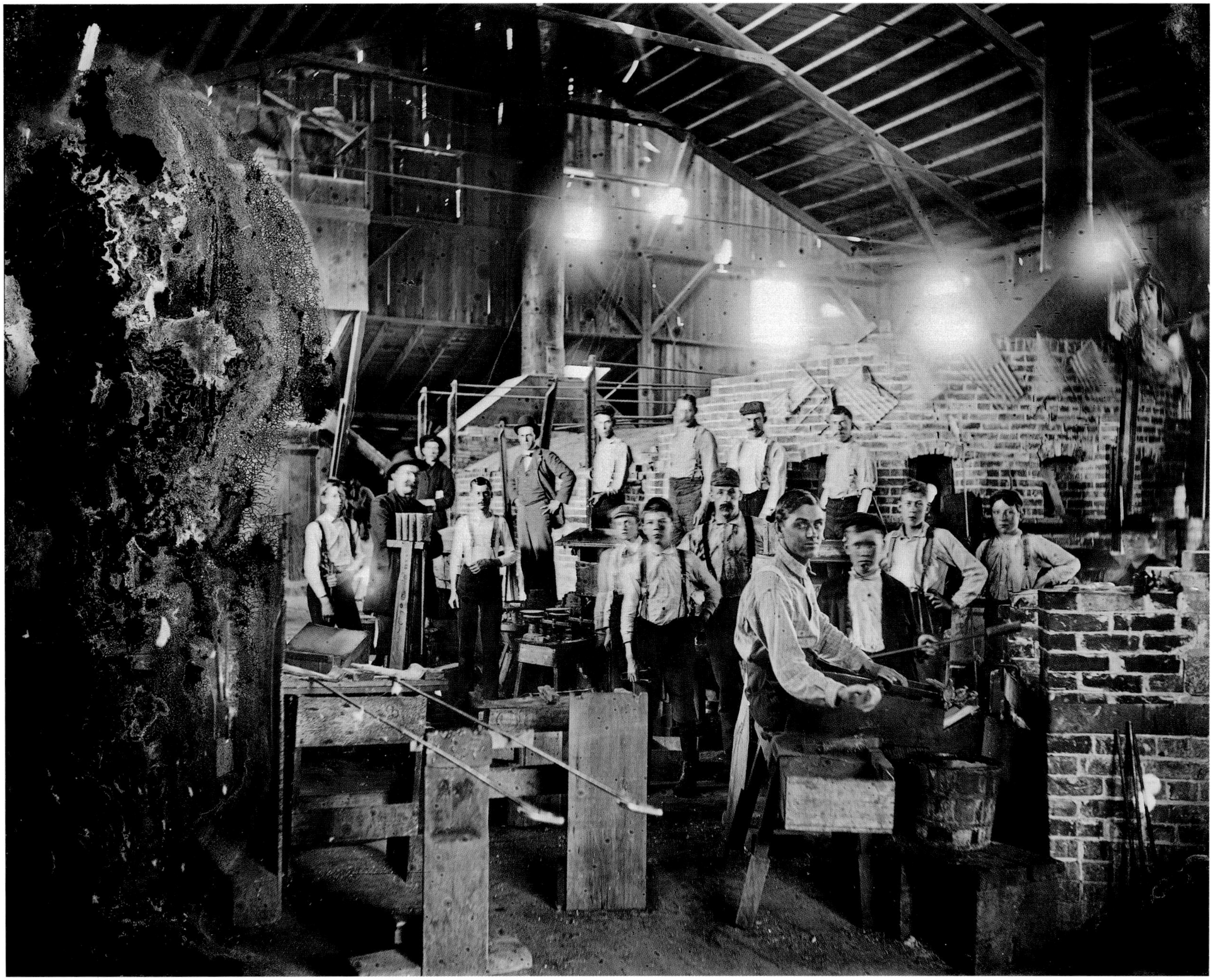

PLATE 85

Glass factory interior

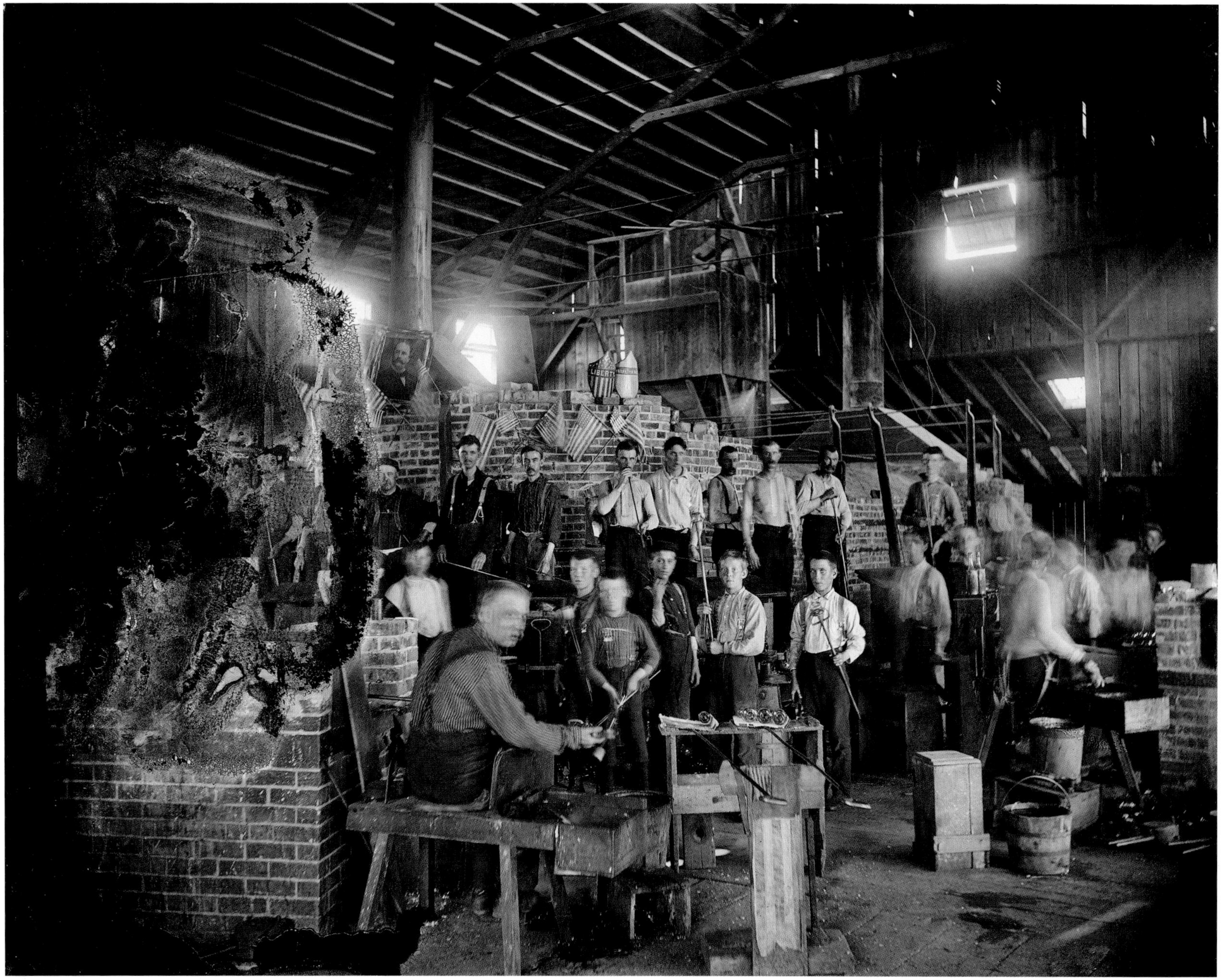

PLATE 86

Glass factory view

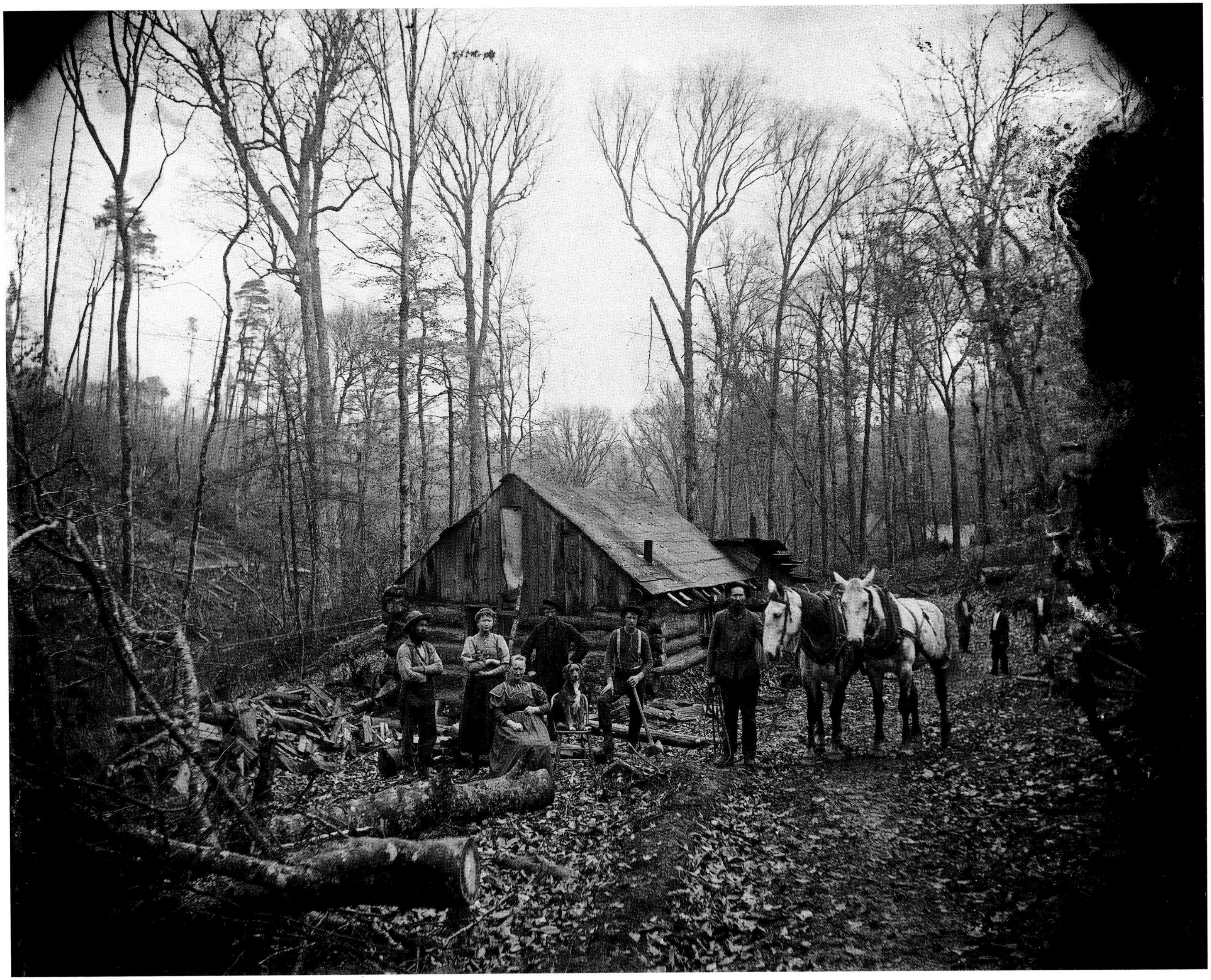

PLATE 87

Can we make a go of farming?

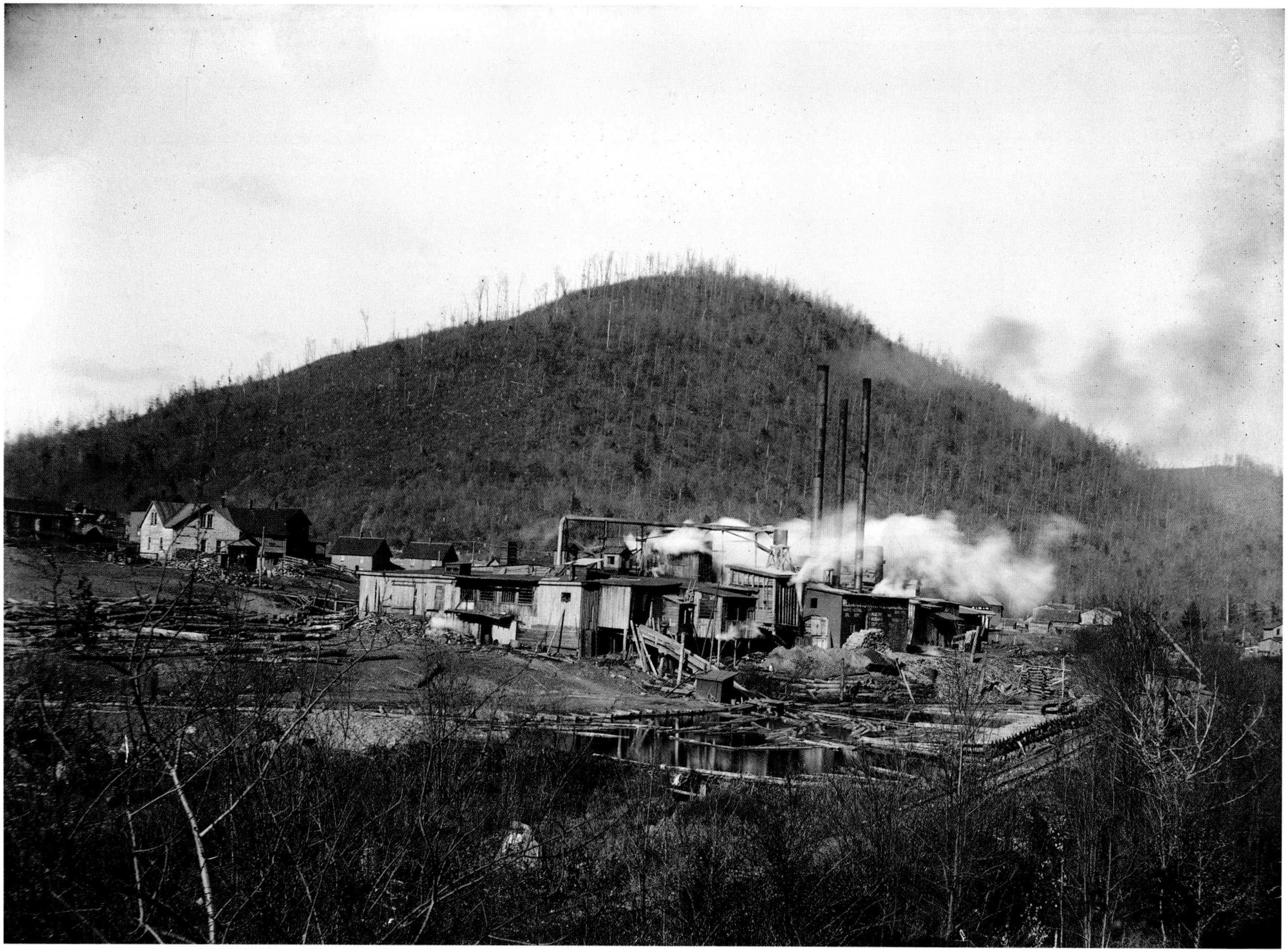

PLATE 88

Heading and stave plant, Hull, Potter County

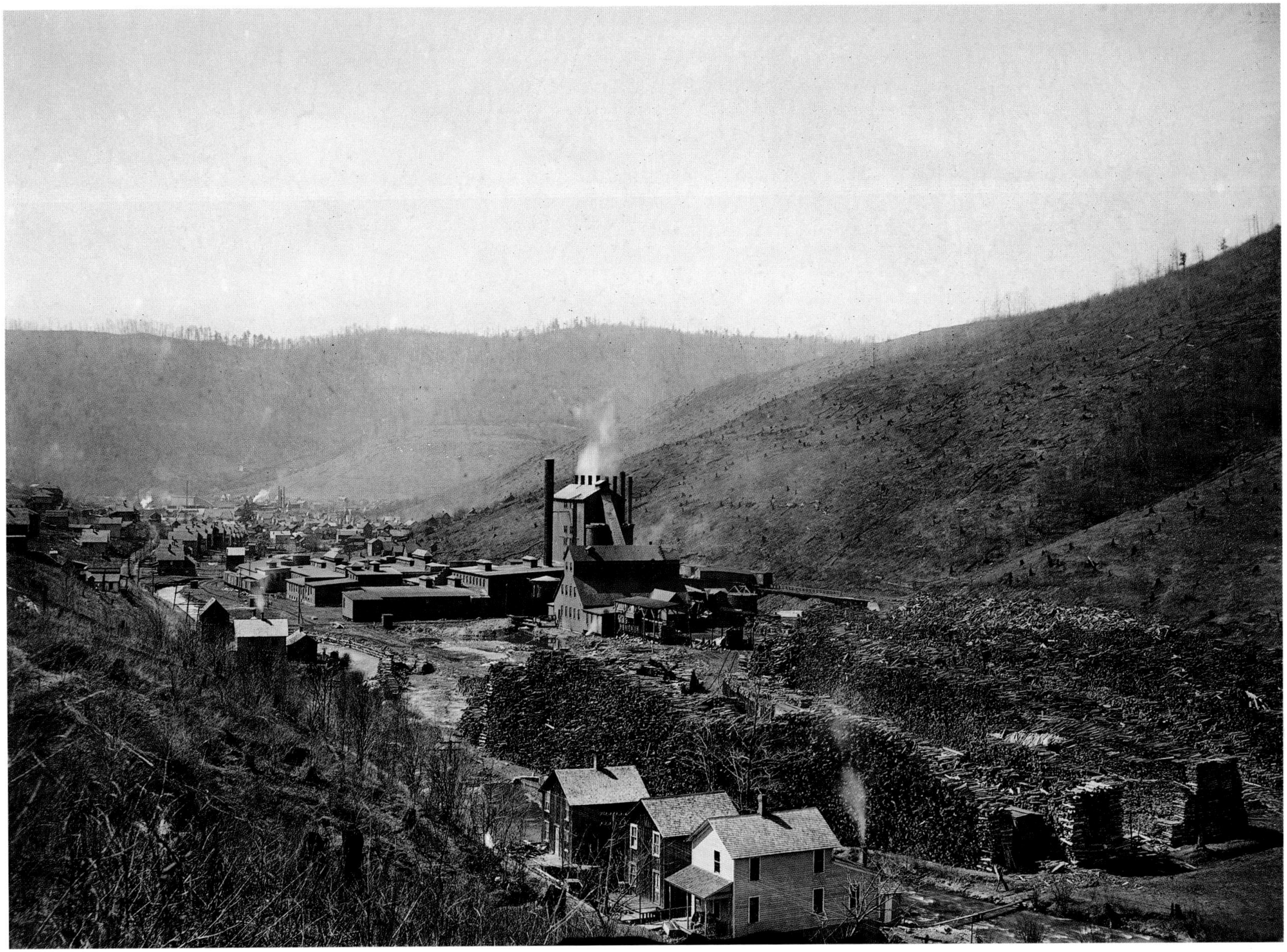

PLATE 89

The Bayless Pulp and Paper Company, mill and pulpwood, Austin, Potter County

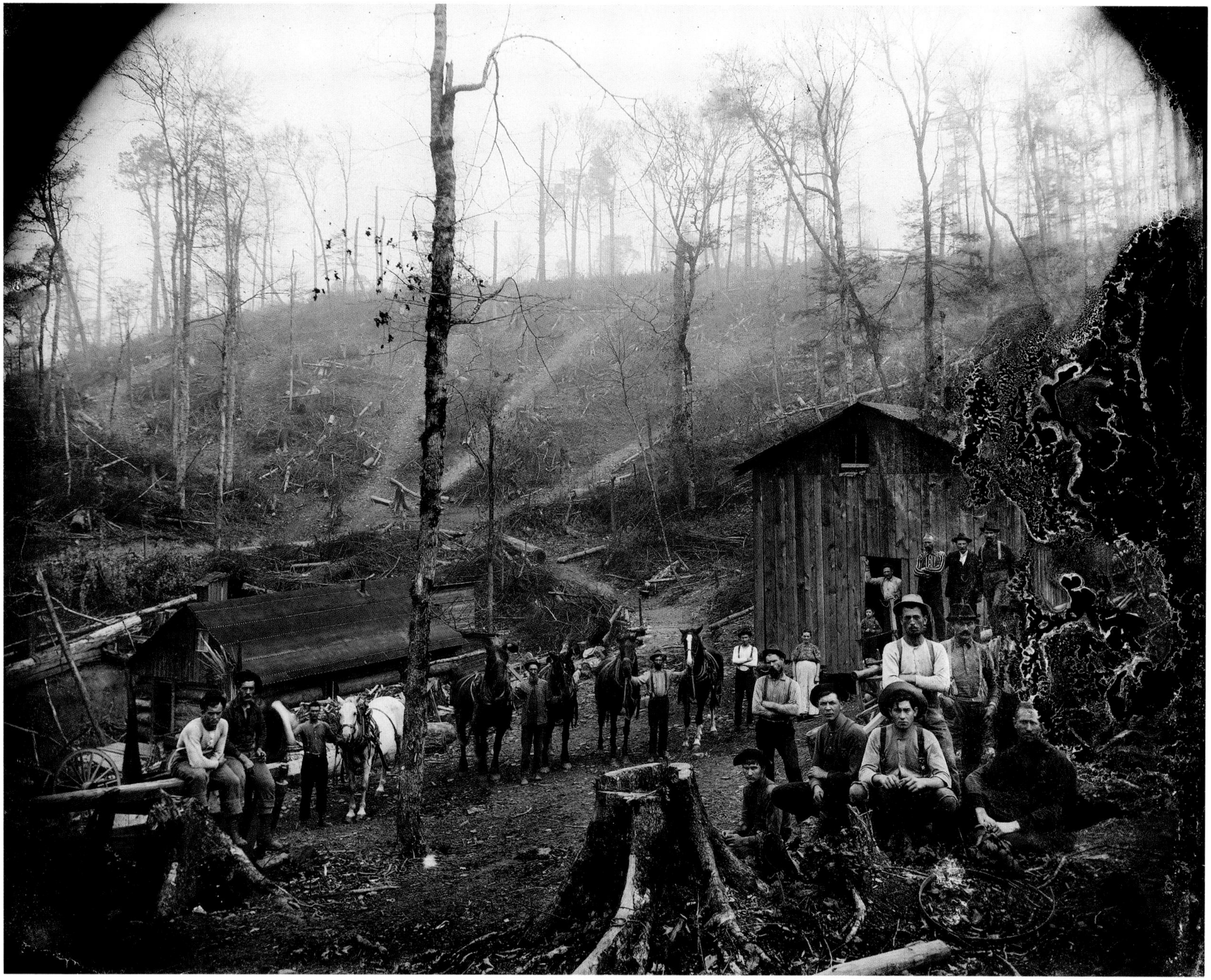

PLATE 90

Roads to ruin

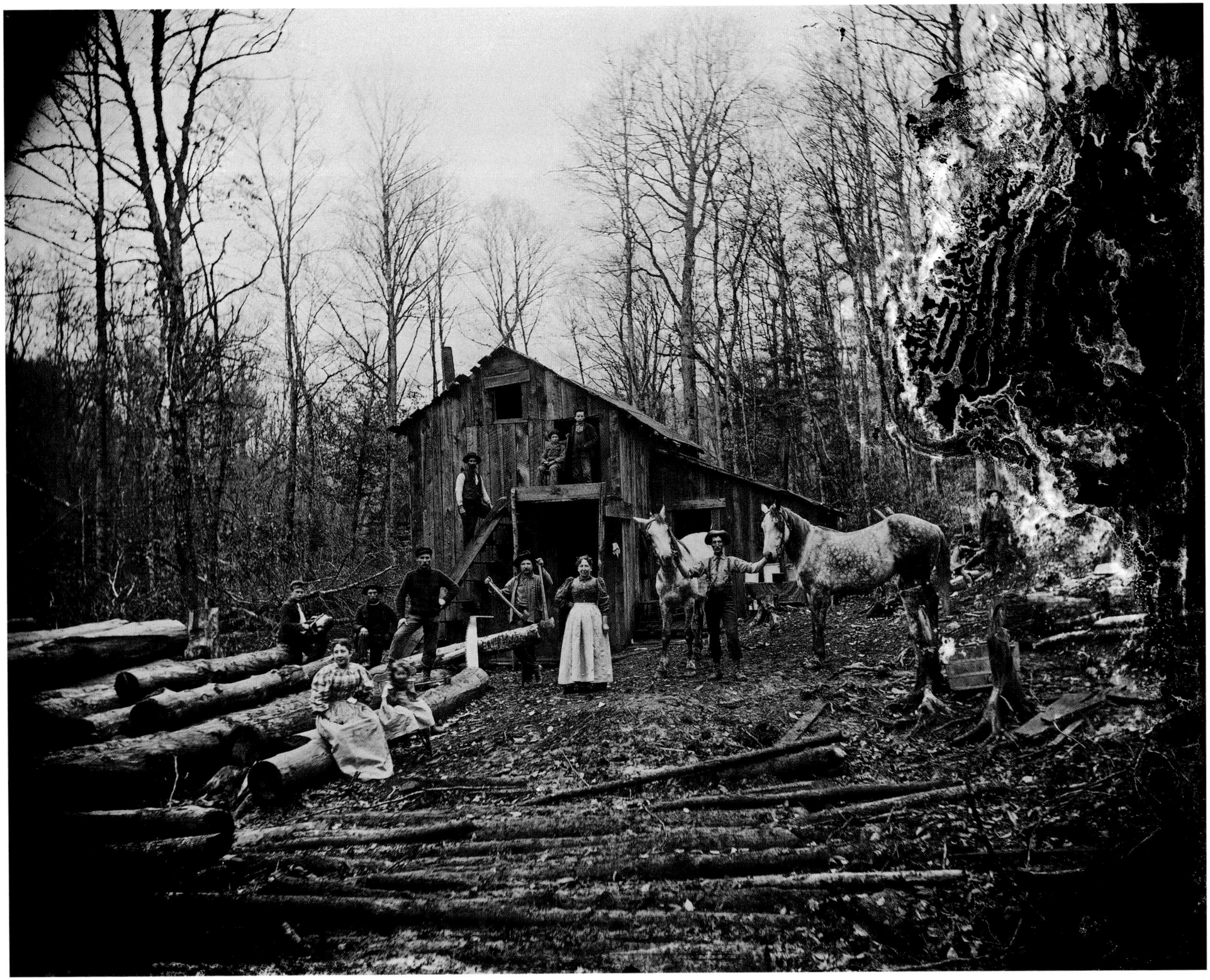

PLATE 91

Perhaps carpentry will work out

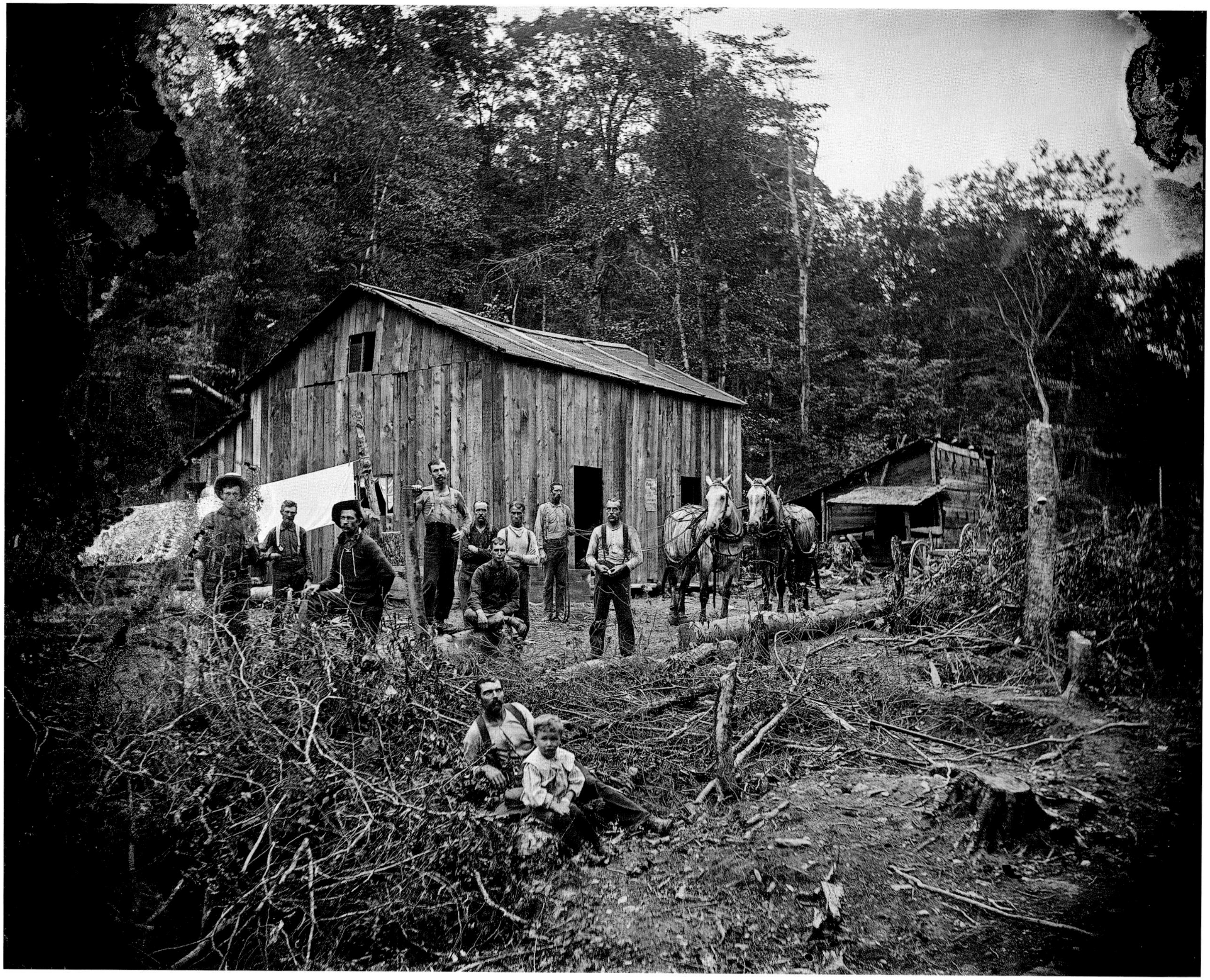

PLATE 92

Clean white sheets and determination

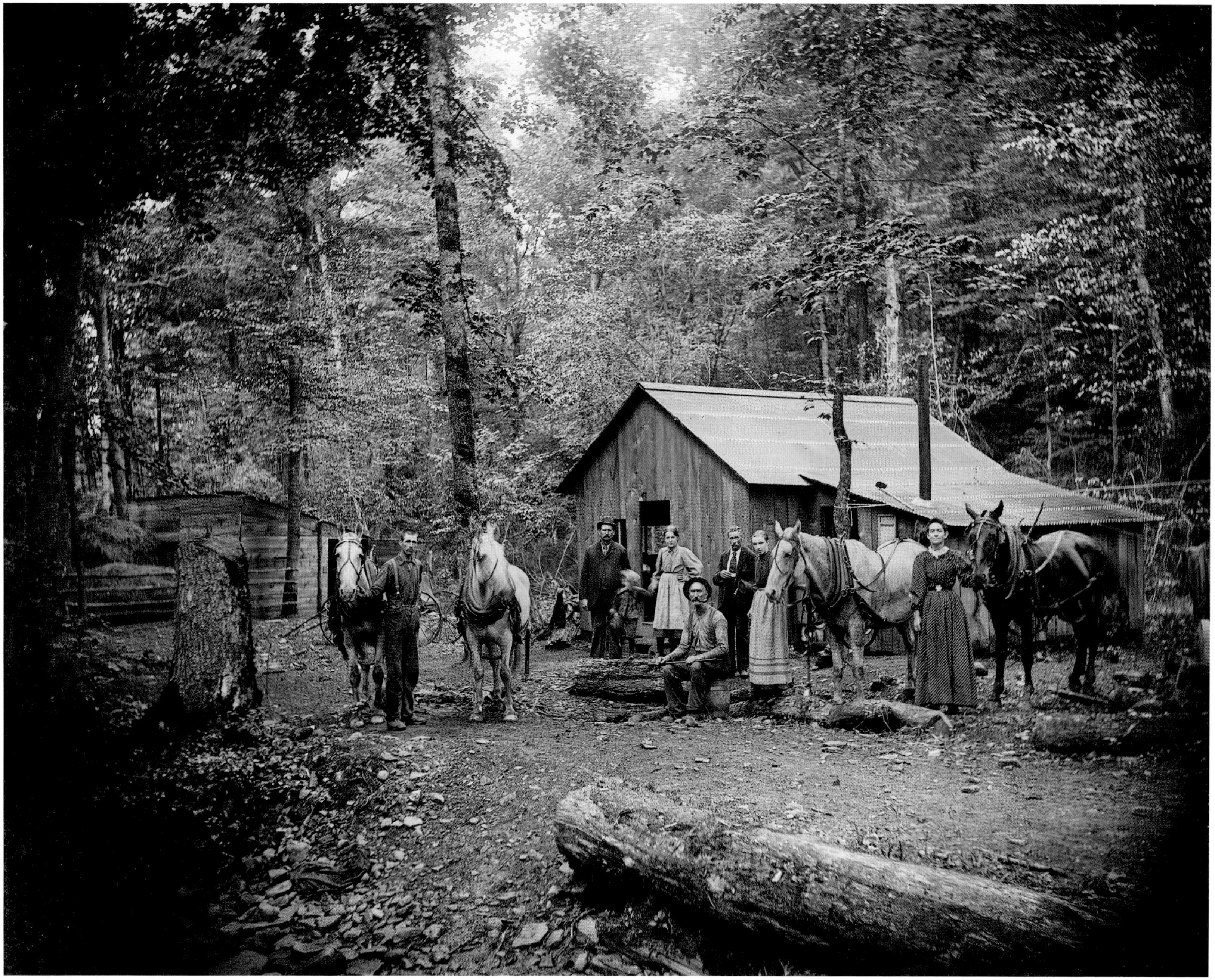

PLATE 93

Grit and dignity

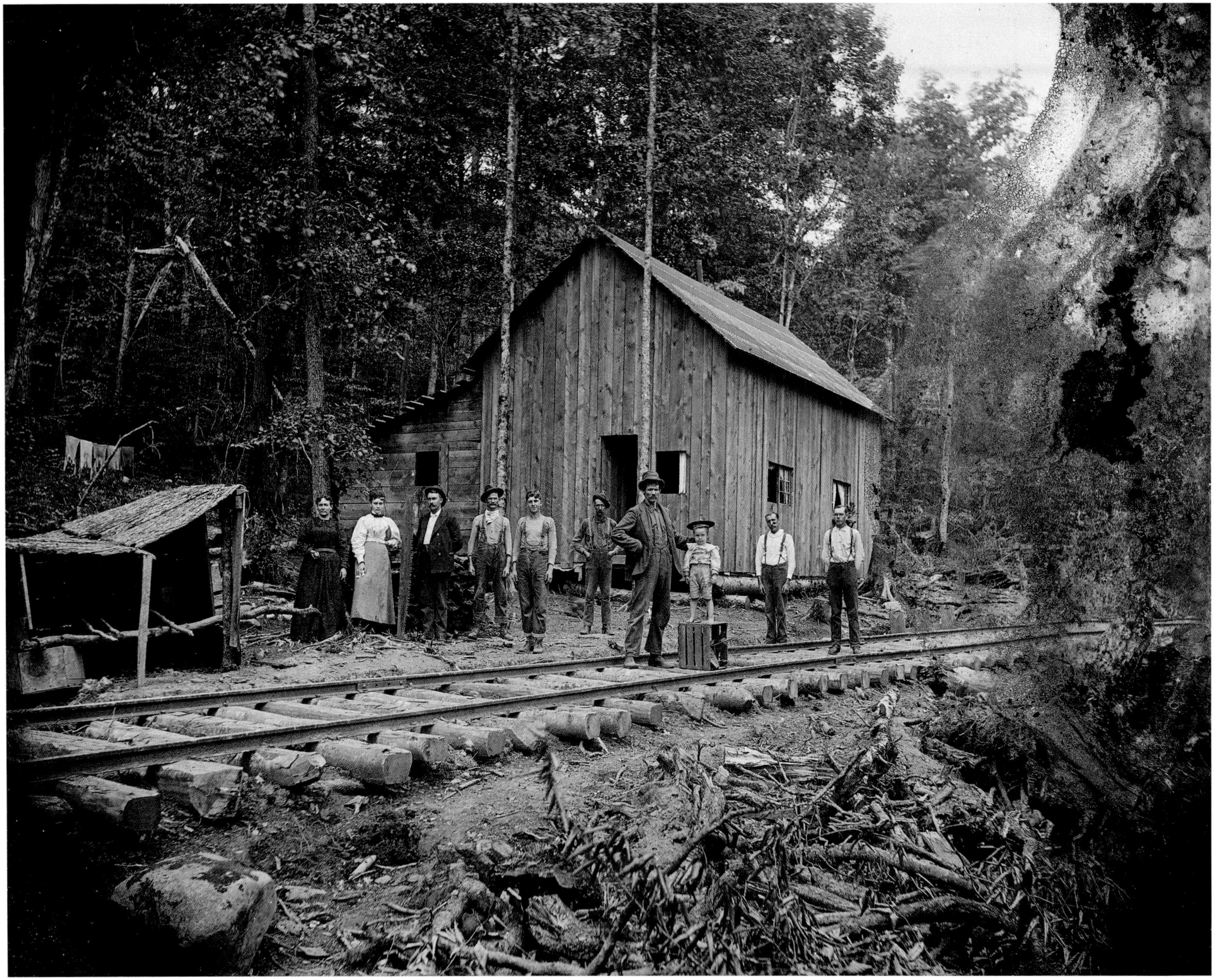

PLATE 94

How secure are we?

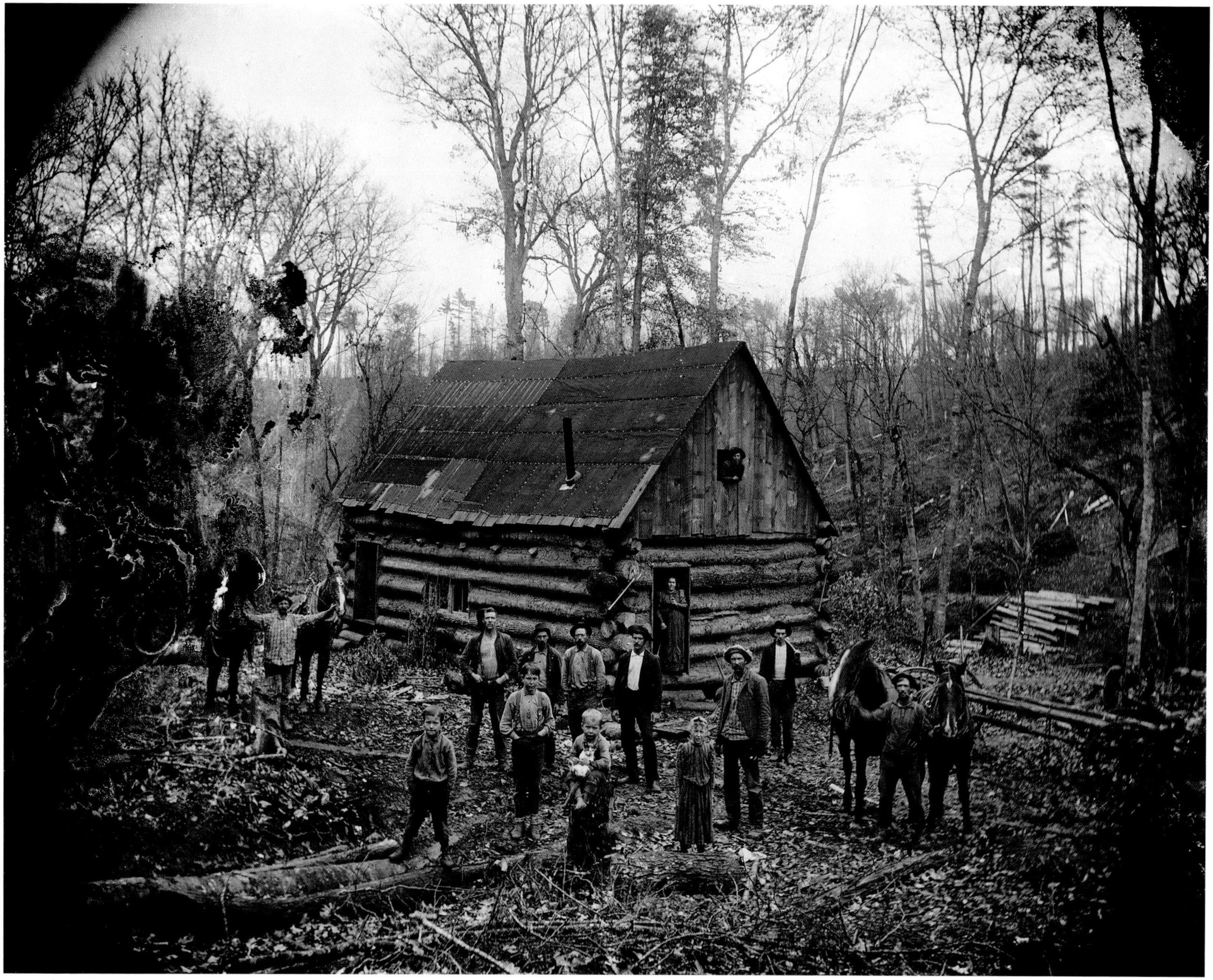

PLATE 95

Lincoln lived with logs, too

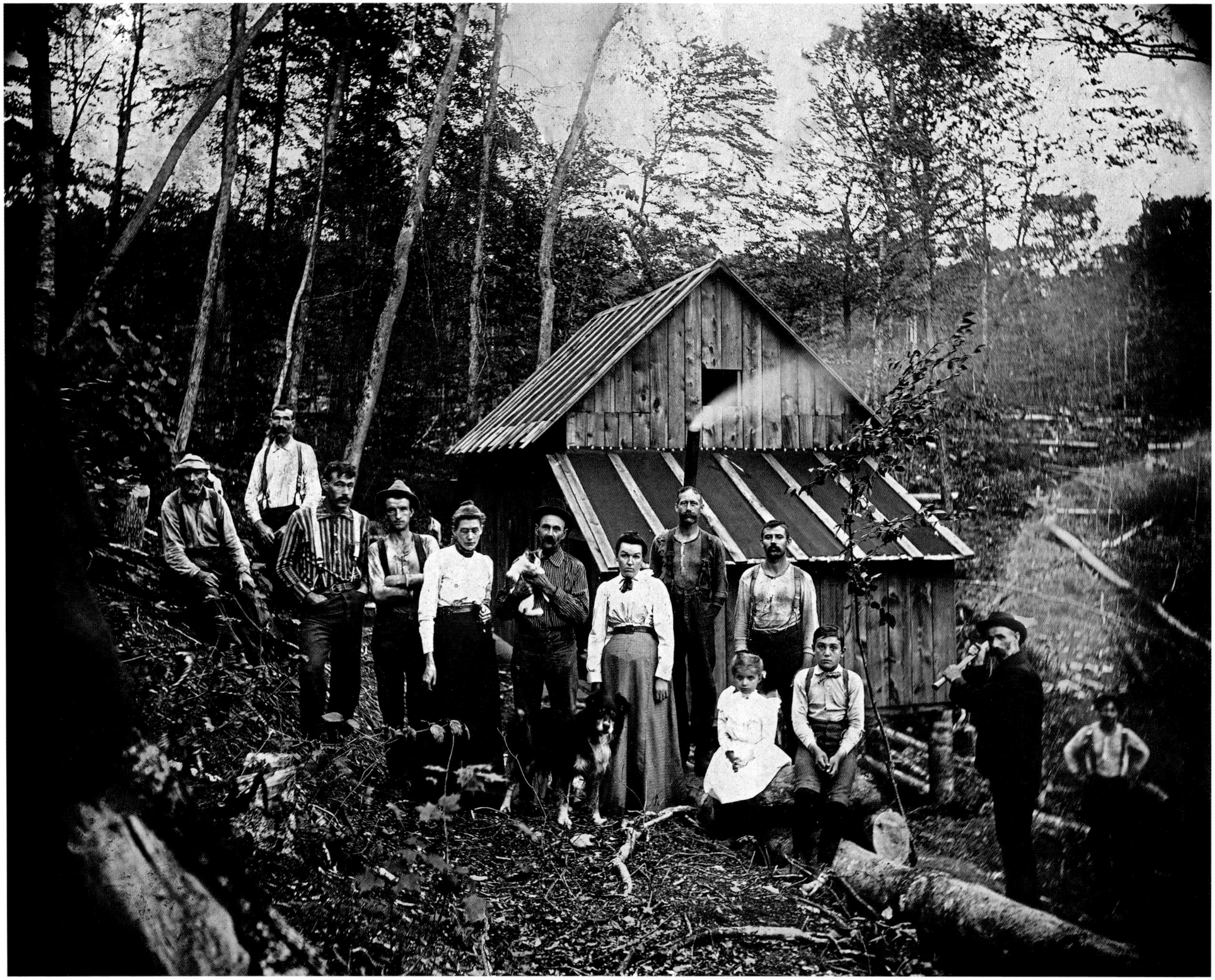

PLATE 96

Coleman Hollow, near Port Allegany, McKean County

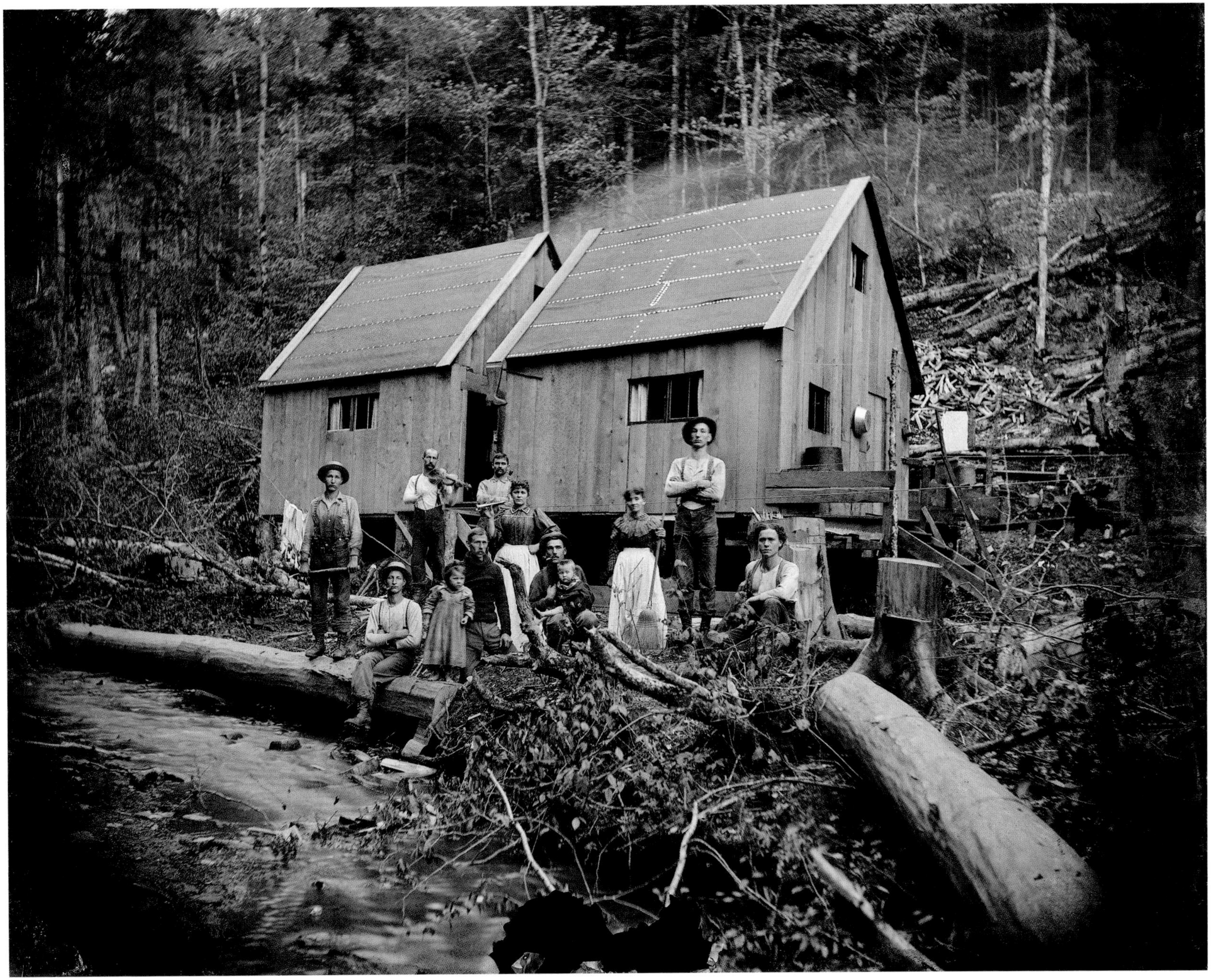

PLATE 97

Music, a boot, and a slight smile

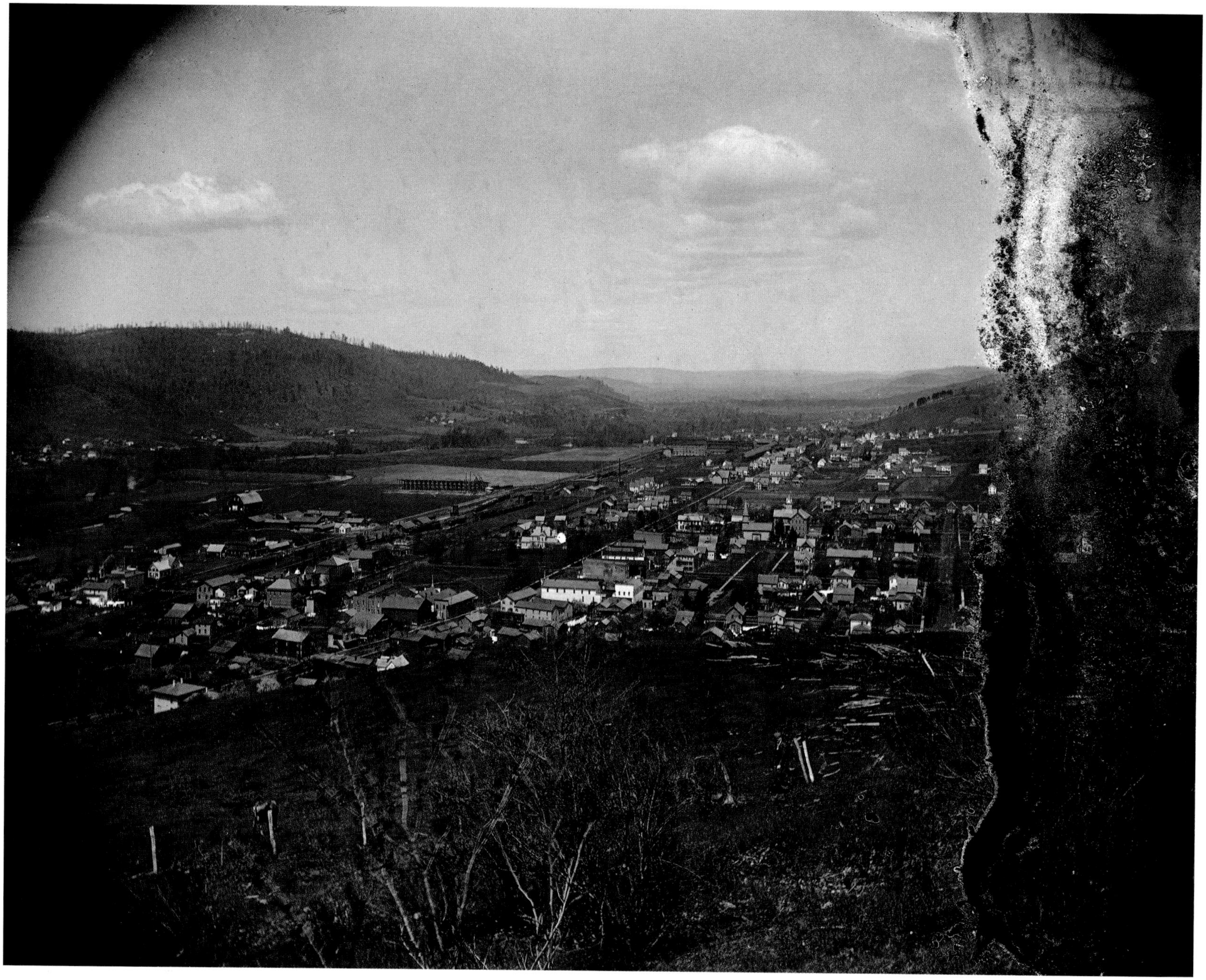

PLATE 98

The attractions of civilization, Port Allegany, McKean County

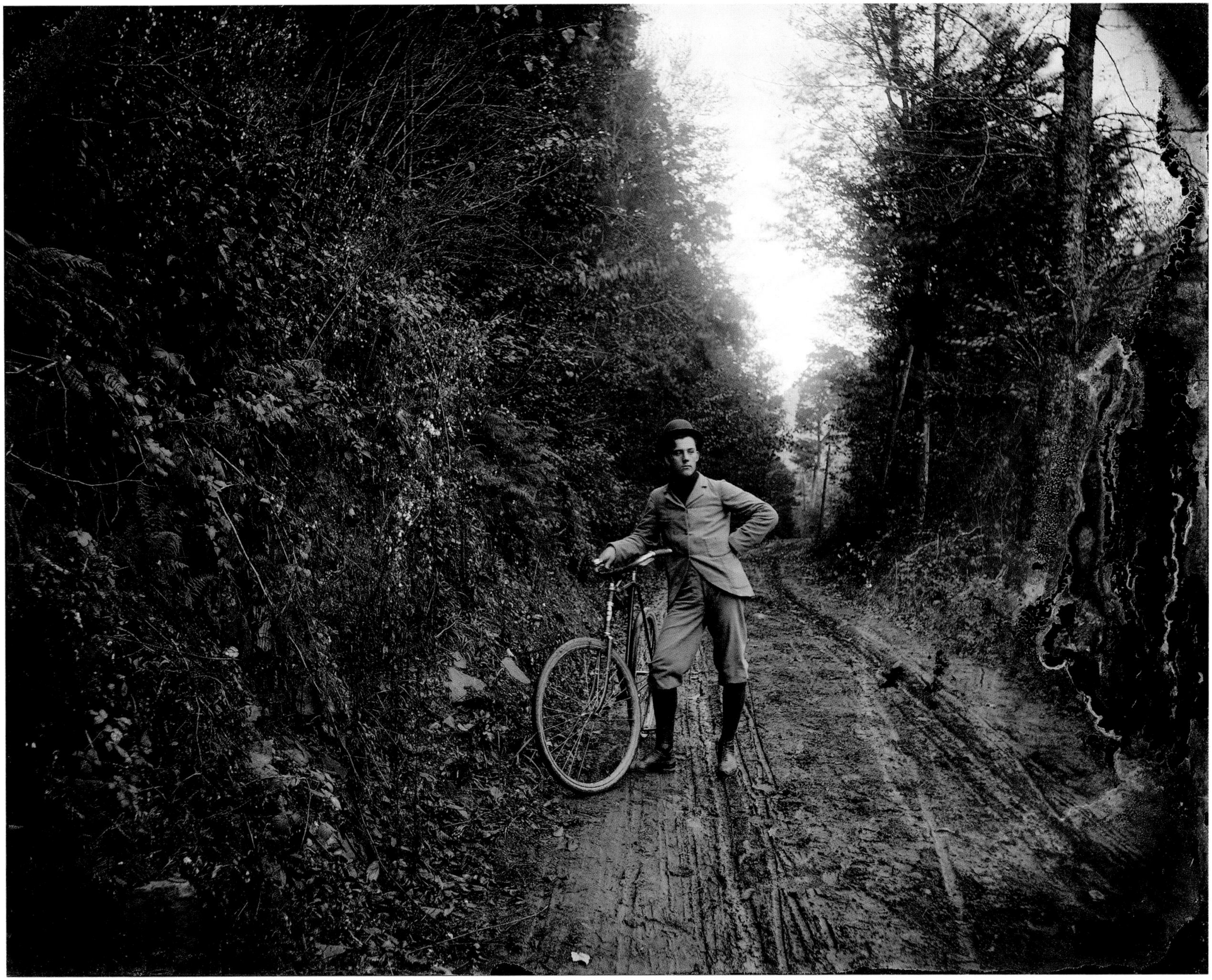

PLATE 99

Road to adventure

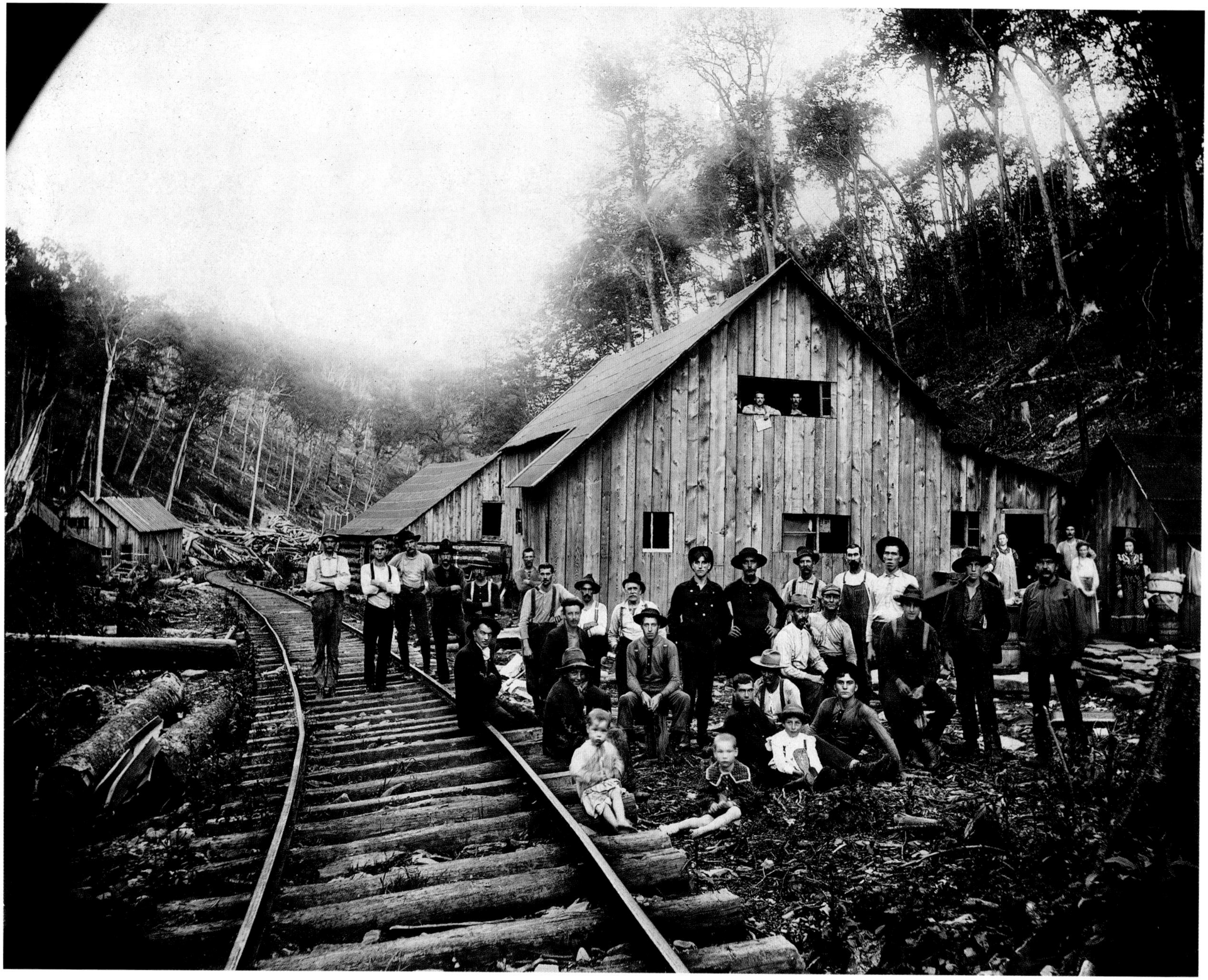

PLATE 100

Road to destiny, Mina, Potter County

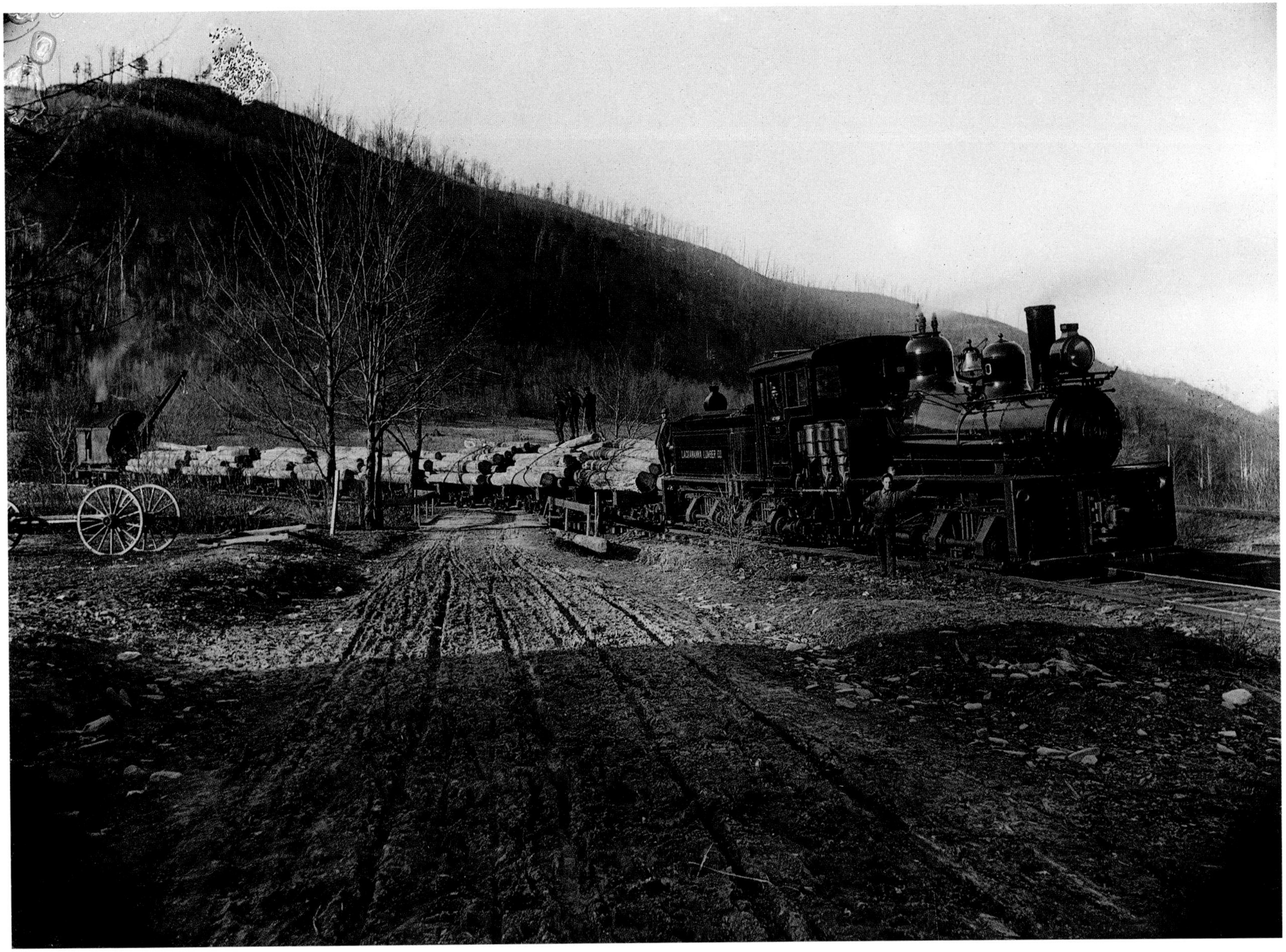

PLATE 101

Lackawanna Lumber Company No. 10, Cross Fork, Potter County

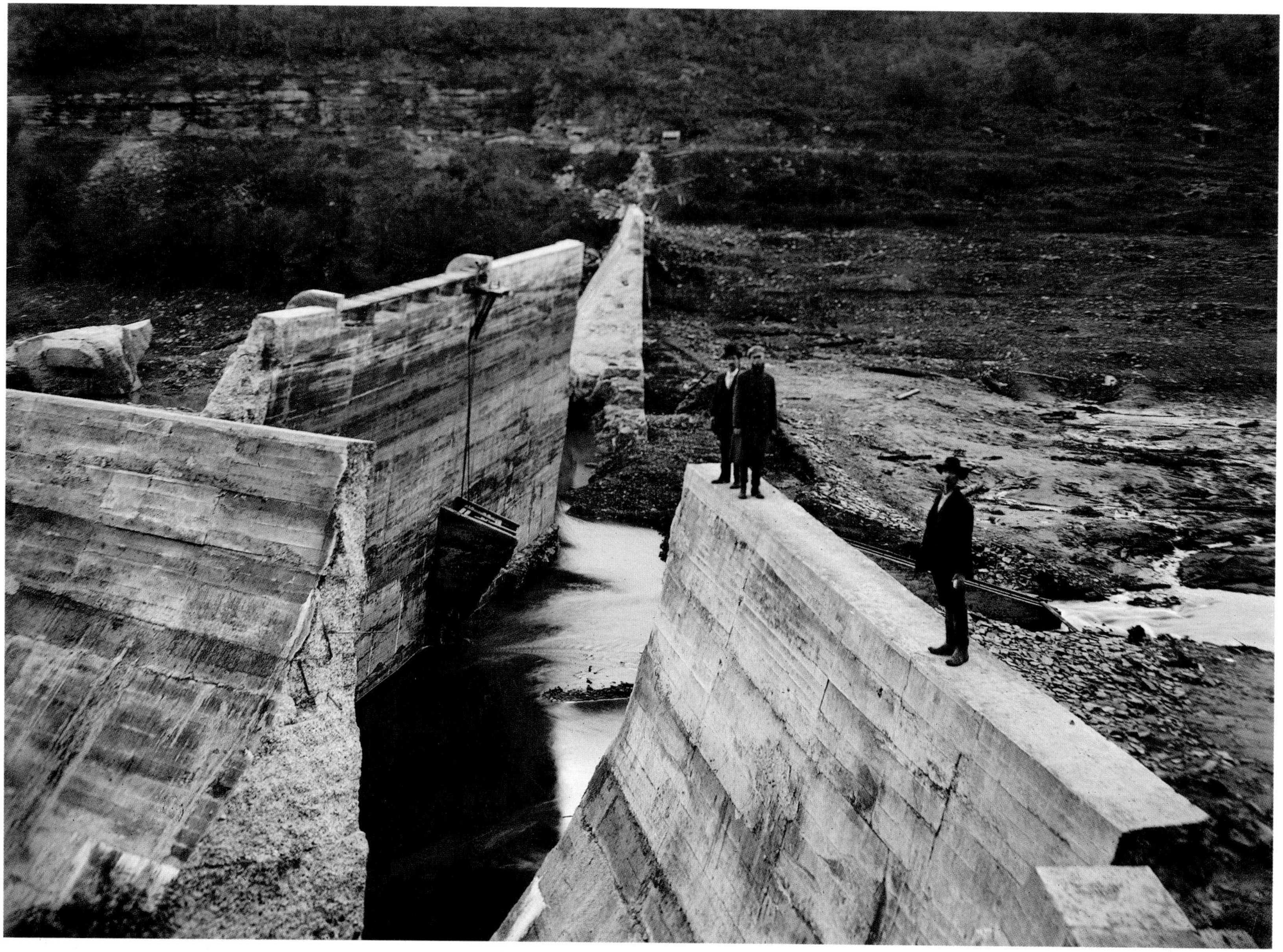

PLATE 102

Austin flood of 1911

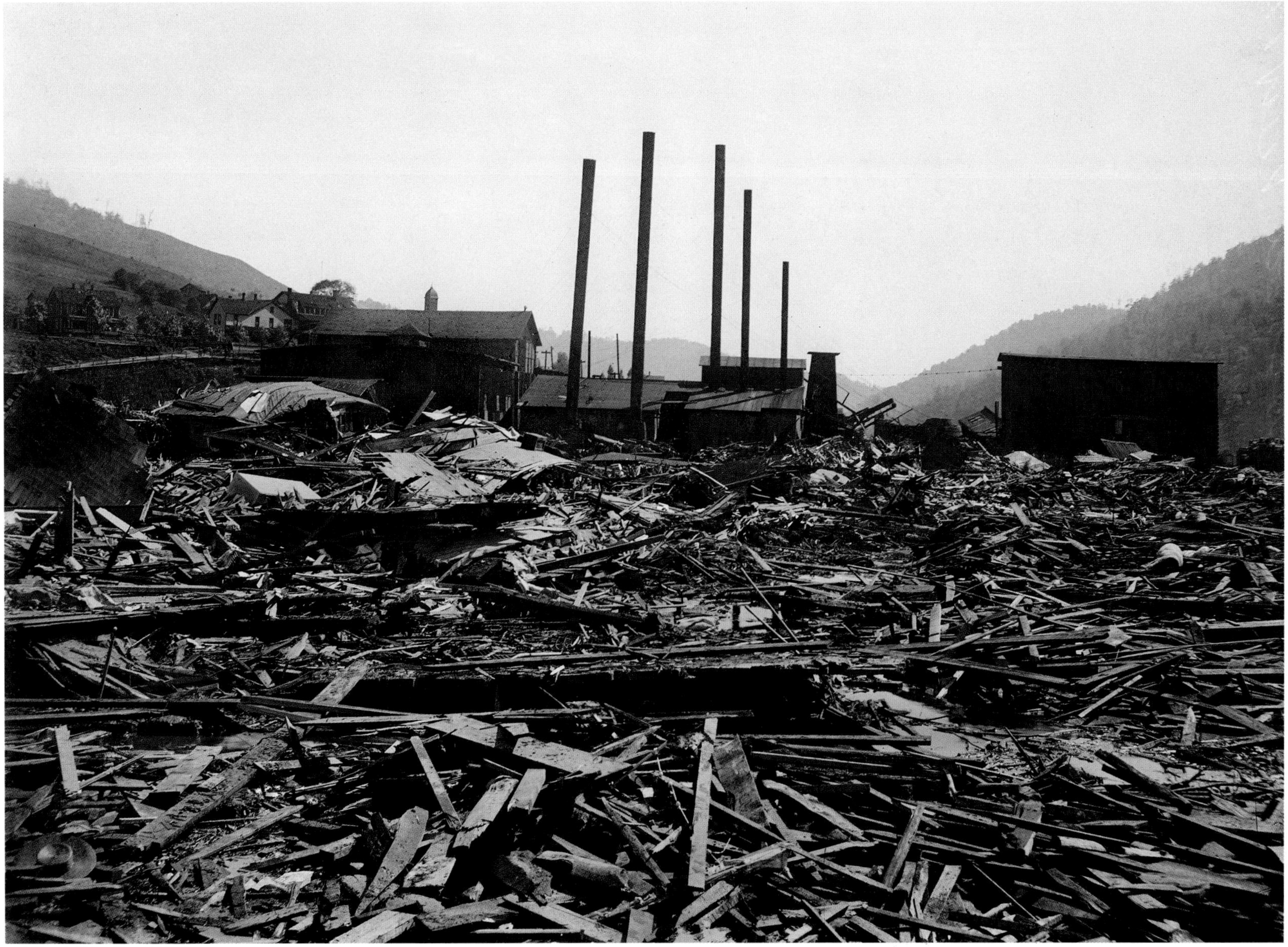

PLATE 103

The pond of Hemlock Mill after a flood

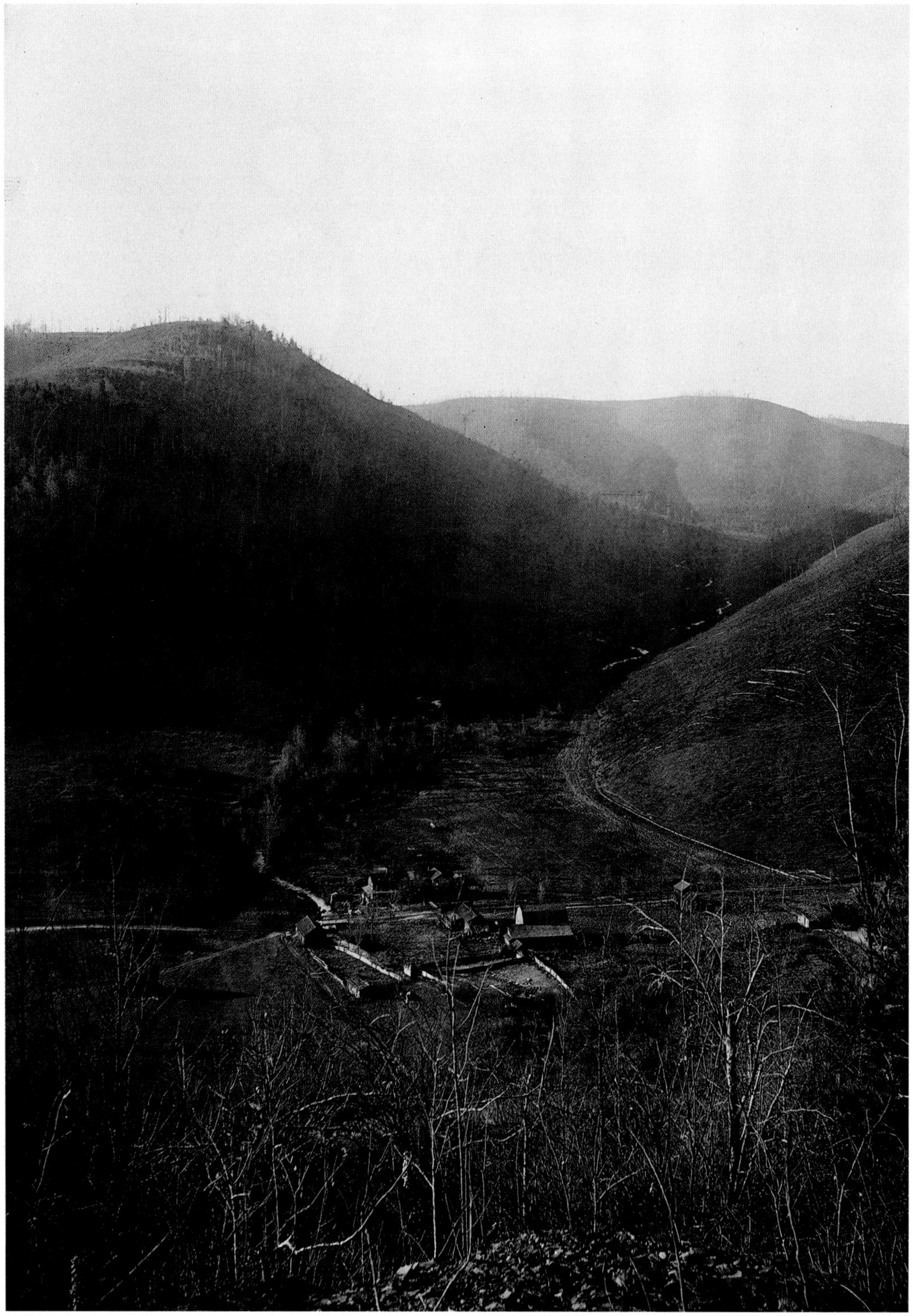

PLATE 104
Bird's-eye view of Logue

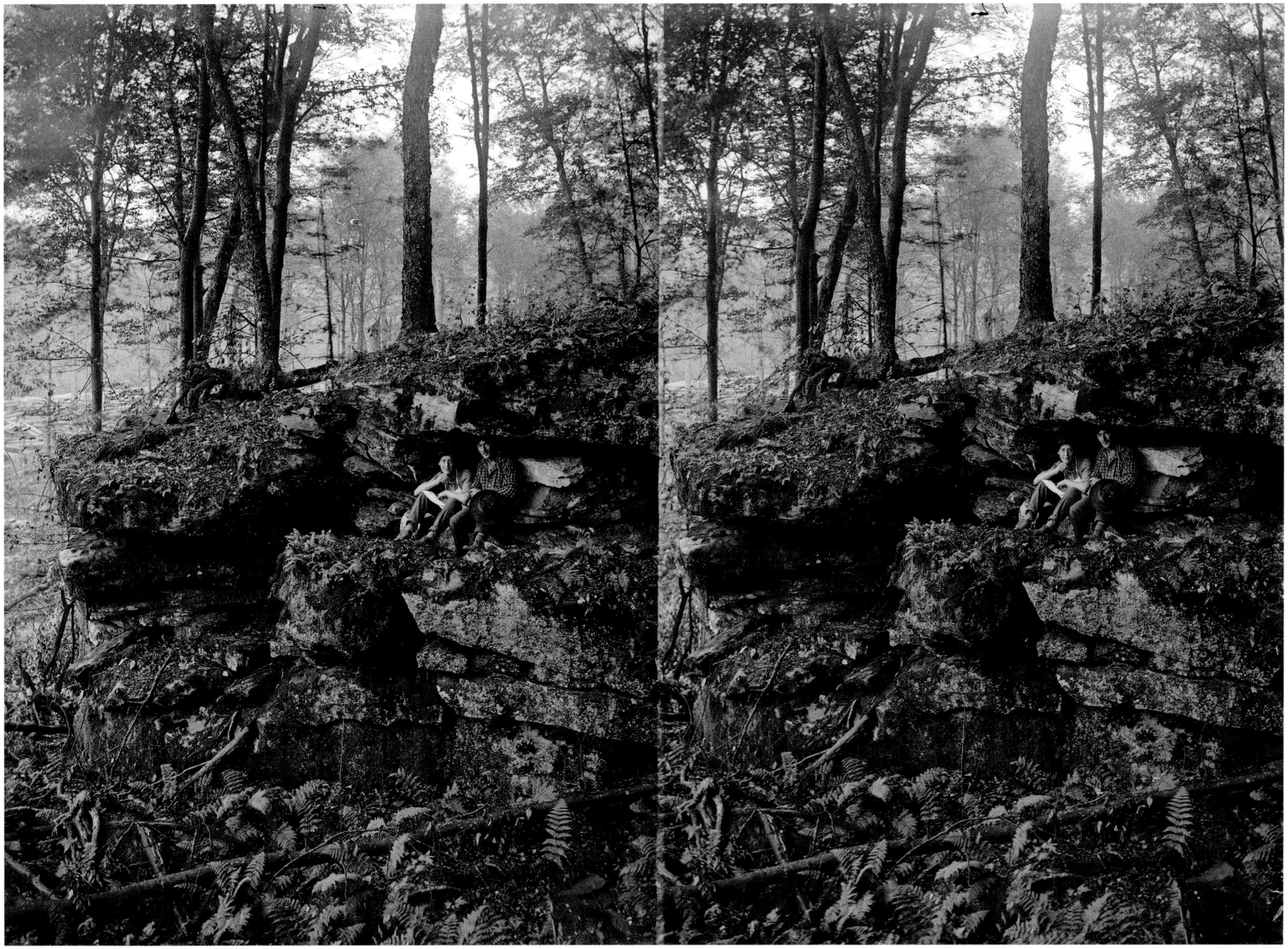

PLATE 105

Under the shelter of the rocks near Betula, McKean County

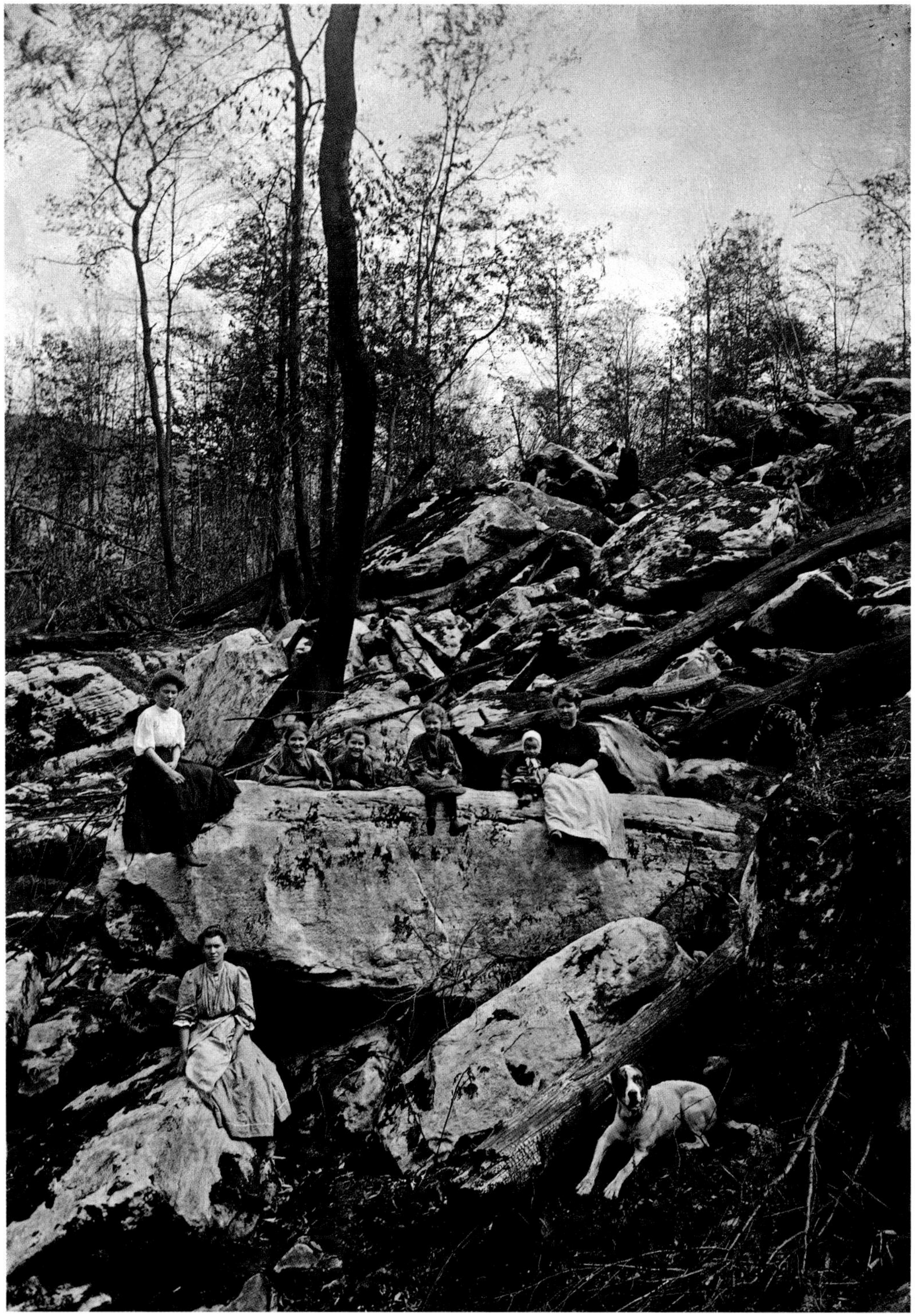

PLATE 106

Happy outdoor hours, Hammersley region, Clinton County

PLATE 107

Having a good time, northwestern Clinton County

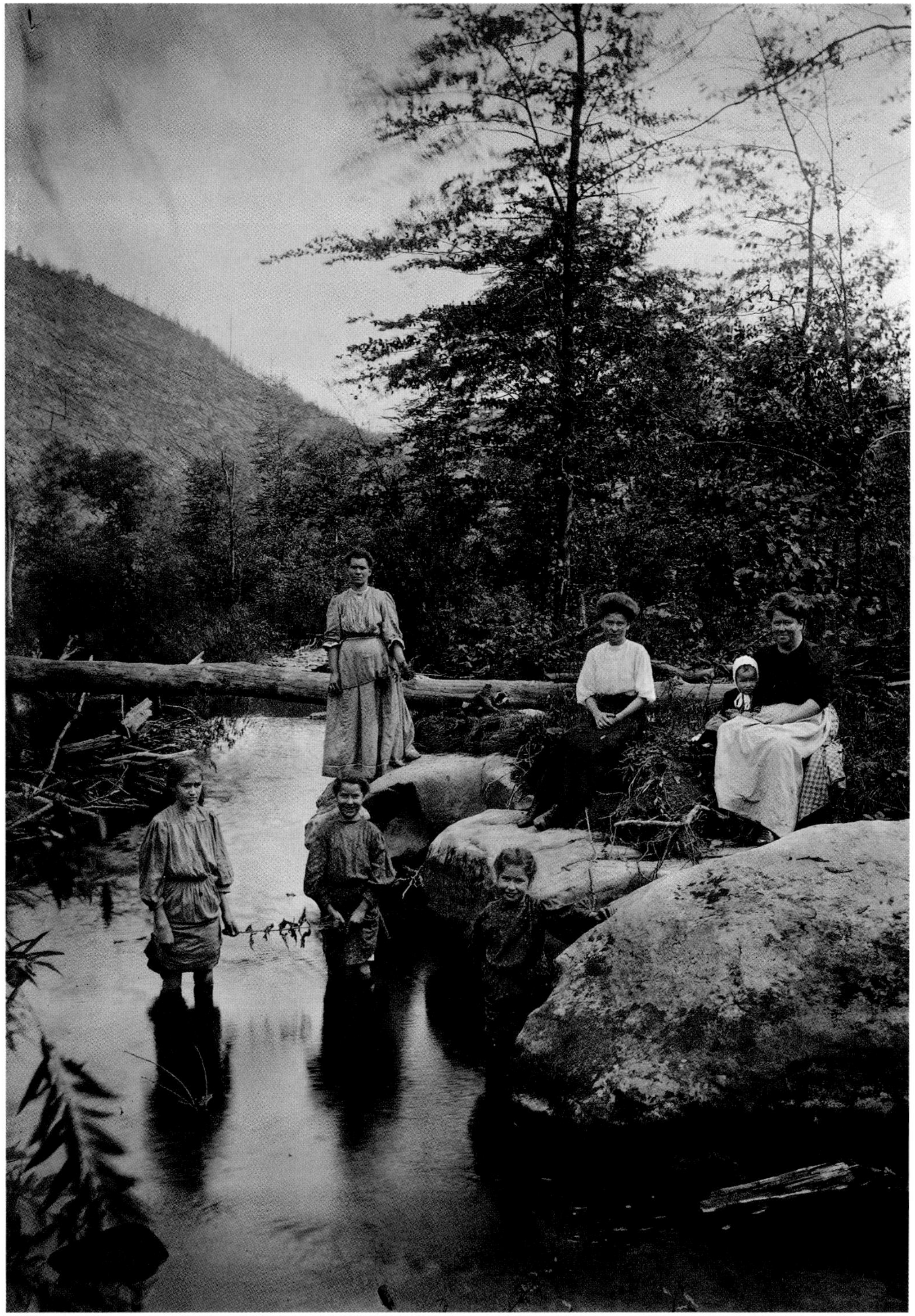

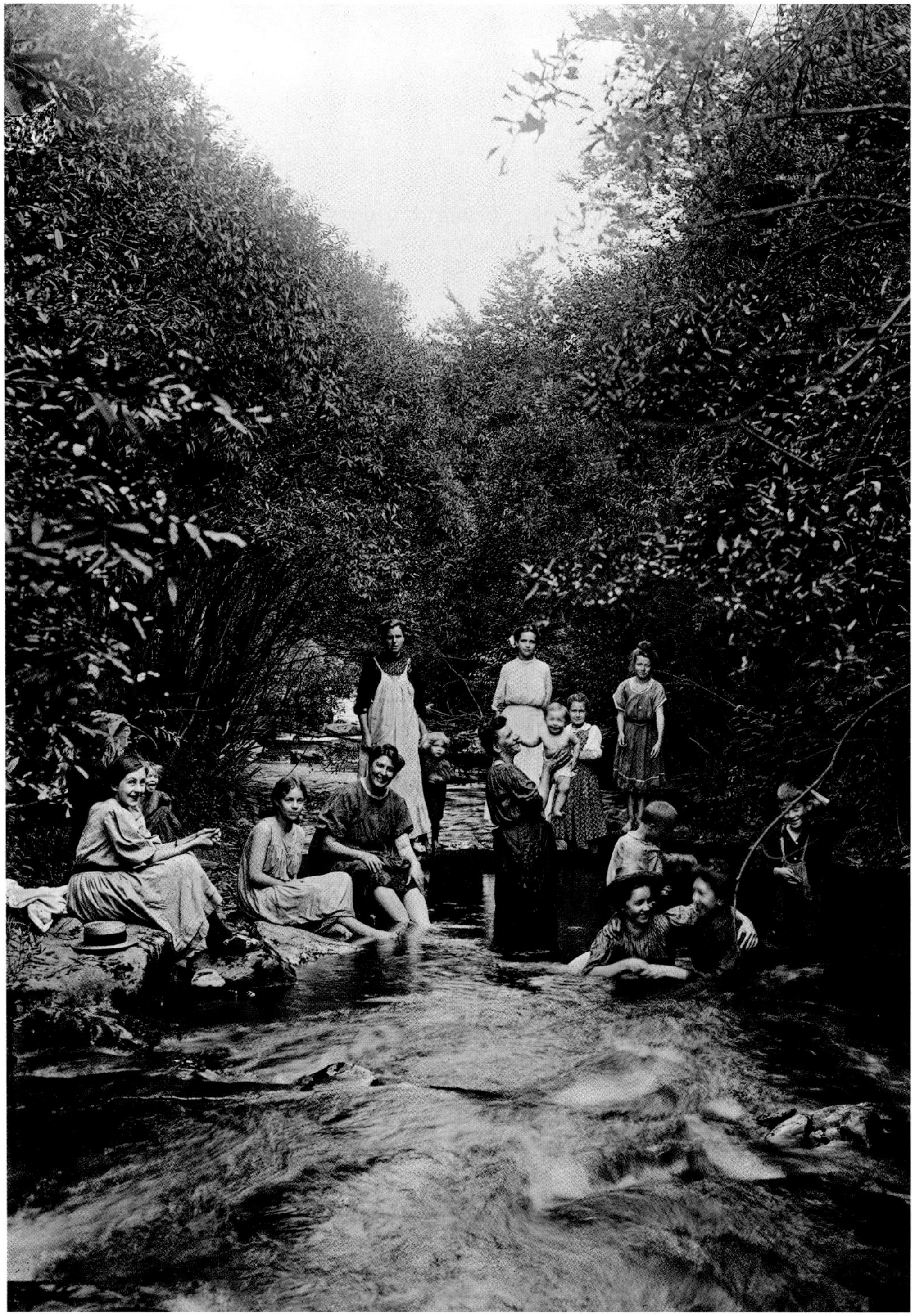

PLATE 108

The water is fine for cooling off on a hot day. Hammersley region, northwestern Clinton County

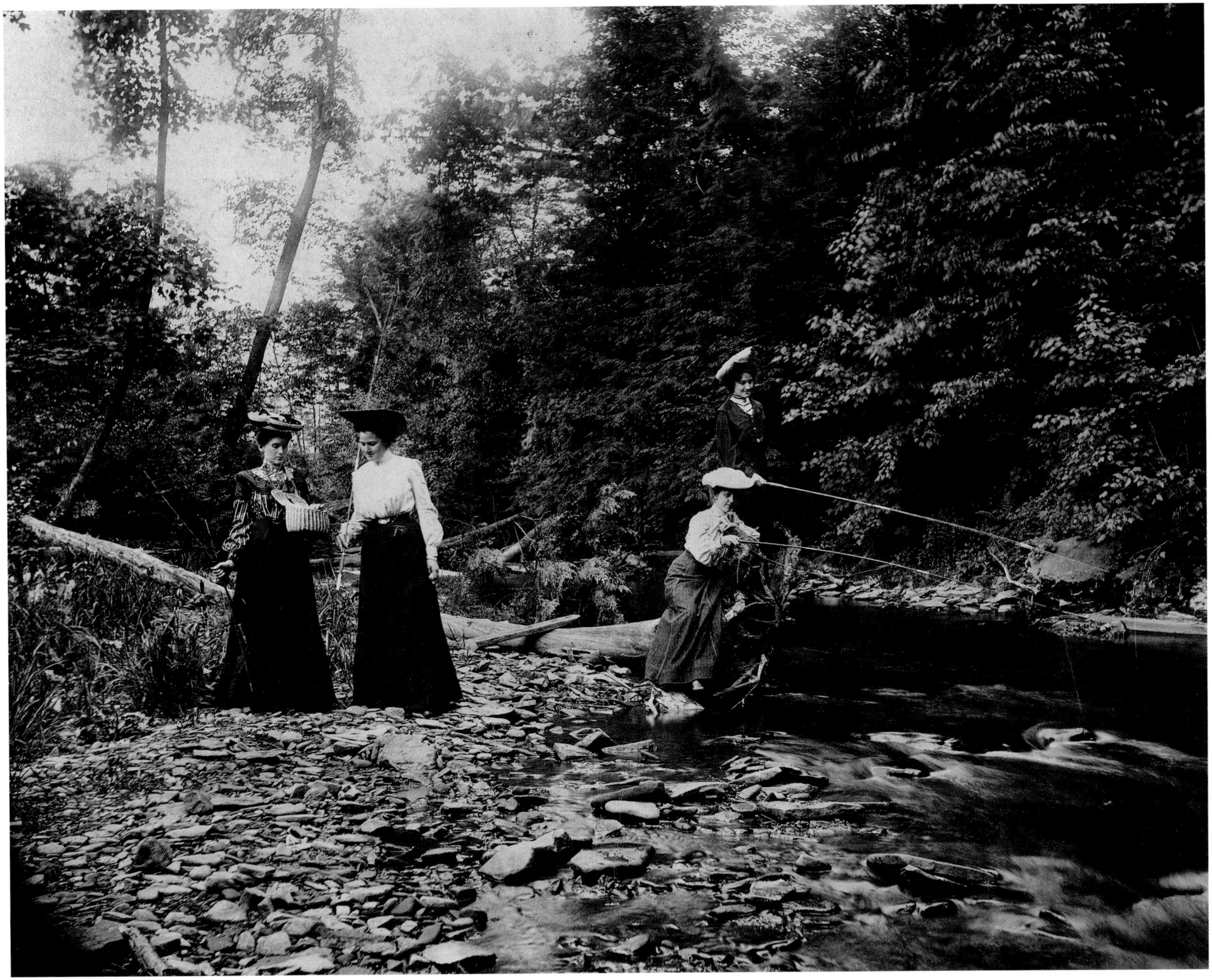

PLATE 109

Women in fine Victorian clothing try their hand at stream fishing

APPENDIX

NOTES ON THE PHOTOGRAPHS

FIGURES

1 Courtesy of the Lois Barden Photograph Collection, Candor, New York, BC 84.

2 Courtesy of the Lois Barden Photograph Collection, BC 03.

3 Courtesy of the Pennsylvania State Archives, Harrisburg, Pennsylvania, Albert F. Zimmerman Collection, MG 482-17.

4 Location unknown. Courtesy of the Pennsylvania State Archives, Records of the Department of Forests and Waters, Public Relations Office, Photographs and Negatives, ca. 1890–1971 (hereafter RG-6.20), Extra Copy Prints, PHMC-21.

5 Courtesy of the Pennsylvania State Archives, RG 6.20, William T. Clarke Glass Plate Negatives (hereafter WTC), no. 3821.

6 Courtesy of the Pennsylvania State Archives, RG 6.20, WTC, no. 3822.

7 Courtesy of the Pennsylvania State Archives, RG 6.20, WTC, no. 3823b.

8 Pattee Library, Pennsylvania State University, University Park. The frontispiece matches the image of an original negative in the Pennsylvania State Archives, RG 6.20, WTC, no. 3700.

9 A ca. 1920s Forestry Department print of this image is located in the Pennsylvania State Archives, RG 6.20, Mounted Prints, no. 331.

10 The image of General of the Armies John Joseph Pershing (1860–1948) of the U.S. Armies was made in 1917. Library of Congress, Washington, D.C. Public domain. Pershing was promoted to this highest military rank after World War I had concluded. George Washington is the only other person to have held that rank before or after Pershing.

11 National Gallery of Art, Washington, D.C. Public domain.

12 Courtesy of the Ronald E. Ostman Collection, Groton, New York.

13 *Locomotives, Jersey City, 1934.* © 2015 Estate of Reginald Marsh / Art Students League, New York / Artists Rights Society (ARS), New York and Smithsonian American Art Museum, Washington, D.C. Photo: Smithsonian American Art Museum, Washington, D.C. / Art Resource, New York.

14 The recumbent man stretched out on the lumber pile has invented the wood hick's version of a recliner. He may be one of the few "lazy boys" of the deep woods. Courtesy of the Pennsylvania Lumber Museum, Ulysses, Pennsylvania, A7 77.4.1.

15 Courtesy of Nicholas A. Tonelli, Lehigh Valley, Pennsylvania. Tonelli's original is in color; see it and other examples of his gorgeous landscapes and rural scenes at www.flickr.com/photos/nicholas_t/.

PLATES

1 A giant hemlock dominates the foreground. Progressive Era advocates would later be inspired by images such as this and display them to advocate for wise forest management (see introduction). Clarke captured this image at the beginning of timbering to create the town of Norwich. Courtesy of the Pennsylvania State Archives, RG-6.20, WTC, no. 3700.

2 Possibly taken in the same location as that shown in fig. 1. Courtesy of the Lois Barden Photograph Collection, BC 116.

3 Compare this late 1800s Clarke image with a contemporary view of the same area (fig. 15). Courtesy of the Potter County Historical

Society, Coudersport, Pennsylvania, Lumbering Pictures, LP 33.

4 Courtesy of the Pennsylvania State Archives, RG-6.20, WTC, no. 4111.

5 Courtesy of Lois Barden Photograph Collection, William T. Clarke Pictures, WTCP e.

6 Courtesy of the Potter County Historical Society, LP 35.

7 The original caption on the negative sleeve also referred to "a helping hand." This wreck probably was the same derailment that Thomas T. Taber III referred to: "Engine No. 19 is hoisted up the bank after running away down Long Hollow. No one was killed when wet rails caused the train to get away from engineer Fred VanGilder" (Taber, *Goodyears*, 534). Courtesy of the Pennsylvania State Archives, RG-6.20, WTC, no. 3936.

8 Courtesy of the Lois Barden Photograph Collection, BC 44.

9 Courtesy of the Pennsylvania State Archives, RG-6.20, WTC, no. 4121.

10 Given that at least five men can be seen holding peaveys, the logs probably were loaded by hand, not machine. Courtesy of the Lois Barden Photograph Collection, BC 94.

11 Businesspeople were eager to link Pennsylvania lumber with national markets and were prepared to expend huge sums to become part of that booming commerce. Railroad information courtesy Kurt Bell, an archivist in the Arrangement and Description Section of the Pennsylvania State Archives, in a text message, May 28, 2015, to Linda A. Ries, Pennsylvania Historical and Museum Commission, Harrisburg; Bell noted that Frank H. and Charles W. Goodyear bought the locomotive for their north-central Pennsylvania lumbering operations. This, the Goodyears' second No. 5, which weighed sixty-five to seventy tons, was standard gauge, equipped with fourteen-by-twelve-inch cylinders and thirty-three-inch drivers. In this photo the locomotive was pristine, without the dings, dents, or modifications that were inevitable results of lumbering operations. The brothers later sold it to the Potato Creek Railroad at Keating Summit to become the new owner's No. 5. When the timber was depleted, the locomotive veteran of thirty-seven years was scrapped in October 1919. Courtesy of the Lois Barden Photograph Collection, BC 04.

12 "Brothers" in the original caption may have referred to a regional nickname for the Goodyear brothers' logging enterprises and company equipment, or it may suggest that the engineer and locomotive had a special bond. This Goodyear Shay locomotive No. 15 may have been purchased from the James E. Strong Lumber Company in 1902 or 1903. Courtesy of the Pennsylvania State Archives, RG-6.20, WTC, no. 3930.

13 Thomas T. Taber III wrote of this image, "Crews and shop force pose for their picture. The locomotives are No. 11, Emporium Lumber Company; No. 5, unknown; and Climax No. 20. The Emporium locomotive was leased because of a shortage of motive power. However, Emporium wouldn't allow a Goodyear man to run the engine. John Swanson and Frank Dollinger took turns working one week at Hammersley and one week at Emporium Lumber Company's Keating Summit mill. Left to right are Lee Mallory, brakeman; Pat Lee, Dan Conway, brakeman; Joe Repolino, mechanic; John Swanson, Joe Ruggles (hands in pocket); unknown, Fred Green; Fred VanGilder, left on running board; Jim Lilley, and the others unknown" (Taber, *Goodyears*, 530). Courtesy of the Pennsylvania State Archives, RG-6.20, WTC, no. 3940.

14 Locomotive and mechanical log loader pick up logs from a "rough and tumble" skidway. Swanson's crew had been cutting hemlock in that area from at least 1900, according to a story in the *Potter County Journal* of June 27, 1900. Courtesy of the Potter County Historical Society, LP 02.

15 Freeman Run 01. Courtesy of the Potter County Historical Society.

16 Taber noted that portable railroad camps were common in the West, but not in Pennsylvania (Taber, *Whining Saws*, 654). Courtesy of the Potter County Historical Society, LP 64.

17 Taber commented, "The band saw and carriage on the short side of the mill. Logs were brought up a jack slip behind the photographer, and the lumber passed out the far end of the mill. Two men rode the carriage. On the near end was the dogger who operated the log and board dogs; on the far end was the setter who positioned a new log on the carriage in

preparation for the first cut. The projection sticking up in front of the band saw wheel is the 'nigger,' used to roll the log on the carriage and operated by two steam cylinders. The nigger and the carriage were operated by the head sawyer who stood on the floor beside the saw and operated and determined how the log would be cut" (Taber, *Goodyears*, 570). This likely is one of the last images Clarke took in Pennsylvania before returning to Rochester, New York. Courtesy of the Pennsylvania State Archives, RG-6.20, WTC, no. 3719.

18 Courtesy of the Pennsylvania State Archives, RG-6.20, WTC, no. 4005.

19 A bird's-eye view of Betula in McKean County shows the Pennsylvania Stave Company plant, drying sheds, and company houses. Norwich appears in the distance, at right. Wooden staves were stored inside sheds to season thoroughly before being assembled in the heading and stave plant, after which they were shipped to market. This plant and the nearby Keystone Chemical plant were the economic basis for the company town's existence. Betula is still on the map; Norwich is not. Courtesy of the Pennsylvania State Archives, RG-6.20, WTC, no. 3755a.

20 Taber commented that the plant ran only six or seven years, "unusually short for a chemical factory" (Taber, *Whining Saws*, 611). The denuded hills behind the plant may be a clue to why the factory closed. Perhaps they ran out of trees, a raw material for making chemicals. Courtesy of the Pennsylvania State Archives, RG-6.20, WTC, no. 4021.

21 Courtesy of the Pennsylvania State Archives, RG-6.20, WTC, no. 4117.

22 Courtesy of the Lois Barden Photograph Collection, BC 08.

23 A paper label glued to the original five-by-seven-inch negative identifies this person as C. E. McGuire, possibly Chris McGuire, fifty-two, of Costello, an Irish immigrant who listed his occupation in the 1910 census as "woodsman." Courtesy of the Pennsylvania State Archives, RG-6.20, WTC, no. 3820.

24 Courtesy of the Pennsylvania State Archives, RG-6.20, WTC, no. 4009.

25 One of the first trees cut at Norwich—clearing the forest prior to building the town in 1910. Courtesy of the Pennsylvania State Archives, RG-6.20, WTC, no. 3703b.

26 Victor L. Beebe noted that this photo was taken "on Will Hartwick's job at Lyman Run." A slightly different pose, from a Clarke stereoview at the Potter County Historical Society, further describes the scene as "on the Cross Fork Branch of Lyman Run" (Beebe, *History of Potter County*, 188). Courtesy of the Pennsylvania State Archives, RG-6.20, WTC, no. 4012.

27 Courtesy of the Potter County Historical Society, LP 60.

28 Courtesy of the Potter County Historical Society, LP 63, and the Pennsylvania State Archives, RG-6.20, Mounted Prints, no. 666.

29 The buckers cut the trunk to moveable lengths on a steep slope, using a wedge at the top of their cut. Courtesy of the Pennsylvania State Archives, RG-6.20, WTC, no. 3825.

30 Henry W. Shoemaker used this image in *Black Forest Souvenirs*. His caption noted that the location was near Corbett, a place no longer on the Pennsylvania map. Courtesy of the Potter County Historical Society, LP 62, and the Pennsylvania State Archives, RG-6.20, Mounted Prints, no. 100.

31 Courtesy of the Lois Barden Photograph Collection, BC 41.

32 Courtesy of the Lois Barden Photograph Collection, BC 108.

33 The man in the center holds a long peavey. Railroad bark cars are barely visible at the upper left of the image. Courtesy of the Lois Barden Photograph Collection, BC 39.

34 Men on the logs hold peaveys, center rear. The woman and probably the dog (front center) are visitors to the site. Courtesy of the Lois Barden Photograph Collection, BC 48.

35 Courtesy of the Lois Barden Photograph Collection, BC 92.

36 This image includes a man who eventually would leave Pennsylvania for a job as a logger in another state. Laura Ishler of Curwensville, Pennsylvania, writes, "This photo from the Coudersport area of Potter County came from my grandparents' collection, the late Philip and May Wolfe Lecker of Saint Marys. The fifth man from the left in the dark shirt is John Lecker (1880–1964), my grandfather's brother. The second man from the right (standing and wearing dark shirt) is Clarence

Michael Hafer (1862–1935), husband of Mollie Lecher, Philip and John's sister. They were three of 12 children of George and Mary Simbeck Lecher of Saint Marys. George and Mary came to the U.S. from Bavaria in 1869. John joined two more of my grandfather's brothers, left Saint Marys in 1907, and settled in the state of Washington. They found work in the lumber industry there" (personal communication to the authors). The Lechers who moved to Washington were being realistic. The year they left Pennsylvania the logging boom was nearly over, leaving barren hillsides, decaying sawmills, and struggling or doomed wood-dependent industries. For impressive logging photos of the redwoods and the Pacific Northwest, see Bohn and Petschek, *Kinsey Photographer*. Courtesy of the Laura Ishler Collection, Curwensville, Pennsylvania.

37 The cook is standing and wears a tie. The cookee is seated and holds a wash pan. The sign "Pleasant Valley Camp" suggests that this scene was located somewhere in Pleasant Valley Township, Potter County. Courtesy of the Lois Barden Photograph Collection, BC 11.

38 Taber identified this scene as taken at the Hammersley Boarding House in Clinton County. "The dining room had two long tables. The second table is visible on the right. Many of the railroaders were single men who lived and ate there" (Taber, *Goodyears*, 533). Note there is a woman poised to serve the meal. Courtesy of the Pennsylvania State Archives, RG-6.20, WTC, no. 3873.

39 Eight or nine pigs can be seen on or near the railroad tracks parallel to the building. Swanson's camp, one of many in the region, gives an idea of the large scale of lumbering activities. Brute strength and stamina were needed to denude a hillside such as the one behind the camp. Courtesy of the Potter County Historical Society, LP 57.

40 Courtesy of the Potter County Historical Society, LP 04.

41 Courtesy of the Potter County Historical Society, LP 04b.

42 A blacksmith holds a heavy hammer and pinchers (far left), the boy fourth from left holds a peavey, and two Italian immigrants (fifth and sixth from left) hold "spivas," logging tools from the old country that they preferred to the peavey. The location is named after Lewis Stephens, a hermit known as "the wild boy." Stephens lived in the area until the 1870s and died in Shinglehouse in 1900. See Thompson, "History of the Wild Boy." Courtesy of the Potter County Historical Society, LP 41.

43 The long peeled stick on the ground in front of the log where the boy is sitting may be a homemade scaling stick. Courtesy of the Lois Barden Photograph Collection, BC 16.

44 The jobber, his wife, and their two daughters at Allison Lumber Camp on 11 Mile Run, Potter County, near Austin. Courtesy of the Lois Barden Photograph Collection, BC 83.

45 Henry W. Shoemaker included this Clarke image in *Allegheny Episodes* and had Brewer marrying a couple of mountaineers in a story titled "A Misunderstanding." Brewer was committed to Kane Hospital, McKean County, in 1923. A March 17, 1923, story in Shoemaker's *Altoona Tribune* newspaper, headlined "'Sky Pilot' Goes to County Home," reported that the "Sky Pilot of the Alleghenies" was "slightly deranged" (9). William E. Brewer was born in 1845 in Webster, New York, and died in 1928 in Warren State Hospital, Warren County, Pennsylvania; he had suffered from senile psychosis. Courtesy of the Pennsylvania State Archives, RG-6.20, WTC, no. 3879.

46 Courtesy of the Lois Barden Photograph Collection, BC 01.

47 Courtesy of the Lois Barden Photograph Collection, BC 51.

48 Courtesy of the Lois Barden Photograph Collection, BC 06.

49 The man with the ax also holds a newspaper or magazine. The curved black area is the result of a missing broken piece from the glass plate negative. Courtesy of the Pennsylvania State Archives, RG-6.20, WTC, no. 4017.

50 Courtesy of the Lois Barden Photograph Collection, BC 27.

51 Courtesy of the Lois Barden Photograph Collection, BC 07.

52 Courtesy of the Lois Barden Photograph Collection, BC 104.

53 Courtesy of the Pennsylvania Lumber Museum, A7 76.15.

54 Courtesy of the Lois Barden Photograph Collection, BC 102.

55 The older man with white hair and beard is a likely candidate for camp lobby hog. Courtesy of the Lois Barden Photograph Collection, detail from BC 27.

56 Independence Day was an important community celebration and also informally marked the end of the season's bark peeling. Hartsburg is no longer on the map. Courtesy of the Potter County Historical Society, LP 49.

57 Many consumer items are evident, among them baby carriages, straw hats, pots, dishes, brooms, and mops. Women and children dominate the foreground, while men are in the background. Courtesy of the Lois Barden Photograph Collection, BC 20.

58 The railroad and a road fill the photograph's foreground, with people in the middle and buildings in the background. Men, women, and children tended to be segregated. The corduroy log road in the foreground appears heavily used. Courtesy of the Lois Barden Photograph Collection, BC 49.

59 These people are of varied ages. There is a possibility the building under construction will become a home for an extended family. Courtesy of the Lois Barden Photograph Collection, BC 50.

60 The way these people stand hints that there is a personal relationship between the young man and woman in the center. The building's log construction and the framing of the windows and doorway suggest a commitment to staying in that location for the long haul. Courtesy of the Lois Barden Photograph Collection, BC 55.

61 Also known as Short Run, the town shows its pride with flowers, gardens, fences, houses with trimmed double-hung windows, clapboard siding, shingled roofs, and brick chimneys. The small community was located in southwestern Abbott township, near Cross Fork Creek, some nine miles north of the larger town of Cross Fork. Baileyville contained a butcher shop, general store, pool hall, barbershop, livery stable, schoolhouse, and pig's ear. The post office closed in 1908, when the timber was almost completely clear-cut. Courtesy of the Lois Barden Photograph Collection, BC 40.

62 Courtesy of the Pennsylvania State Archives, RG-6.20, WTC, no. 3971.

63 Courtesy of the Lois Barden Photograph Collection, BC 91.

64 Unlike the the interior inhabited by Whistler's mother, with its stark simplicity, this interior is highly decorated. A woman sits in a comfortable rocking chair with her hands folded and her needlework nearby, surrounded by items that denote her economic, cultural, and social status. Among them are a piano, a marble statue on a stand, a large book with a cloth cover and hinges on a stand, five pieces of fancy glassware, and a large portrait of a man in an ornate frame on an easel. There are four other photographs in ornate frames and one on a cabinet card (these five photos show two adolescent women, two of a younger boy, and one of a child). Fancy cloth draperies (lambrequins) or crocheted lace textiles cover much of the furniture. The floor is carpeted. Floral designs accentuate the wallpaper. A hanging glass lamp extends from the ceiling. This probably is a gaslight, fed by a pipe visible along the baseboard and up the wall at the left side of the image. Fancy lace curtains dress a large window. Courtesy of the Lois Barden Photograph Collection, BC 28.

65 Clarke photographed the woman before a painted backdrop with her head turned slightly as she gazes to her side. She has prepared carefully for her portrait. Her hair is styled elaborately in cascades of curls on either side of a central part. She showcases her attractive face with a stiff, snowy-white collar. A black ribbon at the nape of her neck presumably holds her hair back. Her dress is dark, trimmed at the neck below the collar and vertically across the shoulders with what look like two black velvet stripes that end in points, each dotted with three white buttons. Her dress has puffed sleeves at the shoulders, leading to tightly gathered cuffs, in the high-Victorian style known as "leg o' mutton." This probably was not a simple mail order catalog costume. Courtesy of the Lois Barden Photograph Collection, BC 121.

66 Courtesy of the Lois Barden Photograph Collection, BC 125.

67 The woman to the right appears to be wringing her hands. Like the child, she looks directly at the camera. However, she probably

did not know her picture was being taken at the same time as the child's. Clarke probably cropped her out of the final print. This image shows the same canvas backdrop as that in plate 65. Courtesy of the Lois Barden Photograph Collection, BC 62.

68 Forty children of various ages, as well as several adults, are in attendance. As suggested by the frilly collars and ruffles worn by the boys, gender-neutral ruffled collars, gathered sleeves, and lace-trimmed cuffs were fashionable for nineteenth-century boys (see McShannock, "Boys' Clothes"). A male teacher, in a dark suit and bowtie, stands slightly in profile and holds an open book. Perhaps he was a scholar, or perhaps not—according to Leo Bailey, "in those days many of the teachers lacked the equivalent of a high school education; in fact, after passing the Eighth Grade one could take the County Superintendent's Examination and, if successful, get a license to teach the first eight grades." Courtesy of the Lois Barden Photograph Collection, BC 30.

69 This United Methodist Church building still houses an active congregation. Courtesy of the Pennsylvania State Archives, RG-6.20, WTC, no. 3844.

70 Lackawanna Lumber Company locomotive No. 3 is shown under steam, pulling a huge load of logs. It was built by the Baldwin Locomotive Company in 1885. It's a 2-6-0 "Mogul" type, shown here running astride what appears to be a 1906 Pierce Great Arrow touring car (Kimes and Clark, *Standard Catalog*). The locomotive was formerly owned by Buffalo and Susquehanna as No. 101, acquired in 1893. The Pierce Arrow Company was in Buffalo, New York, where the Goodyear brothers had their headquarters and lived in palatial mansions on Delaware Avenue, known as Millionaire's Row. Charles W. Goodyear had spent, in 1902, five hundred thousand dollars (equivalent to nearly $13 million in 2014 dollars) to build his residence, carriage house, and power plant building at 888 Delaware Avenue. Frank finished his new Delaware Avenue manse at number 762 in 1906; it was modeled on a house on the Champs-Élysées in Paris. F. H. lived there for only a short while before he died in 1907. The car in this image contains nattily dressed gentlemen, suggesting they might be Goodyear executives or directors visiting their Pennsylvania operations. Courtesy of the Pennsylvania State Archives, RG-6.20, WTC, no. 4047.

71 The caption on the negative envelope noted that "the company-built houses are based on a common structure plan." Courtesy of the Pennsylvania State Archives, RG-6.20, WTC, no. 3746.

72 While some forest (glimpsed in the background) remained, the stumps in the foreground give a preview of what was in store for 5 million north-central Pennsylvania acres. Courtesy of the Potter County Historical Society, LP 03.

73 In addition to the Goodyear operation, the Rich Lumber Company and a dozen others in the region were in the same business for three decades and more. This letterhead image was adapted from Clarke's photograph of a Lackawanna Lumber Company train (plate 10). Clarke photographed the Rich operation at Moore's Camp in September 1892, according to Taber, *Sawmills Among the Derricks*, 794. Pennsylvania State Archives / public domain.

74 Tall timber from Austin's barren and grim hillsides had been cut much earlier. The Bayless Pulp and Paper Company had to rely on logs hauled in from considerable distances. Courtesy of the Pennsylvania State Archives, RG-6.20, WTC, no. 4109.

75 A man, possibly William T. Clarke, in one of the last remaining remnants of the original forest in north-central Pennsylvania. This 1910 Clarke image was perhaps made as a remembrance of what had been. Courtesy of the Pennsylvania State Archives, RG-6.20, WTC, no. 3701.

76 Along the Buffalo and Susquehanna Railroad, a few skeletal trunks remain to show how high the former forest canopy reached before clear-cutting. Courtesy of the Pennsylvania State Archives, RG-6.20, WTC, no. 4066.

77 Courtesy of the Pennsylvania State Archives, RG-6.20, WTC, no. 4071.

78 Courtesy of the Pennsylvania State Archives, RG-6.20, WTC, no. 3814.

79 Courtesy of the Pennsylvania State Archives, RG-6.20, WTC, no. 3800.

80 Courstesy of the Potter County Historical Society.

81 Courtesy of the Pennsylvania State Archives, RG-6.20, WTC, no. 3810.

82 Courtesy of the Pennsylvania State Archives, RG-6.20, WTC, no. 3808.

83 Courtesy of the Pennsylvania State Archives, RG-6.20, WTC, no. 3817.

84 Owners of small-scale coal mining operations in the Pennsylvania Desert era found it advantageous to send coal to larger industrial collection points for further processing and disposition. This Clarke image depicts a locomotive poised to move a modest load of locally mined coal in Cameron County. Courtesy of the Pennsylvania State Archives, RG-6.20, WTC, no. 3992.

85 Courtesy of the Lois Barden Photograph Collection, BC 15.

86 Courtesy of the Lois Barden Photograph Collection, BC 98.

87 This family probably built their modest dwelling with enough care that they wouldn't just leave it when the trees were harvested nearby. Once the trees were gone, they'd have had to consider what to do next if they were to stay. With two horses, they might have been tempted to try farming. However, the roots and stumps of trees would need to be removed, rocks cleared from the intended field, and a lot of work and time invested before cash crops could be grown, harvested, and marketed. Meanwhile, probably enough could be grown for family consumption and barter with nearby neighbors. Self-sufficiency has been a common model of existence throughout American history and these people would not have thought such a life was unusual. Their aspirations would have included farm animals for eggs, milk, cheese, and meat. However, additional buildings for animal husbandry and storage of hay, fair weather pastures, and fences would have been required. Courtesy of the Lois Barden Photograph Collection, BC 12.

88 While logging was in full swing, small industrial centers formed to manufacture various wood products. Railroads brought the raw materials to the centers and transported whatever was manufactured to market. As the raw materials became scarce, transportation costs increased. At some point, many small industries were no longer economically viable. Owners considered if their plants could be modified to manufacture something else at a profit. Employees were dependent upon the owner's decisions. If something else was to be manufactured, new investments would have to be made, and employees would need to learn new skills and techniques in order to be retained. If the owners decided to leave and couldn't sell their plant, they would strip the buildings of anything of value before they left. The employees would be out of work. If they could not figure out a different way of making a living, they would try to sell their houses and land or, failing that, gather up their possessions and abandon their homes. In *The Goodyears*, Taber published a Clarke photo of the same plant but from a different perspective. Taber gave a description of the plant, the "Weidmann Stave and Heading Mill at Hull about 1905. Built in 1899, it was sold to the Emporium Lumber Company in 1906 and continued to operate to about 1914. Thirty thousand feet of hardwood logs were cut each day into barrel staves and heads" (528). Courtesy of the Pennsylvania State Archives, RG-6.20, WTC, no. 3787.

89 The Bayless Pulp and Paper Company in Austin, Potter County, founded in 1900, produced fifty tons of paper daily. To encourage the mill's establishment, money was donated by 125 businesses and citizens. Also, 90 percent of taxpayers agreed to fund a dam that would supply the growing needs of Austin and the paper mill, according to Heimel (*Austin Flood*, 25–28). The mill became a mainstay of economic prosperity in north-central Pennsylvania along with the Emporium Lumber Company and the Goodyear operations. The surrounding area had grown into a sizeable settlement. However, the forests being logged for raw materials were increasingly distant, necessitating longer railroad hauls to keep the mill's vast pulpwood piles replenished. While paper could be made from trees other than hemlock or white pine, those other trees were also in decline. Courtesy of the Pennsylvania State Archives, RG-6.20, WTC, no. 4099.

90 The lumber industry was focused. Cut the trees, transport them, use them; repeat. In

reality, cutting, transporting, and using went on simultaneously, not in separate steps. Such a pragmatic approach obviously worked and was very successful for all parties concerned—for a while, that is. Little thought went into what would happen to the country and inhabitants once the trees were gone. It probably seemed logical that the trees would just grow back, although it would take time. As a result, industrialists and workers focused on finding the easiest solutions to day-to-day problems in order to get the immediate job done. Thus, makeshift roads went straight up and down the hills without detours, if possible. Trees were logged and removed as the road was built. There are very few curves in the converging roads pictured here. Once the forests were removed, rain and the soil it carried would run straight down those paths of least resistance. The more water and loose soil that ran down, the deeper the former roads would become. Soon they would better be described as big ditches. Eventually they would become deep gullies. Nothing would hold the water back until vegetation was reestablished. Courtesy of the Lois Barden Photograph Collection, BC 17.

91 The person with the carpentry skills who built the stairs to the second floor of this house would be in demand if the economy of north-central Pennsylvania improved. Courtesy of the Lois Barden Photograph Collection, BC 14.

92 The clean white sheets on the clothesline and the people's apparent determination suggest a human asset that would be important to whatever new future would be faced once the forests were gone. Courtesy of the Lois Barden Photograph Collection, BC 26.

93 The family's demeanor suggests grit and dignity. One man's suit and tie were unusual in the remote wilderness, but people were anxious to put a good face on their situation for future viewers of photographs. The family had achieved much as working-class people. Their house is plain but nicely finished. They had animals that appear well kept. A sturdy outbuilding and fence are evidence of their intention to stay and make a go of their lives. Courtesy of the Lois Barden Photograph Collection, BC 87.

94 With laundry on the line, children, and a well-constructed large building, these adult residents probably had feelings of security and contact with the outside world, given the important rail bed nearby. However, when some railroad companies started leaving or going bankrupt, the residents' confidence no doubt evaporated along with rail traffic. If this was a spur built to get at and transport logs, the rails and ties would also disappear after the tracts had been clear-cut. Courtesy of the Lois Barden Photograph Collection, BC 89.

95 With Abraham Lincoln as a model, children may have felt that anything was possible in their future, including becoming president of the United States. After all, they had log cabin credentials too! But children everywhere in north-central Pennsylvania with similar aspirations at that time would find the path to making their dreams materialize long, arduous, and nearly impossible. Courtesy of the Lois Barden Photograph Collection, BC 99.

96 At the last minute, the cat was caught and included in the photograph. The cat and the dog probably were the only mammals in the photograph without a worry about the future. Courtesy of the Potter County Historical Society, LP 58.

97 Some residents of north-central Pennsylvania had talents and entrepreneurial ambitions that could supplement or replace their dependence upon logging. This family has a violin player who could, if so inclined, earn money playing for special occasions such as community dances, family reunions, holiday celebrations, and weddings. An intriguing sign of a boot hangs from one of the buildings, suggesting footwear repair or manufacture and retail sales to neighbors. Economically, enough money might be generated to sustain a small group such as this. Is that a slight smile on the woman's lips? For the most part, people of Clarke's era were of somber countenance. Searchers for a true smile among the many people who posed for his photographs will be frustrated. Photography came to the fore in 1839 and from then until the second decade of the twentieth century,

dour faces were the norm. Why? Generally, smiles showing teeth from that era are from actresses, performers, and prostitutes. It was believed that an open mouth, showing teeth, was vulgar, disreputable, obscene. Simply put, the belief was that strangers' smiles solicited and sold sex. Smiles intimated "come-ons." It wasn't until motion pictures and early modern advertisements led the way that attitudes changed. About 1920, the adoption of a widening toothy smile became the proper response across the American and European middle class when cheap roll-film cameras began to be used by ordinary people in family and small-group situations (Veronica Horwell, "That Certain Smile," *Le monde diplomatique*, English version, May 2004). Hence, the "Say cheese!" directive from photographers came *after* Clarke's north-central Pennsylvania photography (but see plates 107 and 108). Courtesy of the Lois Barden Photograph Collection, BC 05.

98 For many used to living in the deep forest, a trip to an established community such as Port Allegany, pictured, would have seemed exciting and filled with activities and sights that were exhilarating and perhaps dangerous. Here was organization, with named streets and sidewalks laid out in a grid, a business district, a major rail line, factory jobs, and hundreds of people all engaged in busy activities that didn't exist at home. The trappings of "civilization" portended the future for more and more backwoods residents. Courtesy of the Lois Barden Photograph Collection, BC 45.

99 Roads, even muddy ones, went somewhere. This young man was committed to a modern future, with his genteel clothing and passion for the latest craze, bicycling. Courtesy of the Lois Barden Photograph Collection, BC 79.

100 Large buildings and rail as far as the eye could see were part of Mina, a growing community near Coudersport, the county seat in Potter County. In such a place, there probably was a sense of joint destiny and companionship that would lead to discussion, planning, decision, and cooperation. Courtesy of the Potter County Historical Society, Mina 01.

101 No. 10 was a seventy- to eighty-ton three-truck Shay built by the Lima Locomotive and Machine Works of Lima, Ohio. It was ready for business on January 24, 1902. As the expensive Shay locomotives and mechanical log loaders made up the front and rear of heavy haul after heavy haul of logs, the Goodyear brothers and other lumber barons must have contemplated a vexing question: What next? They knew that the hills and valleys of the once-rich north-central Pennsylvania timberland were nearly depleted. Part of the answer to this question is sensed in the Goodyears' decision to sell No. 10 to the Emporium Lumber Company, Galeton, Pennsylvania, as their No. 10. Emporium operated it from 1908 to 1910. About 1917, the Shay became the Grasse River Railroad's No. 40 at Cranberry Lake, New York. About 1950, it switched numbers once again and became Elk River Coal and Lumber Company's No. 12 at Swandale, West Virginia, where it was scrapped in the late 1950s. Thus, the choices urged by Gifford Pinchot and President Theodore Roosevelt—to stay put, to replant trees, and to adopt selective, managed cutting in the future—apparently weren't appealing. Instead, the timber barons opted to find new purposes for their equipment; to move assets to vast forests elsewhere; or to sell off aging machinery or scrap it as no longer needed, worn out, or obsolete. The Goodyear brothers chose to move their empire to the pine forests of the American South. Most lumber barons eventually sold their north-central land for pennies an acre or let it go to tax forfeit. Courtesy of the Pennsylvania State Archives, RG-6.20, WTC, no. 4152.

102 This photograph shows the broken dam above Austin, the immediate cause of the 1911 flood (for an important prior cause, see plate 90). At that time, former itinerant photographers were making economic adjustments, branching out into different aspects of business entrepreneurship. William T. Clarke opened a studio; took institutional photographs of schools, churches, and main streets; documented communities and industries; made "reality" or "real photo" postcards; and purchased additional equipment and supplies in order to make stereographs. When the opportunity arose, he also participated in

"disaster tourism," a form of photojournalism (Godbey, "Disaster Tourism"). All those expanded activities led to new markets. People wanted images of their community; their businesses; their institutions; and important or unusual events that were entertaining, newsworthy, or momentous. The Austin flood of September 30, 1911, fit this last category. Clarke had an advantage over his competitors, because he also had "before" pictures to sell along with "after" images (see plate 89). The disaster prompted the State of Pennsylvania to enact legislation regulating dams in 1913. Courtesy of the Pennsylvania State Archives, RG-6.20, WTC, no. 4131.

103 This image also has been titled *Austin Flood View 1911*. Nearly everything in the path of the powerful flood was flattened. The importance of wood in building a community is apparent in this photograph. A frantic warning by telephone from a woman living upstream led to the activation of an emergency alarm siren that gave many people advance warning to evacuate and no doubt saved many lives. Courtesy of the Pennsylvania State Archives, RG-6.20, WTC, no. 4148.

104 Fortunately for some residents, the ancient valley floodplains provided rich soil and relatively level land for farming. Under those conditions, agriculture could provide a very good living. However, in the early twentieth century, the copious runoff from rains on the denuded hills was a problem, so a life of farming then wasn't carefree—as it is not now. Farmers gambled then, as now, and hoped the odds against weather disaster and other threats would be in their favor. Logue is in the East Fork Valley of the Sinnemahoning, just downstream from Conrad. Logue contains a small collection of houses today. Courtesy of the Pennsylvania State Archives, RG-6.20, WTC, no. 4033.

105 The geology of north-central Pennsylvania often is rugged, but it holds many pleasures and surprises for hikers and explorers, as these men discovered in a William T. Clarke stereograph made in McKean County. Courtesy of the Pennsylvania State Archives, RG-6.20, WTC, no. 3773.

106 These women and children and a dog find a few hours for pleasure on a boulder-strewn hillside in northwestern Clinton County. Courtesy of the Pennsylvania State Archives, RG-6.20, WTC, no. 3881.

107 The same Hammersley region group shown in plate 108 enjoy moments of carefree abandon in tranquil waters. Courtesy of the Pennsylvania State Archives, RG-6.20, WTC, no. 3886.

108 This photograph is quite different from many Clarke images in its candid quality, even though it almost certainly was thought out and posed. The people held still during the camera exposure (except for some of the children, who appear blurred). A daring young woman exposes her thighs, and a partially hidden child (left) sticks out her tongue. The smiles of this group and in plate 107 appear genuine and spontaneous. Courtesy of the Pennsylvania State Archives, RG-6.20, WTC, no. 3885.

109 This image is another of Clarke's deliberately posed, and well-composed, photographs. Courtesy of the Potter County Historical Society, William T. Clarke Pictures, WTCP a.

NOTES

PROLOGUE

1. Hesch, Pieper, and Littell, *Ithaca Then and Now*.
2. Koch, *Shay Locomotive*, 333, 337; Taber, *Whining Saws*, 642–54, 695.
3. Ostman and Littell, *Cornell Then and Now*.
4. In the 1930 U.S. census, the birthplace of Clarke's parents is given as the Irish Free State. Irish immigrants were plentiful in Rochester. Irish laborers were "well respected as the backbone" of digging and constructing the Erie Canal, extending some 363 miles from Albany to Buffalo, built from 1817 to 1825; see Wyld, *Low Bridge!*, 55. Irish immigrants continued to arrive in Rochester ("the Flour City") during the era of industrial transition (1855–80) as traditional wheat farming and milling moved westward. The Irish provided labor for new industries, working in carriage factories, in breweries, on railroad lines, and in plant nurseries. Irish politicians and voters also became a significant voice in politics and government. See McKelvey, *Panoramic History*, 65–69.
5. The children were Robert J., age 15; Mary, 13; William T., 10; Bessie A., 4; and Isabelle E., 1.
6. Monroe County Library System, "Rochester City Directories." A sales publication for 1882 noted that Union View and Copying was promoting stereoviews of "10,000 negatives, comprising city and country views in all parts of the world" ("Descriptive Price List").
7. Henry Wharton Shoemaker, "Women's Part in Pennsylvania Lumbering Days," undated, unpublished manuscript, Ben Yarosz Collection, Carmen D. Valentino, Rare Books and Manuscripts, Philadelphia. The Yarosz Collection contains original Shoemaker material.
8. Clarke's father died of "congestion of the brain" at age fifty-four on April 28, 1879, and his mother died of nephritis at eighty on January 4, 1916. Isabelle died, also of nephritis, February 24, 1928. William T. Clarke died at the Madison Square address and was interred July 26, 1930, in Rochester's Mount Hope Cemetery. There may be some truth to Shoemaker's statement that Clarke moved to Pennsylvania for health reasons. He may well have had a delicate constitution, for his burial record states he died of chronic heart valve disease at the age of seventy years, ten months, and sixteen days. Monroe County Library System, "Rochester City Directories"; University of Rochester Library, Mount Hope Cemetery records (select 22, 83, 102 Clark).
9. Ronald E. Ostman and Harry Littell, "Forgotten Forest," *Smithsonian*, February 2006, 50–51, 54–59.
10. Munson, "Gelatin Chloride Printing-Out Paper." Positive prints, as well as negatives, have been affected by the passage of years. Fading, color changes, and damage caused by handling are not uncommon. We believe the images in this book are a close approximation of the original POP prints.
11. Kendal at Ithaca, New York, exhibit, *Forgotten Forest: The Photography of William T. Clarke*, October 2006–February 2007; History Center in Tompkins County, Ithaca, New York, exhibit, *One Thousand Words: A Lost Era Captured on Glass*, May 2005–April 2006; Colgate University, Clifford Gallery of the Art and Art History Department, Hamilton, New York, exhibit, *Land of the Free: Contemporary Prints from Nineteenth Century Glass Plate Negatives*, November 3–December 10, 2004.
12. Ries, Ostman, and Littell, "Wood on Glass"; see, for example, Donald Gilliland, "Exhibit

Showcases Long-Forgotten Logging Images: The State Museum of Pennsylvania," *Sunday Patriot-News* (Mechanicsburg, Pa.), November 21, 2010, D1, D7; "Lumber Heritage Region Presents 'Wood on Glass' (Cameron County Historical Society Annual Meeting)," *Cameron County Echo* (Emporium, Pa.), May 2, 2012.

13. Laura Reindl, "Window on Another Time: A Lost Treasure Trove of Glass Photo Negatives Brings a Bygone World to Life," *Mountain Home: Pennsylvania and the New York Finger Lakes*, September 2010, 1, 3, 10–13, 38, 40, 42.
14. Smith, "On the Importance of a Man," 158–66.
15. Wheeler, "Interview with Henry Ford," 362.
16. See http://www.thehenryford.org/village/historicdistricts.aspx; http://www.thehenryford.org/museum/exhibits.aspx (accessed June 5, 2015). Bruner, "Why Study Business and Economic History?" http://www.goodreads.com/quote/5382-history-doesnt-repeat-itself-but-it-does-rhyme/; http:/www.quoteinvestigator.com/2014/01/12/history-rhymes/ (accessed May 26, 2015).

INTRODUCTION

1. Biographical information regarding William T. Clarke and his family is found in the prologue. The author would like to thank the following for their assistance: Nicolas Graver, Karl S. Kabelac, Simon Bronner, Kay Brownell Reed, Dolores Buchsen, Kurt Bell, and Michael Wennin.
2. See, for example, Bajac, *Invention of Photography*.
3. *Potter County Journal*, February 12, 1891, p. 3, col. 1. This is definitely W. T. Clarke, not the later photographer George Clark of Austin, who would have been ten years old at the time.
4. Two of J. W. C. Floyd's images exist as copy glass negatives and are part of MG-482, the Albert Zimmerman Collection at the Pennsylvania State Archives, Harrisburg, Pa. The Zimmerman Collection contains 116 plates by Zimmerman and Hess. About five hundred of Francis Edwin Bradley's eight-by-ten-inch glass plates are in the Bradley-Hahn Collection at the Railroad Museum of Pennsylvania, Strasburg, and are featured in Kaseman, *Story of the Susquehanna*. For Nelson Caulkins, see Stephenson, *History of Little Pine Valley*. For George Clark, see Pippin, *Wood Hick Pigs-Ear*, 142. A number of C. H. Krumm's glass negatives are now preserved at the Bradford County Historical Society.
5. Previously printers had to first produce a copy of an image through lithography or wood engravings to reproduce for mass media. The halftone broke a photographic image into a series of black dots on a metal plate, widely separated for light tones, densely packed for dark tones. The plate then was set with other type, inked, and put through a press.
6. Heck, *Black Forest of Pennsylvania*.
7. Rosenberger, *Adventures and Philosophy*, 278, 281. The definitive study on Shoemaker is Bronner, *Popularizing Pennsylvania*.
8. These books include Shoemaker, *Indian Steps*; *Black Forest Souvenirs*; *Eldorado Found*; and *Allegheny Episodes*.
9. The letter is part of the Yarosz Collection, now in private hands. Shoemaker credits Clarke in the front matter of *Black Forest Souvenirs*: "Illustrations from photographs by W. T. Clarke, Betula, Pa." Clarke's letter is reproduced with his usage, spelling, punctuation, and grammar intact.
10. Henry W. Shoemaker papers, Special Collections Library, The Pennsylvania State University Libraries.
11. Scott, *Old Mortality*.
12. "Women's Part in Pennsylvania Lumbering Days," undated, unpublished manuscript by Shoemaker, Yarosz Collection. Shoemaker also has Clarke traveling the world in a sailboat on a photographic expedition after the Great War before settling down in Rochester in the 1920s.
13. Yarosz Collection.
14. Shoemaker, "An Antique Dealer's Romance," 65–81. Joseph S. Illick (1884–1967) wrote an early and definitive work, *Pennsylvania Trees*. He was also a photographer and the chief of research at the time of the Clarke negatives' retrieval. He became state forester in 1927.
15. "Search for Photographer Successful," *Potter Enterprise* (Coudersport, Pa.), February 8, 1923, and an undated, unlabeled news clipping titled "A Valuable Find" at the Potter County Historical Society, Coudersport, Pa.

16. Department of Forests and Waters, Record Group 6, Series 6.20 (WTC 3700–WTC 4163), Pennsylvania State Archives. The Potter County Historical Society, however, suffered a disastrous flood in 1942 and "many valuable records and relics were lost." It is possible some Clarke materials were lost at that time. Mary Welfling, supplement to Beebe, *History of Potter County*, 308.

17. The Department of Forests and Waters was established in 1923, replacing the Department of Forestry. It was abolished in 1970 and became part of the Department of Environmental Resources. Today it is part of the Department of Conservation and Natural Resources.

18. Pinchot, "The Blazed Trail of Forest Depletion." The Clarke images used by Pinchot were "Norwich Pond and Sawmill" (WTC 3713), created November 7, 1917, and the image of the giant hemlock (WTC 3700) used in the 1916 pamphlet by Albert Heck, available online in the Biodiversity Heritage Library.

19. Beebe, *History of Potter County*. In this work they are credited to the *Potter Enterprise*, indicating that the newspaper also had a set of images. See also Potter County Historical Society, *Historical Sketches of Potter County*.

20. Taber, *Goodyears*, has the largest number of Clarke images in this series. Taber stated, "Of the many photographers who photographed the logging railroad era of lumbering in Pennsylvania, only Clarke's glass plates are known to have been preserved" (543). While that may have been true in 1971, we do know that negatives from Albert Zimmerman, Francis Edwin Bradley, and C. H. Krumm survive today (see n. 4). Clarke images also were published in Taber, *Whining Saws*.

CHAPTER 1

1. A map of America and areas of virgin forest in 1620 shows Pennsylvania almost completely covered by trees. Greeley, "Relation of Geography," 174. On Pennsylvania's attractiveness to persecuted religious minorities of many European countries, see www.explorepahistory.com (accessed June 30, 2015).

2. Where agricultural crops were grown by aboriginal populations, the forests sometimes were burned or trees were girdled to die and then removed to clear land. Some aboriginal populations also used fire to create favorable habitat for certain game animals, thus improving hunting (Fletcher, *Pennsylvania Agriculture*, 1). However, lightning also can create natural fires. Today, there is widespread support for the premise that natural fires have beneficial effects if human life or significant property losses are not threatened (Pyne, Andrews, and Laven, *Introduction to Wildland Fire*; "Bipartisan Wildfire Disaster Funding").

3. The lumber industry traditionally has referred to processed trees in units of "board feet." A board foot is one foot square and an inch thick.

4. Cochran, *Pennsylvania*, 5–6.

5. Fletcher, *Pennsylvania Agriculture*, 377.

6. There are no known historical references to support Shoemaker's claim. His use of the "Black Forest" moniker stuck in people's memories, however. By 1923, for example, it was being used in professional language. See Bishop, "Pines of Hearts Content."

7. Michael Williams provides a study of the history of the American forest and the differing ways forest history has been framed. He discusses pre-1600 geography, economics, culture, symbols, and the Indian, but the main thrust of his book concerns forest history from 1600 to the later twentieth century. Williams notes increasing forest changes from 1600 to 1860, resulting from pioneer life, agricultural clearing, commercial uses of the forest, and the impact of industry. Those changes had striking regional and national effects after 1860. His work pays particular attention to the lake states (1860–90), the southern forest (1880–1920), and the Pacific Northwest (1880–1940). He considers political and social questions, inquiry, and debate focused on the forest from 1870 to the present, revolving around preservation, management, ownership, supply, protection, use, and gains and losses, as well as forest rebirth from the early 1930s and after. Williams, *Americans and Their Forests*.

8. Riesenman, *History of Northwestern Pennsylvania*, 343.

9. Korson, *Pennsylvania Songs and Legends*.

10. Horst and Smith, *Logging in the Pennsylvania North Woods*.
11. La Gorce, "Penn's Land."
12. E. W. Miller, *Pennsylvania Keystone*; Eastern hemlock sometimes grew to nine feet in circumference and could yield two hundred thousand board feet an acre. See http://en.wikipedia.org/wiki/Pine_Creek_Gorge (accessed June 11, 2015).
13. Williams, *Americans and Their Forests*, 186.
14. Hotchkiss, *History of the Lumber and Forest Industry*, 378; Wilhelm, "History of the Lumber Industry," 9–10.
15. Albertson, *In Memory's Lane*, 34. This work is in the possession of Anne E. Wenger, great-granddaughter of Alonzo Baldwin Albertson; quoted with permission.
16. Whitney, *From Coastal Wilderness*, 144, citing Michaux, *North American Silva*, 240–43.
17. Whitney, *From Coastal Wilderness*, 186. See also Pringle, "Report upon the Principal Lumber Producing Regions."
18. Robert K. Currin, personal interview with the authors, Potter County Historical Society, Coudersport, Pa., May 22, 2006. See also Currin, *Our Forestlands*; http://en.wikipedia.org/wiki/Nail(fastener)#Forged_nails (accessed June 11, 2015).

CHAPTER 2

1. The preferred method was to cut the tree with the blade at a slant, but someone had to back up as the saw moved through the log.
2. Peaveys were known as "cant dogs" in New England's logging parlance; for greenhorns, a peavey was "a stick with a spike in its end and a hook hanging on it" (Holbrook, *Yankee Loggers*, 11, 120).
3. See, for example, Ambrose, *Nothing Like It*. Taber referred to the third phase as the "logging railroad era." He summarized this era: "No longer was the lumberman dependent upon the river valley and the spring thaws for the movement of his timber, but rather, on steam power and railroads that penetrated everywhere and operated without regard to changes of seasons. The log railroad supplied the wood ingredients for more than 600 saw mills, tanneries, wood chemical companies, and stave plants beginning with the first known railroad operation in 1864 in Jefferson County and ending in 1948 in Elk County. The variety and size of these railroads was extensive; no other state ever approached the diversity in Pennsylvania" (Taber, *Ghost Lumber Towns*, 296).
4. The "garden" in this case recalled the biblical Garden of Eden, paradise before humans came along to spoil things. Marx, *Machine in the Garden*. "And the Lord God planted a garden eastward in Eden; and there he put the man whom he had formed. And out of the ground made the Lord God to grow every tree that is pleasant to the sight, and good for food; the tree of life also in the midst of the garden, and the tree of knowledge of good and evil" (Genesis 2:8–9, King James Version).
5. The Delaware and Hudson Canal Company bought the English-built Stourbridge Lion to replace horses that had been hauling anthracite coal over a crude sixteen-mile rail line from the Carbondale mines to a canal in Honesdale, northeastern Pennsylvania. However, the locomotive was too heavy for the tracks and was retired after a few runs to become a stationary engine. See Hollingsworth and Whitehouse, *North American Railways*, 11–12; J. H. White, *Early American Locomotives*, plate 5; and www.portal.state.pa.us/portal.pt/community/things/4280/railroads/528625 (accessed June 9, 2015).
6. Garratt, *World Encyclopedia of Locomotives*, 23.
7. This prediction was realized with the linking of the Union Pacific and the Central Pacific Railroads at Promontory Point, Utah Territory, on May 10, 1869: the United States was connected by rail from the Atlantic to the Pacific Oceans, a task begun in 1863 (Ambrose, *Nothing Like It*; Mayer and Vose, *Makin' Tracks*).
8. Olney, *Practical System*, 70–71.
9. Some historians refer to this period as the age of affluence. See Rinhart and Rinhart, *America's Affluent Age*.
10. See Pennsylvania Historical and Museum Commission, www.portal.state.pa.us/portal/server.pt/community/railroads_of_pennsylvania/20175/developmental_history/943342 (accessed June 9, 2015).

11. Ambrose, *Nothing Like It*, 21.

12. The painting disappeared for thirty years. Miraculously, the work was found in an old Mexico City curiosity shop. Seeing it again after such a long absence, Inness commented, "There is considerable power of painting in it, and the distance is excellent." This was high self-praise from an artist who continually reworked his paintings, seeking perfection. In fact, he also tried to "touch up" other artists' canvases. At least one artist forbade Inness to visit his studio. *The Lackawanna Valley* is in the National Art Gallery, Washington, D.C. See Freedgood, *Great Artists*, 51–52; Yaeger, 28, 71–75; www.metmuseum.org/toah/hd/inne/hd_inne (accessed June 9, 2015).

13. www.westegg.com/inflation/infl.cgi (accessed June 11, 2015).

14. Later in his career, George Inness evinced misgivings about machines in the garden. Industrial mechanization disappeared from his nature landscapes. He came to believe that humans should experience nature spiritually, because God infused the natural world with continuous energy, permitting those in touch with nature to feel heightened emotional mystical sensibilities.

15. See www.westegg.com/inflation/infl.cgi (accessed June 11, 2015). For Pennsylvania's dubious role in financing the transcontinental railroad through the "Pennsylvania Fiscal Agency" (to become the infamous Crédit Mobilier of America, generator of huge insider stockholder profits), see Ambrose, *Nothing Like It*, 92–93.

16. J. H. White, *Early American Locomotives*, vi–viii.

17. See www.explorepahistory.com (accessed June 30, 2015).

18. Milton W. Brown noted that the vast reservoir of American optimism during the early years of the twentieth century was a continuation of nineteenth-century faith in science and technology. People generally believed that machines opened the door to material abundance and progress, leading toward utopia. After World War I, however, a "period of disillusionment" set in. Ordinary people began to question and rebel against the mechanization that had produced so much so fast. Industrial monopolization grew to the extent that only a very few persons controlled vast industries and fortunes, often with little or no regard for, and even callousness toward, ordinary people (Brown, *American Painting*, 5–6, 183–84).

19. Sasowsky, *Prints of Reginald Marsh*, 87, 92, 95–97, 113, 135–41, 168–69, 173, 192, 228, 263.

20. Ibid., 39.

21. Such as the Addison and Pennsylvania and the Sinnemahoning Valley Railroads.

22. Stevens, "When Timber Was King," 391–95.

23. Cox, *Lumberman's Frontier*, 120.

24. The Williamsport boom used many large stone-filled cribs thirty or more feet high across the river to anchor chains of huge floating logs ("boom sticks"). Thus, sections of boom sticks stretched from crib to crib across the middle part of the Susquehanna River. A sheer boom upstream diverted floating logs into the main boom. Once the log drive was completed, the sheer boom was removed to permit river traffic to resume (ibid., 103–4).

25. In 1866, thirty Williamsport sawmills produced 995 thousand board feet per twelve-hour day, or more than 310 million board feet a year. By 1872, forty sawmills were located in Williamsport, employing four thousand workers (ibid., 101; Taber, *Williamsport*).

26. Despite declines in the practice of floating logs over long distances to central sawmills as railroads became a fundamental element of the lumber industry, the Williamsport boom remained in service until just before World War I. See Tonkin, *Last Raft*.

27. A monumental history of this saga is told in thirteen volumes by Thomas T. Taber III, William C. Casler, and Benjamin F. G. Kline Jr. in a number of books originally published individually in Williamsport by the Lycoming Printing Company or were privately printed—with one exception, Taber and Casler, *Climax*, which was published by Railroadians of America. See also Pietrak, *B&S*.

28. Taber, *Goodyears*, 524; Defebaugh, *History of the Lumber Industry*.

29. By 1896, the Pine Creek Gorge rail bed carried 7 million tons of freight over this route, much of it coal north to New York and hemlock bark to area tanneries for leather

production. In addition, three daily passenger trains went between Wellsboro Junction and Williamsport. Culip, "Pine Creek Gorge."

30. Owlett, *Seasons Along the Tiadaghton*, 53–62.
31. The Pine Creek Gorge is deepest at 1,450 feet at Waterville, near the south end. It is 800 feet deep at Leonard Harrison and Colton Point State Parks, where this image was made before the area was logged and before the parks came into existence. The average width rim to rim is 4,000 feet. Today, the parks are surrounded by more than 160,000 acres of Tioga State Forest.See http://en.wikipedia.org/wiki/Pine_Creek_Gorge; http://visittiogapa.com/canyon.html (accessed June 11, 2015).
32. Taber, *Sunset*, 450.
33. For full explication of details that follow, see Taber, *Short-Line Steam*; *Goodyears*; Taber and Casler, *Climax*; and Pietrak, *B&S*.
34. The manufacturing cost of $3,250 in 1890 is equivalent to about $84,335 in 2014 U.S. dollars. That amount in 1900 equals almost $91,036 in 2014 dollars. See http://www.westegg/inflation/infl.cgi (accessed June 11, 2015). A mechanical log loader's boiler firebox consumed an average of three hundred pounds of coal daily. The Goodyears also owned coal mines that provided fuel to fire B&S locomotives.
35. Oak was a less desirable source of tanbark. Whitney, *From Coastal Wilderness*, 186.
36. Fulling, "Botanical Aspects," cited in Whitney, *From Coastal Wilderness*, 186, 372.
37. Onthank, *Tanning Industry*; Riley, "History of Tanning," 51, 78. Cited in Whitney, *From Coastal Wilderness*, 186. The trend toward consolidation and virtual monopoly in the tanning industry was evident in the great size and scope of operations of the U.S. Leather Company and its successor, the Central Leather Company, for the decades just before and after the turn of the century.
38. Marquis, *Allegheny Hardwood Forests*.
39. See "Buffalo and Susquehanna Railroad Corporation Records, 1882–1930," Manuscript Group 457, Pennsylvania State Archives, Harrisburg, www.phmc.state.pa.us/bah/dam/mg/mg457.htm (accessed June 29, 2015). In addition to his other presidencies, Frank H. Goodyear was president of the Buffalo and Susquehanna Railroad and the Goodyear Lumber Company and was a director of the U.S. Leather Company, which the Goodyears supplied with bark from harvested hemlock trees to use in making leather. He was a director of the Marine National Bank and the Rogers Brown Iron Company. Later, he was president of the New Orleans Great Northern Railroad and the Great Southern Lumber Company in the new Goodyear company town of Bogalusa, Louisiana. His older brother, Charles, a lawyer and former district attorney in New York State, was prominent in many of the above business concerns. He was a close personal friend of Grover Cleveland and was instrumental in securing Cleveland's nomination for governor of New York State, a springboard that led him to the U.S. presidency. Charles also was an organizer of the Pan-American Exposition held in Buffalo in 1901. It is said that he and brother Frank were not always on the best of terms. F. H. installed several sons-in-law in high positions in the Goodyear businesses. One left as soon as his wife died in 1904, because he and F. H. also did not get along. Frank was known to be impatient and did not tolerate indecision. F. H. was about five feet eight inches tall, an overeater who weighed 220 lbs., a lifelong workaholic, chronically unhappy, a perfectionist obsessed with achievement and success, and an avoider of physical exercise. He died of Bright's disease (primarily a malfunction of the kidneys with many additional complications) at age fifty-eight (see Dunn, *Buffalo's Delaware Avenue*). His estate was valued at $10 million (more than a quarter of a billion dollars in 2014 dollars). Clearly, his fortune was huge, particularly since he'd spent so much, just prior to his death, on the Delaware Avenue mansion, on the Great Southern Lumber Company, on many buildings in the new company town (such as a commissary, a depot, and employee housing), on Louisiana and Mississippi land acquisitions of three hundred thousand acres in the Pearl River area, and on his two railroads (for example, the New Orleans Great Northern Railroad required a new ninety-mile extension from Bogalusa to link with main line

networks). Also see Matthews, *Men of New York* and Goodyear, *Bogalusa Story*.

40. However, the expansion of the B&S to interstate commerce proved an Achilles' heel and represented one of the few Goodyear businesses to flounder rather than flourish. Expansion and construction of the line some 90 miles from Wellsville to Buffalo proved very expensive and duplicated competing existing services. In financial trouble, the B&S Railroad was rechristened and leased as the B&S Railway, with F. H. Goodyear assuming a lesser leadership role. However, financial default, litigation, and receivership followed. Later, under Baltimore and Ohio Railroad ownership, the B&S (by then known as the Wellsville and Buffalo) finally came to an end, the last straw being massive flooding south of Galeton, Pa., in 1942.
41. Cranmer, "Harvesting the Hemlock," 124, cited in Cox, *Lumberman's Frontier*, 110, 408, 411.
42. The first mobile gasoline-powered chainsaw ("the tree-felling machine") operated by an individual was invented by a German, Andreas Stihl, in 1929. See http://inventors.about.com/od/cstartinventions/a/Chainsaws.htm (accessed June 11, 2015).

CHAPTER 3

1. Bogdan, *Adirondack Vernacular*; Carter, *Solomon D. Butcher*; Lucey, *Photographing Montana*; McKeen, "Erwin Evans Smith," 156–62.
2. Bailey, *Lumber Activity*.
3. Double-bitted axes generally were kept sharp on only one end of the blade. The dull end was used to chop roots or when the ax blow was in danger of striking dirt or rocks, which would ruin a sharp edge.
4. A "schoolmarm," a risqué way to refer to a crotched tree with two trunks branching off from the base trunk (thus, a double-topped tree), was particularly unpredictable when cut and falling.
5. "Logging was the nation's most dangerous occupation, ahead even of coal mining" (Cox, *Lumberman's Frontier*, 321).
6. Peeled trees on the ground were bucked into standard moveable lengths by their being sawed from top to bottom, with the wedge being pounded from the top of the cut, following the path of the saw. However, trees didn't always fall on level or solid ground and cutting lengths under those circumstances approached an art. The various jobs—chopping, sawing, trimming, bark peeling, and bucking—sometimes were performed by the same individuals in that sequence.
7. A gravity slide was constructed from trees split in half lengthwise and assembled in a continuous V configuration (plate 30). The troughlike gravity slides carried logs down steep slopes to skidways (also known as "landings"). Slides sometimes were oiled or iced to facilitate log movement, depending upon the weather.
8. Bailey, *Lumber Activity*.
9. Pennsylvania Department of Agriculture, *Report No. 6, 1897*, 255–67.
10. The sum of $1.50 was equivalent to slightly more than $42 in 2014 dollars. See www.westegg.com/inflation/infl.cgi (accessed June 16, 2015).
11. These were also known as "calked" boots. In the 1880s, boot calks cost a penny each. Typically, there were forty or more calks to a boot. Holbrook, *Yankee Loggers*, 50. The calks for a pair of boots, not installed, in 1885 would have cost the 2014 equivalent of almost twenty-two dollars. Clearly, they were an important requirement for the job. See www.westegg.com/inflation/infl.cgi (accessed October 10, 2015).
12. Hanlon, *Ball-Hooter*, 111–12; see also Huntley, *Story of the Sinnamahone*.
13. "Life Expectancy by Age." See also Shrestha, *Life Expectancy in the United States*.
14. E. W. Miller, *Pennsylvania Keystone*. The woods workers had a fatalistic attitude about the dangers. They thought risks were unavoidable but could be thwarted if they were careful. They reasoned that they'd accepted the risks when they took the job. Also, there was a streak of machismo ("I'm one tough hombre") associated with logging, mixed with a dash of optimistic bias ("others may get hurt, but not me"). See Cox, *Lumberman's Frontier*, 321–22.
15. Jobbers were more likely to hire employees who could fulfill more than one of these job descriptions, depending upon need and

season. The roles of desk workers, storekeepers, blacksmiths, saw sharpeners (also called saw "filers"), cooks and cookees, lobby hogs, and scalers will be discussed in chapter 4, on camp life.

16. Candid action photography was not an option, given the bulky equipment, slow shutter speeds, and slow film speeds of the era.

CHAPTER 4

1. Cochran, *Pennsylvania*, 162–70.
2. In 2014 dollars, $3.50 a day would have been about $85 a day in 1880, $91 in 1890, and $98 in 1900. See http://www.westegg.com/inflation/infl.cgi (accessed June 16, 2015).
3. Cooks and cookees almost always were men. Occasionally, women did live in lumber camps, but they usually were wives, daughters, or relatives of the camp jobber or manager, in which case they might serve as cooks, cookees, store managers, clerks, bookkeepers, or a combination. Otherwise, very few women visited lumbering camps, which were considered coarse. On rare occasions, when women were in camp, the men were on their best polite, proper, and respectable behavior. Most lumber camps were men's preserves rarely visited by females. Huntley, *Story of the Sinnamahone*.
4. Holbrook, *Yankee Loggers*; Huntley, *Story of the Sinnamahone*; Korson, *Pennsylvania Songs and Legends*; Nuschke, "Hicks, Fighters, and Clog Dancers."
5. The lobby was a lounge area where men could take their ease. The term "lobby hog" could be construed as implying that this man was a slacker.
6. As a security measure, a knowledgeable jobber or woods worker might well have kept his own informal, unofficial estimate of the camp's yield that could be used to challenge a scaler's estimate that seemed low. An alternative was if the contract called for clear-cutting a given parcel of land, the need for scaling as the men cut trees would have been unnecessary. All that would be required would be a final inspection of the parcel prior to payment to determine that it was, in fact, denuded and that the final quantity of wood available to the owner was reasonably close to the scaler's original estimate when the timber was standing.
7. One dollar in 1900 is equivalent to about $28 in 2014 dollars, $2.50 is $70, and $3.50 is about $98. See http://www.westegg.com/inflation (accessed June 16, 2015).
8. Reed, *Lumberjack Sky Pilot*. Reed was a sky pilot who preached to loggers throughout the Adirondacks area in New York State beginning in 1917. He remained engaged with the lumber industry and its personnel throughout the Northeast and upper Midwest until 1966. His counterpart in north-central Pennsylvania was William E. Brewer.
9. For a list of specific maladies and herbs, roots, and leaves used for home medical treatments, see Nuschke, *Pennsylvania's Freeman Run Valley*, vol. 1, 56–58; see also Korson, *Pennsylvania Songs and Legends*.
10. Bailey, *Lumber Activity*, 23.
11. *Grit* was an inexpensive weekly newspaper that began in Williamsport in 1892 and quickly expanded to national prominence with a primarily rural circulation of twenty thousand in 1887. Circulation soared to three hundred thousand in 1916 and seven hundred thousand in the 1950s. It featured positive, heartwarming human interest items and news, comic strips, puzzles, and serial fiction. "America's Greatest Family Newspaper" was still available after September 2006, when it became a bimonthly in magazine format with print runs of 150,000, sold at selected general newsstands, bookstores, and farm feed/supply stores. See https://en.wikipedia.org/wiki/Grit_(newspaper) (accessed June 17, 2015).
12. Bailey, *Lumber Activity*, 26–27.
13. Good camp books might include such titles as James Fenimore Cooper's *The Pathfinders* (1840), Harriet Beecher Stowe's *Uncle Tom's Cabin* (1852); Horatio Alger Jr.'s *Ben's Nugget; Or, A Boy's Search for Fortune* (1882) or *The Erie Train Boy* (1890); Jack London's *Call of the Wild* (1903); Mark Twain's *A Tramp Abroad* (1880); Theodore Roosevelt's *The Wilderness Hunter* (1893); Ernest Thompson Seton's *Biography of a Grizzly* (1903); and Robert Louis Stevenson's *Treasure Island* (1883).
14. Kuhns, "Loleta in 1910," 452–60.
15. Lynch, "Jobbing on Skinners Creek," 799.
16. Holbrook, *Yankee Loggers*, 55.

17. Nuschke, "Hicks, Fighters, and Clog Dancers" and *Pennsylvania's Freeman Run Valley*, vol. 2, 77–89.
18. Longwell, *Steichen*; Newhall, "Pictorial Photography," 140–65; Newhall and Newhall, "Alfred Stieglitz," 60–75; "Edward Steichen," 76–91.

CHAPTER 5

1. Montgomery Ward and Sears, Roebuck led the way in mail order retail consumer shopping in 1872 and 1886, respectively. See http://en.wikipedia.org/wiki/Montgomery_Ward.htm; www.searsarchives.com/history 1886.htm (accessed June 24, 2015).
2. Fox, *Victorian Woman*, 4; Schroeder, *Sears, Roebuck*, 417.
3. A letter by Walter E. Spencer regarding the Hamlet of Baileyville, mailed to Maybel Swanson, secretary, Potter County Historical Society, Coudersport, from Grapeview, Washington, and postmarked December 23, 1976, gave details and reminiscences that were helpful in developing this chapter. Baileyville is no longer on the map.
4. Nuschke, "Industry Comes to Freeman Run," 85–88.
5. *Godey's Lady's Book*, whose publishing house was located in Philadelphia, was the most popular women's periodical from 1830 to just prior to the U.S. Civil War in the early 1860s. It stressed fashion with fashion, plates of women's dress styles that were hand-colored by more than one hundred women workers. Almost every issue contained a pattern for a garment that could be sewn at home, as well as piano sheet music for the latest popular music (waltzes, polkas, galops). *Godey's* also contained original American stories and poetry by well-known authors and occasional food descriptions and recipes, and its articles were sympathetic toward women's education and greater workforce participation. Large national circulations (up to 150,000 copies an issue in 1860) were possible because of increased visual content via engravings, subscription selling (at three dollars a year, compared with two dollars for the *Saturday Evening Post*), manufacturers' advertising quest for mass markets, railroad shipping, and the establishment of post offices in most villages, burgs, and crossroad centers. *Godey's* ceased publication in 1898; https://en.wikipedia.org/wiki/Godey's_Lady's_Book (accessed June 23, 2015).
6. "Conrad," *Potter County Journal*, June 27, 1900, 8.
7. The well-coiffed woman in plate 64 wears a fancy floor-length dress and assumes a demeanor like that of Whistler's mother in *Arrangement in Grey and Black: Portrait of the Painter's Mother*; see Dorment and MacDonald, *James McNeill Whistler*, 140–43.
8. Palmer's method became popular after the publication of *Palmer's Guide to Business Writing*.
9. Nuschke, *Pennsylvania's Freeman Run Valley*, vol. 2, 76.
10. Tichenor, Donohue, and Olien, *Community Conflict*.
11. Thomas T. Taber III gives a thorough account of the life and death of Pennsylvania towns to the east of Potter County, several of which are summarized here. A pattern of boom-to-bubble-to-bust was characteristic, just as it was for Cross Fork. The town of Ricketts straddled Wyoming and Sullivan Counties. It existed from 1891, when a sawmill was established, followed by a planing mill, allowing both hemlock and hardwoods to be harvested and processed. Later, in 1898, a stave mill was established. By 1914, the last log was cut and the company town was abandoned to return to the wild, leaving only crumbling concrete foundations of the stave mill and sawmill a few decades later, along with a few foundation stones where houses had existed and "here and there a large, rambling lilac bush." Laquin in Bradford County lasted from 1902 to 1944 (including some ten years as a Civilian Conservation Corps [CCC] camp), producing lumber, bark for tanneries, wood-based chemicals (acetate of lime, wood alcohol), charcoal, barrel staves, kindling wood, wagon wheel hubs, beech chips for brewing beer, and veneer. Sonestown in Sullivan County was a lumber town from 1902 to 1922, in much of that time a base for logging that sent hemlock bark and logs to other communities to be cut into timber. Masten was one of the towns

that received logs from Sonestown. It was in existence from 1905 through 1930, when the last log was cut into lumber. Masten remained barely alive, with a few souls remaining to sell lumber stockpiled there and to ship it out, a job that was completed in 1933. Thereafter, a CCC camp took over the site and, like their counterparts based in Laquin at about the same time, members of the CCC converted old rail beds into forest roads, a job they finished in 1940. Taber, *Ghost Lumber Towns*.

CHAPTER 6

1. Graves, "Forests," 227.
2. Longfellow's *Evangeline: A Tale of Acadie* (1847) begins, "This is the forest primeval. The murmuring pines and the hemlocks. . . ."
3. Duggar, *Plant Physiology*, 1–8.
4. Spurr and Barnes, *Forest Ecology*, 475–93.
5. Havemeyer, introduction to *Conservation of Our Natural Resources*, 3–14.
6. Many sources allude to the pastoral setting of Pennsylvania as part of the new world. See, for example, Graeff, *History of Pennsylvania*; La Gorce, "Penn's Land," 6; E. W. Miller, *Pennsylvania Keystone*; Stevens, Cordier, and Benjamin, *Exploring Pennsylvania*, 16–23, 40–43; Wallace, *Pennsylvania*.
7. Trego, *Geography of Pennsylvania*.
8. Stevens, "When Timber Was King."
9. Lutz, "Original Forest Composition"; and "Vegetation of Heart's Content," summarized, cited in Spurr and Barnes, *Forest Ecology*, 481.
10. Spurr and Barnes, *Forest Ecology*, 490–91.
11. Maass, *Forestry Pioneer*. "By the end of the nineteenth century, a vast belt across northern and western Pennsylvania had been logged and re-logged, burned and re-burned so many times—and so devastatingly—that significant regeneration no longer took place. Scrub growth dominated where giant pines and hemlocks once grew." Cox, *Lumberman's Frontier*, 121.
12. McCormick, "Exhaustion of the Lumber Supply," 8–11.
13. Corrigan, *Calked Boots*; Gibbs, *We Went a Loggin'*; Harrison, *Steel Rails and Iron Men*; King, *Minnesota Logging Railroads*; www.trestlewalker.com/GoodyearLCo. (accessed October 18, 2015).
14. Beebe, *History of Potter County*; Currin, *Our Forestlands*; Pennsylvania Department of Agriculture, "Operations," 255–64; "Logging Operations," 26–27; Potter County Historical Society, *Historical Sketches*; "Stewardson Township," 3–4; "Report to Dr. Joseph Trimble Rothrock"; DeCoster, *Legacy of Penn's Woods*, 18–21; Taber, *Sawmills Among the Derricks*; *Sunset*; *Tanbark*; *Goodyears*; *Whining Saws*.
15. Leo Bailey remembered that a group of Italian-born woodsmen moved into an old lumber camp about half a mile above his family's camp and introduced a new form of lumbering. "They gathered limbs and discarded parts of hemlock trees not suitable for lumber. These were cut into four-foot lengths of wood and were loaded for shipment a considerable distance to the nearest pulp and paper mill." Bailey, *Lumber Activity*.
16. Some people who were out of work would set fires and then hire on as firefighters.
17. Gifford Pinchot was not a shrinking violet. He stated the facts as he saw them. See McGeary, *Gifford Pinchot*.
18. *National Geographic* magazine summarized the sad outcome: "Forest fires completed the destruction, and millions of blackened, barren acres stood as mute witnesses of the profligacy of man in wasting one of the Commonwealth's principal assets. Floods became more frequent, since barren lands cannot hold back water and give it a chance to soak the ground [and] the lumberman had . . . rendered a thousand streams unfit for fish life. . . . Fish by the millions perished when streams were transformed for long periods into dry river and creek beds." La Gorce, "Penn's Land," 6–7, 9.
19. Among those moving to the gulf were Frank and Charles Goodyear, who began making plans to abandon Pennsylvania as soon as it was apparent that their Commonwealth empire would no longer be profitable in a few years. They widened their business interests beginning in 1890. The Goodyears sized up the gulf states, saw that the trees were straight and tall, that the logging conditions were easy compared to those of New York State and Pennsylvania, and that cheap labor could be obtained readily. They began purchasing land filled with yellow pine stands in the 1890s

when timberland was being offered in Mississippi and Louisiana for the ridiculously low price of $1.25 an acre (about $34 an acre in 2014 dollars, using years 1890 and 1900 as the basis for inflation calculation). By 1905, timbered land was selling for $5 an acre and in 1915, for $15 an acre. Ultimately, they invested $9 million of their own capital and attracted another $6 million from others to fund their new enterprises (equivalent to $389,213,787 in 2014 dollars, assuming $15 million raised by 1905). In 1906 and 1907, they built a huge sawmill and houses for employees at the new town of Bogalusa, Louisiana. Their Great Southern Lumber Company was heralded as "the largest lumber manufacturing proposition in the world." The Goodyears and their investors anticipated huge profits from the venture, and they weren't disappointed. Bogalusa, created around their mill, grew rapidly from an initial population of three thousand workers in 1907 to fifteen thousand residents in a few years. The Goodyears were instrumental in building and operating the New Orleans Great Northern Railroad (the "nogan"), which ran from Slidell, Louisiana, to Jackson, Mississippi, through the heart of their land. Bogalusa was a principal terminus for 150 miles of mainline and 75 miles of sidetrack that radiated out from the town. Logs were shipped in to the Great Southern Lumber Company at the rate of forty to one hundred cleared acres a day and were soon shipped out, converted to lumber. By 1916, the company produced a record amount of lumber, a quarter of a billion board feet in a single year. See Cox, *Lumberman's Frontier*, 244–48; Curtis, "Early Development," 347–68; Williams, *Americans and Their Forests*, 255, 262, 280; www.msrailroads.com/NOGN.htm (accessed July 1, 2014); www.acmeme.com/gmo/chapter_xxi.htm (accessed July 1, 2014); www.westegg.com/inflation/infl.cgi (accessed June 22, 2015). Elsewhere, a rosy picture of lumbering prospects was touted. The Upper Midwest states of Michigan, Wisconsin, and Minnesota (then referred to as "the northwest") were painted in glowing colors in stories that appeared in national media telling of untold riches to be made from an endless supply of trees (e.g. J. E. Jones, "Lumbering in the Northwest," *Cosmopolitan*, May 1893, 67–71). Those claims fueled migrations from Pennsylvania and other tree-depleted states.

20. The Goodyear Lumber Company bought the Robinson Logging Company, headquartered in Sekiu, Clallam County, Washington, in 1914. There, the Goodyears logged out the Blowder drainage and expanded up the Clallam River into the hills. They sold their stake in the Pacific Northwest in October 1923 (www.trestlewalker.com/GoodyearLCo [accessed October 18, 2015]). A promotional booklet published in the same year by three railroad companies promoted the Pacific Northwest to loggers and logging companies. "Timber! The world demands more of it to build and grow. The Pacific Northwest can supply it. In the slopes of the Cascades, the Coast Range and the Olympics of Oregon and Washington, in the Coeur d'Alenes, the Bitterroots, and Rockies of Idaho, Montana, and Wyoming is half the timber in the United States. A vast storehouse of national wealth measuring 970 billion board feet!" The booklet promised that the trees would not run out for hundreds of years. "And cutting must shortly increase to two or three times the present amount, in the opinion of the United States Forest Service" (Anonymous, *Timber Billions of the Pacific Northwest*, 1–32). See also Andrews, *This Was Logging!*; *Timber!*; *Timber*; Ripley, *Green Timber on the Flood Tide*.
21. Krebs, *Map of the Oil Region*.
22. Inevitably, production began to slow down. Then, new methods of secondary recovery were advanced. After 1905, the Bradford Field production increased once again and a second production peak was achieved in 1937 with 17 million barrels.
23. See Sisler, *Oil and Gas Fields*.
24. Sisler, *Map of the Coal Fields*.
25. Beebe, *History of Potter County*; Sisler, *Oil and Gas Fields*. The boom-to-bust industry was summarized in "Bottle Plant Was Thriving Industry for 15 Years" and "Building of Big Glass Plant Boomed Shingle House 1901–1902. Erection of Huge Glass Plant Buildings Was Tremendous Job," both in *Oswayo Valley Mail* (Shinglehouse, Potter County, Pa.), June 28, 1956, http://homepages.rootsweb

.com/~mhender/1956.html (accessed June 23, 2015).

CHAPTER 7

1. Three ways of examining the dilemma were operative. An early way of looking at the relationship between environment and humans was to see nature as dominant, sinister, even repugnant. This was possible during the early history of European entry into "Penn's Forest." When there were few humans facing the wild, untamed land and forests, it was natural to see formidable obstacles that the environment placed in the way of human survival and adaptation. Nature seemed forbidding, alien, paramount, supernatural, and almost insurmountable. This was an *environmental determinism* point of view. By 1920, however, when the vast white pine and hemlock forests had been largely decimated by 240 years of successive waves of cutting and other use of the forest, emphasis shifted to stress the huge influence of human beings upon nature, a form of *cultural determinism*. It became common to see environment as a pawn, under the domination of humans with their fires, machines, tools, motivations, and ambitions of exploitation. Cultural deterministic thinking gradually would shift, guided by pioneering conservationists, to consider the reciprocal effects of humans upon nature ("we can fix these problems caused by humans with our sciences and planned, intelligent interventions"), tempered by a realization of the effects of nature upon humans ("we can't control the weather and our best efforts to create change are dependent upon the environment's natural timetable"). The realization grew that human actions have led to benefits *and* deleterious results, mediated by natural forces, both cooperative and competitive. There has been a growing sense that, in an ecosystem, humans affect nature and nature affects humans. Changes in one aspect of the ecosystem lead to responses from other components of the ecosystem in *dynamic reciprocal causation*. Gordon G. Whitney summarizes the current approach: "North America today is neither a lost paradise nor a land of endless resources. Human activities have all but eliminated some of North America's ecosystems, for example, the tallgrass prairie. Other ecosystems have proven more resilient to human activity, or have changed and taken on new values." For example, Whitney related, "the repeated cutover hemlock-hardwood region of northern Pennsylvania is currently the center of the commercial black cherry industry" (Whitney, *From Coastal Wilderness*, 337). On the earliest views of America's forests, see Williams, *Americans and Their Forests*, 10–12.
2. Ironically, the Bayless Pulp and Paper Mill, the Emporium Lumber Company, the Standard Wood Company, and the Goodyear Lumber Company were left standing and substantially unharmed, but their raw materials and stockpiles had been swept downstream by the raging water (Heimel, *1911*, 124–25).
3. See Dixon, "Report of the Austin Disaster"; Heimel, *1911*; Largey, Worthington, and Ogden, *Austin Disaster*; Nuschke, *Dam That Could Not Break*; Owens and Owens, *Victims of the Austin Flood*; Rich, *Lessons in Social Responsibility*.
4. See Maass, *Forestry Pioneer*, 71–111.
5. Ibid., 79.
6. Governor Pennypacker served in that office from 1903 to 1907. Pennypacker signed legislation creating the Departments of Health, Highways, Mines, and Fisheries. Under his administration, the first school for state foresters was opened at Mont Alto; more than a half million acres of land were preserved; and he appointed Joseph Trimble Rothrock as commissioner of forestry, a fortunate choice, as subsequent events demonstrated. Today, because of a merger with Penn State's forestry program in 1929, Mont Alto continues to offer a forward-looking forest technology associate degree program in addition to degree programs in many other academic disciplines: http://cpc.state.pa.us/history/notable/samuel-whitaker-pennypacker.cfm (accessed June 27, 2015).
7. Illick, "Joseph Trimble Rothrock," 83–94. When the timber was mostly gone, Pennsylvania industry exploited other natural resources, such as coal and oil. The impact of those natural resource extractions, including timber, led to some positive developments.

Business enterprises provided a stable foundation and the wealth necessary to establish a sizeable population and communities, fledgling transportation systems, the industries, and the infrastructure that allowed Pennsylvania to prosper and grow as the Commonwealth headed out of the nineteenth century.

8. While many early twentieth-century observers were ready to abandon north-central Pennsylvania as a hopeless cause, the area was able to rebound through the vision and energy of farsighted individuals like Pennypacker, Rothrock, and Pinchot, in tandem with practical efforts of federal and state government agencies in implementation. Together, those early advocates and visionaries crafted new governmental policies affecting land acquisition, protection of remaining forest pockets, seedling production, reforestation, preservation practices, and forest education. In short, they adopted a progressive conservation philosophy and put it into practice. They accomplished wonders. Joseph S. Illick, Jacob Nolde, Mira Lloyd Dock, and Maurice K. Goddard were among the Pennsylvania leaders. Other prominent environmental heroines and heroes in Pennsylvania conservation, forestry, and environmental activism include Kristine M. Anderson, John James Audubon, John Bartram, Basse A. Beck, Ralph Elwood Brock, Rachel Carson, Rosalie Barrow Edge, Richard James, Otto Emery Jennings, Horace McFarland, Ivan McKeever, Jerome Irving Rodale, and Ned Smith.
9. Maass, *Forestry Pioneer*, 85–86, 93–95, 103, 107–8, 115.
10. Rimby, "Better Housekeeping Out of Doors"; *Mira Lloyd Dock*; McElroy, Stangil, and Druse, *Our Hidden Heritage*; Dock, *Forest and Woodland Reservations*.
11. Dock, "A Summer's Work Abroad"; Dock and Ray, *Nature Study Reference Library*.
12. "Reservation" was the term used for a forest reserve, later simply called a "state forest."
13. The October 1903 issue of *Forest Leaves* referred to her as "probably the most accomplished practical lady forester in North America." Rothrock "trusted her judgment as much as his own" (Maass, *Forestry Pioneer*, 94, 107).
14. Pennsylvania Department of Conservation and Natural Resources, *1895–1995*; DeCoster, *Legacy of Penn's Woods*.
15. See Dock, *Mira Lloyd Dock Forestry Lantern Slides Collection*. The collection includes 468 examples of her 1889–1903 glass lantern slides of trees, a description of the digitization process, and biographical information. See also White, Ma, and Wetzel, "Digitizing Glass Lantern Slides." Dock was a frequent lecturer who spoke simply and sincerely on topics such as "forestry at home and abroad." Her family's summer home was at Graeffenburg, near Mont Alto Forest Academy, and she taught botany there until 1929, when the academy merged with Penn State (Maass, *Forestry Pioneer*, 108).
16. Dock, "Profession of Forestry."
17. Dock, *State Forestry Today*.
18. See D. White, *Gifford Pinchot*.
19. The Pinchots moved in the same social circles as Hudson River School landscape artist Sanford R. Gifford (1823–1880), whose last name was appropriated for their son's first name. The artist was godfather to James and Mary's son. Sanford R. Gifford's art emphasized the paradise and natural beauty of the American outdoors that he so obviously enjoyed and admired. One of Gifford's most famous oil paintings, *Hunter Mountain, Twilight* (1866), was among seven of his works owned by the Pinchot family. Interestingly, this famous landscape painting of the New York Catskills shows a number of tree stumps in the foreground, the result of years of tanbarking, charcoal industry exploitation, and logging. The painting is credited as providing a visual gateway to modern environmentalism and with fostering sentiment for the Catskill and Adirondack Forest Preserves. See Avery and Kelly, *Hudson River School Visions*, 78, 175–78; Yaeger, *Hudson River School*, 28, 61, 63–64, 66, 68–69, 71; https://en.wikipedia.org/wiki/Sanford_Richardson_Gifford (accessed June 29, 2015).
20. Kirk Johnson, "From a Woodland Elegy. A Rhapsody in Green. Hunter Mt. Paintings Spurred Recovery," *New York Times*, June 7, 2001. See also Lewis, "Pinchot Family"; McGeary, *Gifford Pinchot*; Pinchot and Price, *Profession of Forestry*.

21. See Pinchot, *Primer of Forestry*; *Training of a Forester*.
22. See C. Miller, *Gifford Pinchot*; Pinchot, *Fight for Conservation*.
23. Balogh, "Scientific Forestry."
24. See Ponder, "Gifford Pinchot."
25. See, for example, Muir, *Our National Parks*.
26. Pinchot and W. J. McGee defined "conservation" as "the use of the natural resources for the greatest good of the greatest number for the longest time." Implicit in their definition is the understanding that some natural resources are finite, but necessary to human benefit, and therefore to be used wisely. Pinchot, "Birth of Conservation," 326.
27. Steen, *Conservation Diaries*.
28. See Pinchot, "Birth of Conservation." Pinchot became embroiled in a controversy with Richard A. Ballinger, secretary of the Interior under President William Howard Taft, over the disposal of Alaskan coal lands. Pinchot eventually was dismissed by the president for insubordination. See Allen and Sharpe, *Introduction to American Forestry*. The Forest History Society notes that Pinchot is generally regarded as the "Father of American Conservation" (http://www.foresthistory.org [accessed June 27, 2015]).
29. Pinchot noted that early in 1905 in Roosevelt's administration, there was a governmental "mess," with highly fragmented government organizations overlooking mineral resources (three), streams (four or five), and forests (six) and supervising wildlife, soils, erosion, and other elements of the land (a dozen). Pinchot had much to do with the eventual consolidation of those fragmented units into a single organization, which cooperated as a unit and made decisions about how to address interrelated ecological concerns under the umbrella of a common conservation ethic.
30. Lillard, *Great Forest*.
31. Prior to his holding the highest office in the land, Roosevelt had been an outdoorsman who experienced many joys and dangers in the wild. He was a strong advocate of wilderness preserves and, despite their later differences, reveled in several trips he made with John Muir to the Grand Canyon and Yosemite. Roosevelt advocated national park status for those natural treasures of "desolate and awful sublimity" and advocated for expanded wilderness preserves (see Roosevelt, "Wilderness Preserves," 146–62). He called a conference on conservation, held at the White House, in May 1908. The conference hosted a thousand national leaders, including all the state governors, members of the Supreme Court, Cabinet officers, relevant national society representatives, and special guests. In his opening address, almost certainly written for him by Gifford Pinchot and Pinchot's colleague W. J. McGee, the president said the most critical problem facing the nation was the danger that its natural resources would be depleted. Some of those resources, he noted, were not renewable and he advocated wise, efficient, careful, and thoughtful husbandry of coal, oil, gas, and iron and other metals. However, other resources, such as timber, soil, and water, could be renewed and improved. Roosevelt called for an end to selfish, greedy individualism, which he said had characterized past practices. He advocated national intervention by leaders such as those in his audience to rightfully oversee and regulate natural resources in the national interest and for the sake of future generations (see Roosevelt, "Opening Address by the President," 3–12). Theodore Roosevelt did much to protect priceless areas of the country. Congress had set aside 26 million acres of land for preservation prior to his assuming the presidency, but President McKinley had allowed 3 million acres to be sold back into private, exploitative ownerships. Roosevelt stopped those private sales, insisted upon selective cutting of trees rather than clear-cutting, and called for a new study that reorganized 60 million acres of national forest and eventually resulted in regulated use. He saved another 16 million acres before "land grabbers" could gain their control (see Donald, *Lion in the White House*, 164–65, 330–34).
32. A history of the Pennsylvania Bureau of Forestry explained, "With so much deforested land in Pennsylvania, tree planting was crucial. Northeastern forests will reproduce naturally if trees are available to produce seeds and sprouts. But large expanses of the state had been cleared and cropped. In other

areas, all the trees had been burned repeatedly for miles around, leaving no natural source of tree seeds. These areas had to be planted to return them to forest." Robert S. Conklin, Pennsylvania Commissioner of Forestry from 1904 to 1920, oversaw the state's initial reforestation program (see DeCoster, *Legacy of Penn's Woods*, 41–42).

33. Pinchot, "Lumberman and the Forester," 11–14.
34. Pinchot's autobiography, *Breaking New Ground*, was published posthumously in 1947 and has been reprinted. See also Pinkett, *Gifford Pinchot*.
35. Illick's major works include *Check-List of the Woody Plants*; *Outline of General Forestry*; *Common Trees of Pennsylvania*; *Forest Research in Pennsylvania*; *Guide to Forestry*; "Joseph Trimble Rothrock"; *Pennsylvania Trees*; "Rebuilding Our Forest Heritage"; *Boy Scouts First Book of Forestry*; *Forest Situation in Pennsylvania*; *State Forests of Pennsylvania*; *Tree Habits*; and *Woody Plants of Franklin County*; Illick and Aughanbaugh, *Pitch Pine*; Illick and Frontz, *Beech, Birch, Maple Forest Type*; and Illick and Shoemaker, *In Penn's Woods: Guide*; *In Penn's Woods: Handy and Helpful*.
36. Illick praised another environmental hero, Jacob Nolde. Illick's contributions to progress in environmental protection and restoration came about through his educational writing. For Nolde, environmental action occurred as a result of an idea. His dream was to create an estate in Pennsylvania that rivaled the forested estates he'd seen as a youth in Europe. Sound environmental principles and practices were applied and he succeeded through planning and hard physical labor. See "Learn about PA's Natural Resources," www.dcnr.state.pa.us/learn/index.htm (accessed June 28, 2015; click on "State Forests," "history"), which proclaims, "The Commonwealth is the nation's largest producer of hardwood lumber (over 1 billion board feet per year). The State Forest system contains twelve percent of the forested area (about 2.1 million acres). This makes it one of the largest expanses of public forestland in the eastern United States. Other public forests include State Game Lands, State Parks and the Allegheny National Forest."
37. See http://www.dcnr.state.pa.us/stateparks/findapark/noldeforest/index.htm (accessed June 28, 2015).
38. Maurice Goddard taught people that they could make a difference, communicating through direct contacts and speeches. He also had great faith in the impact of pamphlets, such as his *Forest Trees to Plant in Pennsylvania* and *Planning for Parks and Waters in Pennsylvania*.
39. See Goddard's biography: Morrison, Leader, and Forrey, *Walk on the Downhill Side*. See also Barrett, "How to Make History Matter," 550–69.
40. For much more detail, see Pennsylvania Department of Conservation and Natural Resources, *1895–1995*. See also Pennsylvania Department of Conservation and Natural Resources Bureau of Forestry State Forest Resource Management Plan 2007 Update, http://www.dcnr.state.pa.us/forestry/_update_2007.pdf. This will lead to the 2015 update draft.
41. The Goodyear brothers, Frank and Charles, have been cast as archetypal villains in the foregoing narrative. Their greed and grasp were excessive in Pennsylvania, as documented. However, even the Goodyear empire, largely through the actions of their heirs, eventually came to practice an early version of sustainable forestry in the southern gulf states, if not in Pennsylvania. However, their behaviors were enacted more to protect investments than to practice conservation and environmental altruism. Michael Williams briefly traced the history of logging in Louisiana. He noted, "Even the most 'boom-minded' companies realized that reforestation was simply good business. The Goodyears in Bogalusa feared that their investment of over $15 million (before even a board came out of the mill) was in danger of being lost. The cutovers were increasing at the startling rate of nearly 20,000 acres per annum so that the mill was, in effect, getting farther and farther away from its supplies and was in danger of being abandoned [a danger hedged somewhat by Great Southern's use of portable sawmills], Bogalusa facing 'the dismal prospect of becoming a ghost town with a mountain of saw dust as a monument

to its passing'" (Williams, *Americans and Their Forests*, 286–87). The immediate passings that most threatened Bogalusa, however, were the deaths of Frank Henry Goodyear in 1907 and Charles Waterhouse Goodyear in 1911, when, according to an online assessment, "the Goodyear empire began to fall apart" (http://buffaloah.com/a/del/888/hist/ and http://buffaloah.com/h/goodyear_FItc.html [accessed June 29, 2015]).

Although the sawmill had been constructed by 1907, it would not go into operation until the following year, delayed by uncertainties caused by the bankers' panic of 1907 that had also caused the Goodyears to put a temporary halt to building the town of Bogalusa. They also stopped further work on the New Orleans Great Northern Railroad's ninety-mile connection that would link the new town to the national rail network. Eventually, however, the railroad and the Great Southern Lumber Company were successful. The Great Southern sawed its first log in 1908 and became the world's largest sawmill. It remained in business until the last log was sawed in 1988. Family heirs and directors of the Great Southern and the New Orleans Great Northern Railroad were responsible for modernizing and improving business practices. They were persuaded to change policies on conservation and woodlands management beginning with a visit to the sandy Landes district near Bordeaux in southwestern France, where they saw that an enormous area planted with pines could stabilize the dunes. Further investigations made clear that sound forest management and large-scale reforestation could bring practical payoffs. Visits were made in 1920 to review forest conservationist Henry E. Hardtner's mills and forests north of Alexandria, Louisiana. Favorable impressions of Hardtner's philosophy and practices spurred the immediate start of a new Bogalusa reforestation program under the supervision of knowledgeable foresters. The lands were fenced to keep out marauding hogs, breaks were cut to control fires, and pine cones were collected for seeds. Germination fields holding up to 17 million seedings were pampered, the best seedlings were replanted, and soon a trial area of eight hundred acres was under way in the first year. A further fifteen thousand acres were added during the next few years, with plans for more than ten times that amount in the future. The initial planted area was supplemented by a company policy that favored cutting mature forests selectively, thus encouraging continuous growth and production (Goodyear, *Bogalusa Story*; Williams, *Americans and Their Forests*, 286–87).

Another innovation was the closing of the Great Southern's refuse burner in 1924. The huge burner had consumed 560 cords of waste matter (mostly sawdust) every day from October 1, 1908, to July 4, 1924, or a total of 2,688,000 cords, by the company's estimate. That much "waste" or "refuse" was valued at $1,344,000. The damage through air pollution was incalculable. The construction of new mill machines that converted sawdust to paper permitted reuse of the waste by-product. Thus, another source of profit was discovered and a nasty problem was solved (Goodyear, *Bogalusa Story*).

While these developments may sound like cutting-edge practices, most actions taken were among the economic conservation practices advocated almost twenty years earlier by Gifford Pinchot, head of the U.S. Forest Service, and President Theodore Roosevelt. Their suggestions and advocated practices were blithely ignored by the founders of the Goodyear empire when they were logging and lumbering in Pennsylvania.

The prediction, set forth by Pinchot and Roosevelt, of a "forest famine" was helpful in spurring action during the first decades of the twentieth century, but the famine scare never materialized. Forests have regenerated—they are resilient, as long as they are allowed to thrive and are not used for agriculture or urban purposes, such as industry and housing. However, the forests that have returned are not necessarily the same as the forests that were logged or lost to other causes. For a balanced review, see MacCleary, *American Forests*.

42. Tom Byrne, "Ansonia to Gaines," *Sunday Telegram* (Elmira, N.Y.), March 29, 1970.
43. See http://www.lumbermuseum.org/ (accessed June 29, 2015).
44. The complete series of sixteen plates is at the State Archives RG-6.20; WTC 4007–4023.

BIBLIOGRAPHY

The question marks that appear in the bibliography indicate missing information. Some material cited or referenced is "fugitive," that is, lacks facts such as publisher or place or date of publication. The use of question marks follows the authority of and notation used by OCLC WorldCat in these instances. See www.worldcat.org/.

MANUSCRIPT AND ARCHIVAL COLLECTIONS

Lois Barden Photograph Collection, Candor, New York

Laura Ishner Photograph Collection, Curwensville, Pennsylvania

Ronald E. Ostman, Historical Collection, Groton, New York

Pennsylvania Lumber Museum, Ulysses, Pennsylvania

Pennsylvania State Archives, State Archives Division, Pennsylvania Historical and Museum Commission, Harrisburg, Pennsylvania

Potter County Historical Society, Coudersport, Pennsylvania

Margaret B. Redmond Manuscript Archive, Scipio Center, New York

Henry W. Shoemaker Papers, Special Collections Library, The Pennsylvania State University, University Park, Pennsylvania

Anne E. Wenger Manuscript Archive, Coldwater Lake, Michigan

SECONDARY SOURCES

Albertson, Aden Byron. *In Memory's Lane.* [Nordmont, Sullivan County, Pa.]: privately printed, 1969. In the possession of Anne E. Wenger, Coldwater Lake, Michigan.

Allen, Shirley Walter, and Grant William Sharpe. *An Introduction to American Forestry.* New York: McGraw-Hill, 1960.

Ambrose, Stephen E. *Nothing Like It in the World: The Men Who Built the Transcontinental Railroad, 1863–1869.* New York: Simon and Schuster, 2000.

Andrews, Ralph W. *This Was Logging! Selected Photographs of Darius Kinsey.* Atglen, Pa.: Schiffer, 1984.

———. *Timber! Loggers Challenge the Great North-West Forests.* Atglen, Pa.: Schiffer, 2007.

———. *Timber: Toil and Trouble in the Big Woods.* Atglen, Pa.: Schiffer, 1984.

Anonymous. *Timber Billions of the Pacific Northwest.* Chicago and St. Paul: Chicago, Burlington & Qunicy R.R., Northern Pacific Railway, Great Northern Railway, 1923.

Avery, Kevin J., and Franklin Kelly, eds. *Hudson River School Visions: The Landscapes of Sanford R. Gifford.* New Haven: Yale University Press, 2003.

Bailey, Leo J. *Lumber Activity of James (Cap) Bailey in Northern Penna. Before and After the Turn of the Century When Daddy Was a Boy.* N.p: Privately printed, Christmas 1967. In the possession of his daughter, Margaret B. Redmond of Scipio Center, New York. Quoted with permission.

Bajac, Quentin. *The Invention of Photography*. New York: Harry N. Abrams, 2002.

Balogh, Brian. "Scientific Forestry and the Roots of the Modern American State: Gifford Pinchot's Path to Progressive Reform." *Environmental History* 7, no. 2 (2002): 198–225.

Barrett, B. "How to Make History Matter: The Maurice K. Goddard Legacy Project." *Pennsylvania History* 79, no. 4 (2012): 550–69.

Bastian, Jonathan D. *Lumbering: A Way of Life at Pardee; Stripping the Forests of Union, Snyder, and Mifflin Counties Through Prosperity and Turmoil.* N.p.: CreateSpace, 2013.

Beebe, Victor L[lewellyn]. *History of Potter County, Pennsylvania.* Coudersport, Pa.: Potter County Historical Society, 1934.

"Bipartisan Wildfire Disaster Funding Included in President Obama's FY15 Budget." Nature Conservancy. http://www.nature.org/news features/pressreleases/ (accessed June 8, 2015).

Bishop, L. L. "Pines of Hearts Content." *American Forestry* 29 (1923): 361–63.

Bogdan, Robert. *Adirondack Vernacular: The Photography of Henry M. Beach.* Syracuse: Syracuse University Press, 2003.

Bohn, Dave, and Rodolfo Petschek. *Kinsey Photographer: A Half Century of Negatives by Darius and Tabitha Kinsey.* San Francisco: Chronicle Books, 1978.

Bronner, Simon J. *Popularizing Pennsylvania: Henry W. Shoemaker and the Progressive Uses of Folklore and History.* University Park: Pennsylvania State University Press, 1996.

Brown, Milton W. *American Painting from the Armory Show to the Depression.* Princeton: Princeton University Press, 1970.

Bruner, Robert F. "Why Study Business and Economic History?" *Robert F. Bruner: Darden* (blog), November 7, 2013. http://blogs.darden.virginia.edu/deansblog/2013//11/why-study-business-and-economic-history/.

Butcher, Solomon D. *S. D. Butcher's Pioneer History of Custer County, and Short Sketches of Early Days in Nebraska.* Broken Bow, Nebr.: Merchants, 1901.

———. *Sod Houses; or, The Development of the Great American Plains: A Pictorial History of the Men and Means That Have Conquered This Wonderful Country.* Kearney, Nebr.: Western Plains, 1904.

Carter, John E. *Solomon D. Butcher: Photographing the American Dream.* Lincoln: University of Nebraska Press, 1985.

Casler, Walter C. *Logging Railroad Era of Lumbering in Pennsylvania.* Vol. 8, *Tionesta Valley: Sheffield, Brookston, Loleta, Bear Creek, Cherry Grove.* Williamsport, Pa.: Lycoming, 1973.

———. *Logging Railroad Era of Lumbering in Pennsylvania.* Vol. 9, *Teddy Collins Empire: A Century of Lumbering in Forest County; Nebraska, Golinza, Kelletville, Mayburg, Bucks Mills, Iron City, Pigeon, Marienville.* [Corry, Pa.]: privately printed, 1976.

———. *Logging Railroad Era of Lumbering in Pennsylvania.* Vol. 11, *Allegheny Valley Logging Railroads: Locomotives, Sawmills, Pine Timber; Warren, Forest, Venango, Crawford, Erie Counties.* Williamsport, Pa.: Lycoming, 1977.

Casler, Walter C., Benjamin F. G. Kline Jr., and Thomas Townsend Taber III. *Logging Railroad Era of Lumbering in Pennsylvania.* Williamsport, Pa.: Lycoming, 1970–78.

———. *Logging Railroad Era of Lumbering in Pennsylvania, 1970–1976.* 2nd ed. Strasburg, Pa.: Railroad Museum of Pennsylvania, 1999.

Cochran, Thomas C. *Pennsylvania: A Bicentennial History.* New York: W. W. Norton; Nashville: American Association for State and Local History, 1978.

Collins, Herman Le Roy. *When Lumber Was King.* [Lycoming County, Pa.?: n.p.], [1926?].

Corrigan, George. *Calked Boots and Cant Hooks: One Man's Story of Logging the North.* Ashland, Wisc.: Northword, 1976.

Cox, Thomas R., ed. "Harvesting the Hemlock: The Reminiscences of a Pennsylvania Wood-Hick." *Western Pennsylvania Historical Magazine* 67, no. 2 (1984), 109–31.

———. *The Lumberman's Frontier: Three Centuries of Land Use, Society, and Change in America's Forests.* Corvallis: Oregon State University Press, 2010.

———. "Transition in the Woods: Log Drivers, Raftsmen, and the Emergence of Modern Lumbering in Pennsylvania." *Pennsylvania Magazine of History and Biography* 104, no. 3 (1980): 345–64.

Cranmer, Hiram. "Harvesting the Hemlock: The Reminiscences of a Pennsylvania Wood-Hick." *Western Pennsylvania Historical Magazine*, no. 67 (1984): 121–22.

Culip, Kimbra. "Pine Creek Gorge: The Other Grand Canyon." *GEOTIMES: Earth, Energy and Environmental News, Travels in Geology*, October 2007. http://www.agiweb.org/geotimes/oct07/article.html?id=Travels1007.html.

Currin, Robert K. *Our Forestlands.* Coudersport, Pa.: Potter County Historical Society, n.d.

Curtis, Michael. "Early Development of Operations of the Great Southern Lumber

Company." *Louisiana History* 14 (1973): 347–68.

DeCoster, Lester A. *The Legacy of Penn's Woods, 1895 to 1995: A History of the Pennsylvania Bureau of Forestry.* Harrisburg, Pa.: Commonwealth of Pennsylvania, Pennsylvania Historical and Museum Commission for Department of Conservation and Natural Resources, Bureau of Forestry, 1995.

Defebaugh, James Elliott. *History of the Lumber Industry of America.* 2 vols. Chicago: American Lumberman, 1906–7.

———. "Pennsylvania-Forest Conditions." In *History of the Lumber Industry of America.* Vol. 2, 530–55. Chicago: American Lumberman, 1907.

"Descriptive Price List Union View and Copying Co., 1882." *Photographic Times and American Photographer* 11, no. 132, n.s., 12 (December 1881): 462.

Dixon, Samuel Gibson. "Report of the Austin Disaster." Pennsylvania Health Bulletin, no. 36 (1912): 1–10. Harrisburg, Pa.: State Department of Health.

Dock, Mira Lloyd. *Forest and Woodland Reservations.* Philadelphia: Art Department of the Civic Club of Philadelphia, 1899. http://cpc.state.pa.us/history/notable/mira-lloyd-dock.cfm.

———. *Mira Lloyd Dock Forestry Lantern Slides Collection.* Mont Alto: Pennsylvania State University Libraries, Penn State Mont Alto Library, 2004. http://www.libraries.psu.edu/psu/digital/mdfls.html.

———. "The Profession of Forestry." In *Vocations for the Trained Woman: Opportunities Other Than Teaching*, edited by Agnes Frances Perkins, 163–70. Boston: Women's Educational and Industrial Union, 1910.

———. *State Forestry Today*. General Federation of Women's Clubs Bulletin, no. 2 (1924). Washington, D.C.: General Federation of Women's Clubs Forestry Committee, Department of Applied Education, Conservation of Natural Resources Division.

———. "A Summer's Work Abroad, in School Grounds, Home Grounds, Play Grounds, Parks, and Forests." Pennsylvania Department of Agriculture Bulletin, no. 62 (1900): 8–10, 13–15.

Dock, Mira Lloyd, and William Stanley Ray. *Nature Study Reference Library: For Use in Public Schools.* Pennsylvania Department of Agriculture Bulletin, no. 64 (1900).

Donald, Aida D. *Lion in the White House: A Life of Theodore Roosevelt.* New York: Basic Books, 2007.

Donehoo, George P. *Report of the Austin Relief Association Dealing with the Relief Work After the Disaster of September 30, 1911, Containing the Reports of the General Association, and the Reports of the Secretary of the Local Committee at Austin, Pa.: Report of the Secretary, George P. Donehoo.* Coudersport, Pa.: Potter County Journal Print, 1912.

Dorment, Richard, and Margaret F. MacDonald. *James McNeill Whistler*. Washington, D.C.: National Gallery of Art, Tate Gallery, 1995.

Duggar, Benjamin M. *Plant Physiology with Special Reference to Plant Production.* New York: Macmillan, 1912.

Dunn, Edward T. *Buffalo's Delaware Avenue: Mansions and Families.* Buffalo, N.Y.: Canisius College Press, 2003.

Edenborn, Harry M., and Dessie Severson. "Characterization of Waste Tar Associated with Abandoned Wood Chemical Sites in Northwestern Pennsylvania, USA." *Water, Air, and Soil Pollution* 183 (2007): 331–40.

Fletcher, Stevenson Whitcomb. *Pennsylvania Agriculture and Country Life, 1640–1840.* Harrisburg, Pa.: Pennsylvania Historical and Museum Commission, 1950.

Fox, Sally. *The Victorian Woman: A Book of Days.* Boston: Little, Brown, 1987.

Freedgood, Lillian. *Great Artists of America.* New York: Thomas A. Crowell, 1963.

Fulling, E. H. "Botanical Aspects of the Paper-Pulp and Tanning Industries in the United States: An Economic and Historical Survey." *American Journal of Botany* 43 (1956): 621–34.

Garratt, Colin. *The World Encyclopedia of Locomotives: A Complete Guide to the World's Most Fabulous Locomotives.* London: Hermes House, 2002.

Gibbs, Esther. *We Went a Loggin'.* St. Cloud, Minn.: North Star Press, 1974.

Godbey, Emily. "Disaster Tourism and the Melodrama of Authenticity: Revisiting the 1889 Johnstown Flood." *Pennsylvania History* 73, no. 3 (2006): 272–315.

Goddard, Maurice K. *Forest Trees to Plant in Pennsylvania.* Harrisburg, Pa.: Pennsylvania Department of Forests and Waters, 1976.

———. *Planning for Parks and Waters in Pennsylvania.* Temple University Libraries, Urban Archives, Pamphlet Collection, Pennsylvania Planning Association, 1957, 1959.

Goodyear, Charles Waterhouse III. *Bogalusa Story.* Buffalo, N.Y.: Wm. J. Keller, privately printed, 1950.

Graeff, Arthur D. *The History of Pennsylvania.* Chicago: John C. Winston, 1949.

Graves, Henry S. "Forests." In *Conservation of Our Natural Resources*, by Loomis Havemeyer, Gar A. Roush, Frederick H. Newell, Henry S. Graves, George S. Wehrwein, Paul G. Redington, and Elmer Higgins, 227–316. New York: Macmillan, 1937.

Greeley, W. B. "The Relation of Geography to Timber Supply." *Economic Geography* 1 (1925). Reprinted in *From Coastal Wilderness to Fruited Plain: A History of Environmental Change in Temperate North America, 1500 to the Present*, by Gordon G. Whitney, 174–75. Cambridge, U.K.: Cambridge University Press, 1994.

Haddock, Keith. *A History of Marion Power Shovel's First 100 Years; & Memorabilia, Excerpts from Company Records.* Marion, Ohio: Marion Power Shovel Division, Dresser Industries, 1984.

Hanlon, Howard A. *The Ball-Hooter: From the Forests They Felled—Cities Grew.* Prospect, N.Y.: Prospect Books, 1960.

Harrison, Frederick G. *Steel Rails and Iron Men: A Chronicle of Logging Railroads in Northeastern Minnesota and Wisconsin During the Great Logging Era of the 1890's to 1935, and an Account of Logging Railroads Up to 1969, Featuring the Duluth and Northeastern Railroad.* Indian Rocks Beach, Fla.: Books Unlimited, 1970.

Hartley, Harry J. "A History of the Lumber Industry in Pennsylvania to 1900." M.A. thesis, Pennsylvania State College, 1926.

Havemeyer, Loomis. Introduction to *Conservation of Our Natural Resources*, by Loomis Havemeyer, Gar A. Roush, Frederick H. Newell, Henry S. Graves, George S. Wehrwein, Paul G. Redington, and Elmer Higgins, 3–14. New York: Macmillan, 1937.

Heck, Albert S. "An Address by Hon. Albert S. Heck, President Judge, 55th Judicial District of Pennsylvania, to the Sportsmen of the Wild Life League, at Conneaut Lake, September 6, 1916." In *The Black Forest of Pennsylvania: An Address*, compiled by John P. Rieff. [Norristown, Pa.?: n.p.], [1916?]. Biodiversity Heritage Library, http://www.biodiversitylibrary.org.

Heimel, Paul W. *1911: The Austin Flood.* Coudersport, Pa.: Knox Books, 2011.

Hesch, Merrill, Richard Pieper, and Harry Littell. *Ithaca Then and Now.* Ithaca, N.Y.: McBooks Press, 2000.

Holbrook, Stewart H. *Yankee Loggers: A Recollection of Woodsmen, Cooks, and River Drivers.* New York: International Paper, 1961.

Hollingsworth, J. Brian, and Patrick Whitehouse. *North American Railways.* London: Hamlyn/Bison, 1977.

Horst, Melvin J., and Elmer Lewis Smith. *Logging in the Pennsylvania North Woods: A Collection of Turn of the Century Photographs.* Lebanon, Pa.: Applied Arts, 1969.

Hotchkiss, George W. *History of the Lumber and Forest Industry of the Northwest.* Chicago: George W. Hotchkiss, 1898.

Huntley, George William, Jr. *A Story of the Sinnamahone.* Williamsport, Pa.: Williamsport Printing and Binding, 1936.

Illick, Joseph S[imon]. *The Boy Scouts First Book of Forestry.* Pennsylvania Department of Forestry Circular, no. 20, 1920. Harrisburg, Pa.

———. *A Check-List of the Woody Plants of a Portion of the South Mountains near Mont Alto, Pennsylvania.* Harrisburg, Pa.: Pennsylvania Department of Forestry, 1913.

———. *Common Trees of Pennsylvania: How to Know Them, Where to See Them; A Handy Pocket Manual of the Common Native and Introduced Trees of Pennsylvania.* Altoona, Pa.: Times Tribune, 1925.

———. *Forest Research in Pennsylvania.* Harrisburg, Pa.: Department of Forests and Waters, 1930.

———. *The Forest Situation in Pennsylvania.* Pennsylvania Department of Forestry Bulletin, no. 30 (1923). Harrisburg, Pa.

———. *Guide to Forestry: Book One.* Pennsylvania Department of Forestry Bulletin, no. 26 (1922). Harrisburg, Pa.

———. "Joseph Trimble Rothrock, 'Father of Pennsylvania Forestry.'" In *The Pennsylvania-German Society: Proceedings and Addresses, October 5, 1923*, 83–94. Norristown, Pa.: Norristown Press, 1929.

———. *An Outline of General Forestry*. 1935. New York: Barnes and Noble. 2nd ed., 1936; 3rd ed., 1949.

———. *Pennsylvania Trees*. Pennsylvania Department of Forestry Bulletin, no. 11 (1914). Harrisburg, Pa.

———. "Rebuilding Our Forest Heritage." *The Pennsylvania-German Society Proceedings and Addresses, October 6, 1922 at Reading, PA*. Vol. 33, pt. 2 (1922): 3–13.

———. *The State Forests of Pennsylvania*. 1927. Reprint. Pennsylvania Department of Forests and Waters Bulletin, no. 46 (1946). Harrisburg, Pa.

———. *Tree Habits: How to Know the Hardwoods*. Washington, D.C.: American Nature Association, 1924.

———. *Woody Plants of Franklin County, Pennsylvania*. Rev. ed. Pennsylvania Department of Forestry Bulletin, no. 21 (1921). Harrisburg, Pa.

Illick, Joseph S[imon], and John E. Aughanbaugh. *Pitch Pine in Pennsylvania*. Pennsylvania Department of Forests and Waters Research Bulletin, no. 2 (1930). Harrisburg, Pa.

Illick, Joseph S[imon], and LeRoy Frontz. *The Beech, Birch, Maple Forest Type in Pennsylvania*. Harrisburg, Pa.: Pennsylvania Department of Forests and Waters, 1928.

Illick, Joseph S[imon], and Henry W[harton] Shoemaker. *In Penn's Woods: A Guide to Recreational Opportunities in the State Forests of Pennsylvania*. 3rd rev. ed. Department of Forests and Waters Bulletin, no. 31 (1928). Harrisburg, Pa.

———. *In Penn's Woods: A Handy and Helpful Pocket Manual of Natural Wonders and Recreational Facilities of the State Forests of Pennsylvania*. Harrisburg, Pa.: Pennsylvania Department of Forests and Waters, 1925.

Kaseman, Edward L. *Story of the Susquehanna and New York*. 3rd ed. Williamsport, Pa.: Lycoming, 1979.

Kimes, Beverly Rae, and Henry Austin Clark Jr. *Standard Catalog of American Cars, 1805–1942*. 1985. 3rd ed., Iola, Wisc.: Krause, 1996.

King, Frank Alexander. *Minnesota Logging Railroads: A Pictorial History of the Era When White Pine and the Logging Railroad Reigned Supreme*. San Marino, Calif.: Golden West Books, 1981.

Kline, Benjamin F. G., Jr. *Logging Railroad Era of Lumbering in Pennsylvania*. Vol. 1, *Pitch Pine and Prop Timber: The Logging Railroads of South-Central Pennsylvania*. Williamsport, Pa.: Lycoming, 1971.

———. *Logging Railroad Era of Lumbering in Pennsylvania*. Vol. 2, *"Wild Catting" on the Mountain: The William Whitmer & Sons Company and the Whitmer-Steele Company Operations in Cambria, Centre, Clearfield, and Union Counties, Pennsylvania and Cornwall, Rockbridge County, Virginia*. Williamsport, Pa.: Lycoming, 1970.

———. *Logging Railroad Era of Lumbering in Pennsylvania*. Vol. 12, *Dinkies, Dams, and Sawdust: The Logging Railroads of West Central Pennsylvania; DuBois, Clearfield, Brookville, Brockway, Penfield, Ebensburg, Punxsutawney*. [Lancaster, Pa.]: privately printed, 1975.

———. *Logging Railroad Era of Lumbering in Pennsylvania*. Vol. 13, *"Stemwinders" in the Laurel Highlands: The Logging Railroads of South-Western Pennsylvania*. Williamsport, Pa.: Lycoming, 1973.

Kline, Benjamin F. G., Jr., Walter Casler, and Thomas Townsend Taber III. *Logging Railroad Era of Lumbering in Pennsylvania: A History of the Lumber, Chemical Wood, and Tanning Companies Which Used Railroads in Pennsylvania*. 3 vols. Williamsport, Pa.: Lycoming, 1970–78.

Koch, Michael. *The Shay Locomotive: Titan of the Timber*. Denver: World Press, 1971.

Korson, George. *Pennsylvania Songs and Legends*. Baltimore: Johns Hopkins University Press, 1949.

Krebs, Otto. *Map of the Oil Region of Pennsylvania: Meadville, Pennsylvania; Townsend and Dale, 1865*. In *Oil and Gas Maps of Southwestern Pennsylvania, 1915*. Harrisburg, Pa.: W. S. Ray, 1916.

Kuhns, Mrs. John F. "Loleta in 1910." *Pennsylvania History* 19, no. 4 (1952): 452–60.

Kury, Franklin L., and Maurice K. Goddard. *Natural Resources and the Public Estate: A Biography of Article I, Section 27 of the Pennsylvania*

Constitution. Harrisburg, Pa.: Reed, Smith, Shaw, and McClay, 1985.

La Gorce, John Oliver. "Penn's Land of Modern Miracles." *National Geographic* 68, no. 1 (1935): 6–7, 9.

Largey, Gale, Andy Worthington, and Phillip E. Ogden. *The Austin Disaster, 1911, as Reported in the Media Before Radio, Television, the Internet*. Williamsport, Pa.: Reed Hann Litho, 2011.

Lewis, James G. "The Pinchot Family and the Battle to Establish American Forestry." *Pennsylvania History* 66, no. 2 (1999): 143–65.

"Life Expectancy by Age, 1850–2004, for White Males." Infoplease. http://www.infoplease.com/ipa/A0005140.html (accessed June 15, 2015).

Lillard, Richard Gordon. *The Great Forest*. New York: Alfred A. Knopf, 1947.

Longfellow, Henry Wadsworth. *Evangeline: A Tale of Acadie*. Boston: William D. Ticknor, 1847.

Longwell, Dennis. *Steichen: The Master Prints, 1895–1914, the Symbolist Period*. New York: Museum of Modern Art, 1978.

Lucey, Donna M. *Photographing Montana, 1894–1928: The Life and Work of Evelyn Cameron*. New York: Alfred A. Knopf, 1990.

Lutz, H. J. "Original Forest Composition in Northwestern Pennsylvania as Indicated by Early Land Survey Notes." *Journal of Forestry* 28 (1930): 1098–1103.

———. "The Vegetation of Heart's Content, a Virgin Forest in Northwestern Pennsylvania." *Ecology* 11, no. 1 (1930): 1–29.

Lynch, Clyde. "Jobbing on Skinners Creek, Port Allegany." In *Sawmills Among the Derricks: Kane, Kinzua Valley, Bradford, Crosby, Lewis Run*, by Thomas T[ownsend] Taber III. Muncy, Pa.: Reed Hann Litho, 1975.

Maass, Eleanor A. *Forestry Pioneer: The Life of Joseph Trimble Rothrock*. Lebanon, Pa.: Pennsylvania Forestry Association, 2003.

MacCleary, Douglas W. *American Forests: A History of Resiliency and Recovery*. 1992. Rev. ed., Durham, N.C.: Forest History Society, 2011.

Majot, Tulla, Paul A. Heimel, David Richardson, and Curt Weinhold. *Coudersport Glass, 1900–1904*. Marietta, Ohio: Glass Press, 1999.

Marquis, D. A. *The Allegheny Hardwood Forests of Pennsylvania*. USDA Forest Service General Technical Report, no. NE-15. Broomall, Pa.: Northeastern Forest Experiment Station, 1975.

Marx, Leo. *The Machine in the Garden: Technology and the Pastoral Ideal in America*. New York: Oxford University Press, 1964.

Matthews, George E. *Men of New York: A Collection of Biographies and Portraits of Citizens of the Empire State Prominent in Business, Professional, Social, and Political Life During the Last Decade of the Nineteenth Century*. Buffalo, N.Y.: G. E. Matthews, 1898.

Mayer, Lynne Rhodes, and Kenneth E. Vose. *Makin' Tracks: The Story of the Transcontinental Railroad in the Pictures and Words of the Men Who Were There*. New York: Praeger, 1975.

McCormick, R. L. "The Exhaustion of the Lumber Supply." In *Forestry and the Lumber Supply*, edited by Theodore Roosevelt, Robert Laird McCormick, and Gifford Pinchot, 8–11. United States Department of Agriculture, Bureau of Forestry, circular no. 25, June 11, 1903.

McElroy, Janice H., Mary Ann Stangil, and Margaret D. Druse. *Our Hidden Heritage: Pennsylvania Women in History*. Washington, D.C.: American Association of University Women, Pennsylvania Division, 1983.

McGeary, M. Nelson. *Gifford Pinchot: Forester-Politician*. Princeton: Princeton University Press, 1960.

McKeen, Ona Lee. "Erwin Evans Smith, Cowboy Photographer." In *A Century of Photographs, 1846–1946: Selected from the Collection of the Library of Congress*, compiled by Renata V. Shaw, 156–62. Washington, D.C.: Library of Congress, 1980.

McKelvey, Blake. *A Panoramic History of Rochester and Monroe County, New York*. Woodland Hills, Calif.: Windsor, 1979.

McShannock, Linda. "Boys' Clothes." *Minnesota History* 56, no. 8 (1999–2000): 454.

Michaux, F. A. *The North American Silva; or, A Description of the Forest Trees, of the United States, Canada, and Nova Scotia*. Vol. 3. Philadelphia: Thomas Dobson–Solomon Conrad, 1819.

Miller, Char. *Gifford Pinchot and the Making of Modern Environmentalism*. Washington, D.C.: Island Press/Shearwater Books, 2001.

Miller, E. Willard. *Pennsylvania Keystone to Progress: An Illustrated History.* Northridge, Calif.: Windsor, 1986.

Monroe County Library System. "Rochester City Directories by Decade, 1820s–1940s." http://www3.libraryweb.org/lh.aspx?id=1105 (accessed June 2, 2015).

Morrison, Ernest, George M. Leader, and William C. Forrey. *A Walk on the Downhill Side of the Log: The Life of Maurice K. Goddard.* Mechanicsburg, Pa.: Pennsylvania Forestry Association, 2000.

Muir, John. *Our National Parks.* Boston: Houghton Mifflin, 1901.

Munson, Doug. "Gelatin Chloride Printing-Out Paper: Some Historical and Contemporary Notes." http://www.albumenworks.com/CAW_POP_Historical_Notes.pdf (accessed June 4, 2015).

Newhall, Beaumont. "Pictorial Photography." In *The History of Photography from 1839 to the Present*, 140–65. 5th ed. New York: Museum of Modern Art, 1986.

Newhall, Beaumont, and Nancy Newhall. "Alfred Stieglitz." In *Masters of Photography*, 60–75. New York: Castle Books, 1958.

———. "Edward Steichen." In *Masters of Photography*, 76–91. New York: Castle Books, 1958.

Nuschke, Marie Kathern Brisbois. "Community Growth in Freeman Run." In *Pennsylvania's Freeman Run Valley: An Oral History of Freeman Run Valley, Potter County, Pennsylvania*. Vol. 3, 1–117. Coudersport, Pa.: Potter County Enterprise, 1979.

———. *The Dam That Could Not Break.* Camp Hill, Pa.: Plank's Suburban Press, 1960.

———. "Hicks, Fighters, and Clog Dancers: Early Lumber Camps in Freeman Run Valley." *Pennsylvania History* 19, no. 4 (1952): 435–51.

———. "Industry Comes to Freeman Run." In *Pennsylvania's Freeman Run Valley: An Oral History of Freeman Run Valley, Potter County, Pennsylvania*. Vol. 2, 1–111. Coudersport, Pa.: Potter County Enterprise, 1977.

———. "Little Marie of Freedom Run." In *Pennsylvania's Freeman Run Valley: An Oral History of Freeman Run Valley, Potter County, Pennsylvania*. Vol. 4, 1–103. Coudersport, Pa.: Potter County Enterprise, 1980.

———. *Pennsylvania's Freeman Run Valley: An Oral History of Freeman Run Valley, Potter Country, Pennsylvania*. Coudersport, Pa.: Potter County Enterprise, 1979.

———. "White Men Come to Freeman Run." In *Pennsylvania's Freeman Run Valley: An Oral History of Freeman Run Valley, Potter County, Pennsylvania*. Vol. 1, 1–78. Coudersport, Pa.: Potter County Enterprise, 1976.

Olney, J. *A Practical System of Modern Geography; or, A View of the Present State of the World.* Rev. ed. New York: Pratt, Woodford, 1844.

Onthank, A. H. *The Tanning Industry.* Detroit: Basic Industry Library, Robert Morris Club, National Association of Credit Men, 1917.

Ostman, Ronald E., and Harry Littell. *Cornell Then and Now: Historic and Contemporary Views of Cornell University.* Ithaca, N.Y.: McBooks Press, 2003.

Owens, John, and Denise Owens. *The Victims of the Austin Flood of 1911 and Their Families.* Austin, Pa.: E. O. Austin Home/Historical Society, September 30, 2011.

Owlett, Steven E. *Seasons Along the Tiadaghton: An Environmental History of the Pine Creek Gorge.* Petaluma, Calif.: Interprint, 1993. Also privately printed, Wellsboro, Pa.

Palmer, A. N. *Palmer's Guide to Business Writing.* Cedar Rapids, Iowa: Western Penmanship, 1894.

Pennsylvania Bureau of Forestry. *A Chronology of Events Showing Things as They Happened to Penn's Woods from the Landing of the First Europeans on the Shores of the Delaware to the Present Time, with Dates, Illustrations, and Annotations.* Harrisburg, Pa.: Bureau of Forestry, Pennsylvania Department of Environmental Resources, 1975.

Pennsylvania Department of Agriculture. "Logging Operations of the Lackawanna Lumber Company." In *Annual Report of the Pennsylvania Department of Agriculture*, no. 6. Harrisburg, Pa: State Printer, 1897.

———. "Operations of F. H. & C. W. Goodyear." In *Annual Report of the Pennsylvania Department of Agriculture*, no. 6. Harrisburg, Pa.: State Printer, 1897.

Pennsylvania Department of Conservation and Natural Resources. *1895–1995: The Legacy of Penn's Wood; A History of the Pennsylvania Bureau of Forestry*. 1995. http://www.apps.dcnr.state.pa.us/forestry/sfrmp/docs/The%20Legacy%20of%20Penn%27s%20Woods%20

%20The%20History%20of%20the%20Bureau%20of%20Forestry.pdf.

Pennsylvania Department of Environmental Protection. *Pennsylvania's Environmental Heroes*. Harrisburg, Pa., 2001. http://www.elibrary.dep.state.pa.us/dsweb/Get/Document-103003/EEHeroCards.pdf (accessed September 26, 2015).

Pennsylvania Lumber Museum, Ulysses, Potter County. *Timber!* Harrisburg, Pa.: Pennsylvania Historical and Museum Commission, 1985. http://lumbermuseum.org/.

Pietrak, Paul. *The B&S: Buffalo and Susquehanna Railway*. North Boston, N.Y.: privately printed, 1995.

Pinchot, Gifford. *The Adirondack Spruce: A Study of the Forest in Ne-ha-sa-ne Park, with Tables of Volume and Yield and a Working Plan for Conservative Lumbering*. New York: Critic, 1898.

———. *Biltmore Forest, the Property of George W. Vanderbilt: An Account of Its Treatment, and the Results of the First Year's Work*. Chicago: Lakeside Press, 1893.

———. "The Birth of Conservation." In *Breaking New Ground*, 319–26. New York: Harcourt Brace, 1947. Commemorative edition, Washington, D.C.: Island Press, 1998.

———. "The Blazed Trail of Forest Depletion." *American Forestry* 29, no. 354 (1923): 323–28, 374.

———. *The Conservation of Natural Resources*. U.S. Department of Agriculture, Farmer's Bulletin, no. 327 (1908). Washington, D.C.

———. *The Fight for Conservation*. New York: Doubleday, Page, 1910.

———. "The Lumberman and the Forester." *Forestry and the Lumber Supply*. United States Department of Agriculture, Bureau of Forestry, no. 25. Washington, D.C., Government Printing Office (1903): 11–14.

———. *Practical Assistance to Farmers, Lumbermen, and Others in Handling Forest Lands*. Division of Forestry. Washington, D.C.: Government Printing Office, 1898.

———. *Practical Assistance to Tree Planters*. Forest Products Laboratory, Division of Forestry, USDA Forest Service Circular, no. 22. Washington, D.C.: Government Printing Office, 1899.

———. *A Primer of Forestry*. Washington, D.C.: Government Printing Office, 1899.

———. *The Training of a Forester*. Philadelphia: J. B. Lippincott, 1914.

Pinchot, Gifford, and Henry Solon Graves. *The White Pine: A Study, with Tables of Volume and Yield*. New York: Century, 1896.

Pinchot, Gifford, and Overton Westfeldt Price. *The Profession of Forestry*. Washington, D.C.: American Forestry Association, 1901.

Pinkett, Harold T. *Gifford Pinchot, Private and Public Forester*. Urbana: University of Illinois Press, 1970.

Pippin, Bill. *Wood Hick ,Pigs-Ear, and Murphy: The Historical Story of Galeton, Pennsylvania, and Surrounding Area*. N.P.: Jostens, 1989.

Ponder, Stephen. "Gifford Pinchot, Press Agent for Forestry." *Journal of Forest History* 31, no. 1 (1987): 26–35.

Potter County Historical Society. *Historical Sketches of Potter County, 1776–1976*. Coudersport, Pa.: Potter County Historical Society, 1976.

———. "Stewardson Township [southeastern Potter County]: The Lumbering Era." *Quarterly Bulletin* 97 (1990): 3–4.

Pringle, C. G. "Report on the Principal Lumber Producing Regions of the State [of Pennsylvania]." In *Report on the Forests of North America (Exclusive of Mexico)*, vol. 9, *Tenth Census of the United States*, edited by Charles Sprague Sargent. Washington, D.C.: Government Printing Office, 1880.

———. "Report upon the Principal Lumber Producing Regions of Pennsylvania in the Early 1880s." In James Elliott Defebaugh, *History of the Lumber Industry of America*, vol. 2, 540–42. Chicago: American Lumberman, 1907.

Pyne, Stephen J., Patricia L. Andrews, and Richard D. Laven. *Introduction to Wildland Fire*. New York: John Wiley, 1996.

Reed, Frank A. *Lumberjack Sky Pilot*. 1965. 3rd ed., Lakemont, N.Y.: North Country Books, 1976.

"Report to Dr. Joseph Trimble Rothrock, Pennsylvania Commissioner of Forestry, Williamsport, 1897." In *The Legacy of Penn's Woods, 1895–1995: A History of the Pennsylvania Bureau of Forestry*, by Lester A. DeCoster, 11–14. Harrisburg, Pa.: Pennsylvania Historical and Museum Commission, Department of Conservation and Natural Resources, Bureau of Forestry, 1995.

Rich, Thomas P. *Lessons in Social Responsibility from the Austin Dam Failure.* Emmitsburg, Md.: National Emergency Training Center, 2006.

Ries, Linda A., Ronald E. Ostman, and Harry Littell. "Wood on Glass: The Lumber Industry Photographs of William T. Clarke." *Pennsylvania Heritage* 37, no. 1 (2011): 16–25.

Riesenman, Joseph, Jr. *History of Northwestern Pennsylvania.* Vol. 1. New York: Lewis, 1943.

Riley, G. A. "A History of Tanning in the State of Maine." M.A. thesis, University of Maine, 1935.

Rimby, Susan. "Better Housekeeping Out of Doors: Mira Lloyd Dock, the State Federation of Pennsylvania Women, and Progressive Era Conservation." *Journal of Women's History* 17, no. 3 (2005): 9–34.

——. *Mira Lloyd Dock and the Progressive Era Conservation Movement.* University Park: Pennsylvania State University Press, 2012.

Rinhart, Floyd, and Marion Rinhart. *America's Affluent Age.* South Brunswick, N.J.: A. S. Barnes, 1971.

Ripley, Thomas Emerson. *Green Timber on the Flood Tide to Fortune in the Great Northwest.* Palo Alto, Calif.: American West, 1968.

Roosevelt, Theodore. "Opening Address by the President." In *Proceedings of a Conference of Governors in the White House*, edited by Newton C[rain] Blanchard and W. J. McGee, 3–12. Washington, D.C.: U.S. Government Printing Office, 1909.

——. "Wilderness Preserves: Outdoor Pastimes of an American Hunter." In *The Adventures of Theodore Roosevelt*, edited by Anthony Brandt, 146–62. Washington, D.C.: National Geographic Society Adventure Classics, 2005.

Rosenberger, Homer T. *Adventures and Philosophy of a Pennsylvania Dutchman: A Biography in a Broad Setting.* Bellefonte, Pa.: Pennsylvania Heritage, 1971.

Sasowsky, Norman. *The Prints of Reginald Marsh.* New York: Clarkson N. Potter, 1976.

Schlereth, Thomas J. *Victorian America: Transformations in Everyday Life, 1876–1915.* New York: Harper Perennial, 1992.

Schroeder, Joseph J., Jr., ed. *Sears, Roebuck and Co. Incorporated Consumers Guide, Fall 1900.* Northfield, Ill.: DBI, 1970.

Scott, Sir Walter. *Old Mortality.* 1816 reprint. Oxford: Oxford University Press, 2009.

Shoemaker, Henry Wharton. *Allegheny Episodes: Folk Lore and Legends Collected in Northern and Western Pennsylvania.* Altoona, Pa.: Altoona Tribune, 1922.

——. "An Antique Dealer's Romance." In *Some Stories of Old Deserted Houses in the Central Pennsylvania Mountains.* Altoona, Pa.: Times Tribune, 1931.

——. *Black Forest Souvenirs: Collected in Northern Pennsylvania.* Reading, Pa.: Bright-Faust, 1914.

——. *Eldorado Found, the Central Pennsylvania Highlands: A Tourist's Survey.* Altoona, Pa.: Altoona Tribune, 1917.

——. *The Indian Steps and Other Pennsylvania Mountain Stories.* Reading, Pa.: Bright, 1912.

——. "A Misunderstanding." In *Allegheny Episodes: Folk Lore and Legends Collected in Northern and Western Pennsylvania.* Altoona, Pa.: Altoona Tribune, 1922.

——. *Some Stories of Old Deserted Houses in the Central Pennsylvania Mountains.* Altoona, Pa.: Times Tribune, 1931.

Shrestha, Laura B. *Life Expectancy in the United States.* Table 1, "U.S. Life Expectancy in the United States by Sex, in Selected Years." CRS-3, Congressional Research Service, Library of Congress. Order code RL32792. http://aging.senate.gov/crs/aging1.pdf. Updated August 19, 2006.

Sisler, John D., comp. *Map of the Coal Fields of Pennsylvania* (map). Topographic and Geologic Survey, Bulletin M-6 (1929): plate 1, pt. 2. Harrisburg, Pa.: Commonwealth of Pennsylvania, Department of Internal Affairs.

——. *Oil and Gas Fields of Western Pennsylvania* (map). Topographic and Geologic Survey, Bulletin M19 (1932): plate 1. Commonwealth of Pennsylvania, Department of Internal Affairs, Harrisburg, Pa.

Smith, Alexander. "On the Importance of a Man to Himself." In *Dreamthorp, with Selections from "Last Leaves,"* 148–66. London: Oxford University Press, 1934.

Spurr, Stephen H., and Burton V. Barnes. *Forest Ecology.* 2nd ed. New York: Ronald Press, 1973.

Stebbins, Theodore E., Jr., Carol Troyen, and Trevor J. Fairbrother. *A New World: Masterpieces of American Painting, 1760–1910.* Boston: Museum of Fine Arts, 1983.

Steen, Harold K., ed. *The Conservation Diaries of Gifford Pinchot.* Durham, N.C.: Forest History Society, 2001.

Stephenson, Harry. *History of Little Pine Valley: History and Genealogy of Little Pine Valley and Its Residents.* Camp Hill, Pa.: Planks Suburban Press, 1992.

Stevens, S. K. "When Timber Was King in Pennsylvania." *Pennsylvania History* 19, no. 4 (1952): 391–95.

Stevens, Sylvester K., Ralph W. Cordier, and Florence O. Benjamin. *Exploring Pennsylvania. Geography/History/Civics.* 3rd ed. New York: Harcourt, Brace, and World, 1968.

Taber, Thomas T[ownsend] III. *Ghost Lumber Towns of Central Pennsylvania: Laquin, Masten, Ricketts, Grays Run.* Vol. 3 of *Logging Railroad Era of Lumbering in Pennsylvania.* Williamsport, Pa.: Lycoming, 1970.

———. *The Goodyears: An Empire in the Hemlocks.* Vol. 5 of *Logging Railroad Era of Lumbering in Pennsylvania.* Williamsport, Pa.: Lycoming, 1971.

———. *Logging Railroad Era of Lumbering in Pennsylvania: Addenda and Index, Additions and Corrections.* [Williamsport, Pa.?]: N.p., [1978?].

———. *Railroads of Pennsylvania: Encyclopedia and Atlas.* Muncy, Pa.: T. T. Taber, 1987.

———. *Sawmills Among the Derricks: Kane, Kinzua Valley, Bradford, Crosby, Lewis Run.* Vol. 7 of *Logging Railroad Era of Lumbering in Pennsylvania.* Muncy, Pa.: Reed Hann Litho, 1975.

———. *Short-Line Steam in Pennsylvania.* Muncy, Pa.: T. T. Taber III, [1970?].

———. *Sunset Along Susquehanna Waters: Williamsport, Leetonia, Slate Run, Cammal, Glen Union, Gleasonton.* Vol. 4 of *Logging Railroad Era of Lumbering in Pennsylvania.* Williamsport, Pa.: Lycoming, 1972.

———. *Tanbark, Alcohol, and Lumber: The Forest Industries of Ridgway, St. Marys, Hallton, Wilcox, Portland Mills, Straight, Johnsonburg.* Vol. 10 of *Logging Railroad Era of Lumbering in Pennsylvania.* Williamsport, Pa.: Lycoming, 1974.

———. *Whining Saws and Squealing Flanges: Cross Fork, Emporium, Austin, Hicks Run, Galeton.* Vol. 6 of *Logging Railroad Era of Lumbering in Pennsylvania.* Williamsport, Pa.: Lycoming, 1972.

———. *Williamsport: Lumber Capital.* South Williamsport, Pa.: Hilsher Graphics, 1995.

Taber, Thomas T[ownsend], III, and Walter Casler. *Climax: An Unusual Steam Locomotive.* Rahway, N.J.: Railroadians of America, 1960.

Theiss, Lewis E., and Charles Fisher Snyder. *Lumbering and Rafting on the West Branch.* 1957. Sunbury, Pa.: Northumberland County Historical Society, 1999.

Thompson, W. W. "History of the Wild Boy Hermit." *Potter County Historical Society Bulletin* 30 (1973): 4. Coudersport, Pa.

Tichenor, Phillip J., George A. Donohue, and Clarice N. Olien. *Community Conflict and the Press.* Beverly Hills, Calif.: Sage, 1980.

Tonkin, Joseph Dudley. *The Last Raft.* [Harrisburg, Pa.]: privately printed, 1940.

Trego, Charles B. *A Geography of Pennsylvania.* Philadelphia: Edward C. Biddle, 1843.

University of Rochester Library. Mount Hope Cemetery records. http://www.lib.rochester.edu/index.cfm?page=3311 (accessed June 4, 2015).

Walker, J. H. *Rafting Days in Pennsylvania.* University Park: Pennsylvania State University Press, 2006.

Wallace, Paul A. W. *Pennsylvania: Seed of a Nation.* New York: Harper and Row, 1962.

Wheeler, Charles N. "Interview with Henry Ford." *Chicago Tribune*, May 25, 1916. In *Random House Webster's Quotationary*, edited by Leonard Roy Frank, 362. New York: Random House, 2001.

White, Alicia, Jin Ma, and Lawrence Wetzel. "Digitizing Glass Lantern Slides: The Mira Lloyd Dock Project at the Pennsylvania State University." *Microform and Imaging Review* 34 (2005): 81–90.

White, Dale. *Gifford Pinchot: The Man Who Saved the Forests.* New York: J. Messner, 1957.

White, John H. *Early American Locomotives with 147 Engravings.* New York: Dover, 1972.

Whitney, Gordon G. *From Coastal Wilderness to Fruited Plain: A History of Environmental Change in Temperate North America, 1500 to the Present.* Cambridge, U.K.: Cambridge University Press, 1994.

Wilhelm, S. A. "History of the Lumber Industry of the Upper Allegheny River Basin During the Nineteenth Century." Ph.D. diss., University of Pittsburgh, 1953.

Williams, Michael. *Americans and Their Forests: A Historical Geography.* Cambridge, U.K.: Cambridge University Press, 1989.

Wyld, Lionell D. *Low Bridge! Folklore and the Erie Canal.* Syracuse: Syracuse University Press, 1962.

Yaeger, Bert D. *The Hudson River School: American Landscape Artists.* New York: Todtri, 1996.

INDEX

Page numbers in *italics* refer to figures; plates are referred to by their number; page numbers followed by *a* refer to a numbered appendix entry; page numbers followed by *n* refer to notes.